PHOTO BIOLOGY

PHOTO BIOLOGY

Elli Kohen

Department of Biology
University of Miami
Miami, Florida

René Santus

Laboratoire de
Physico-Chimie de L'Adaptation Biologique
Museum National d'Histoire Naturelle
Paris, France

Joseph G. Hirschberg

Department of Physics
University of Miami
Miami, Florida

Academic Press

San Diego   New York   Boston   London   Sydney   Tokyo   Toronto

Copyright © 1995 by ACADEMIC PRESS, INC.

Academic Press, Inc.
A Division of Harcourt Brace & Company
525 B Street, Suite 1900, San Diego, California 92101-4495

United Kingdom Edition published by
Academic Press Limited
24-28 Oval Road, London NW1 7DX

Library of Congress Cataloging-in-Publication Data

Kohen, Elli.
    Photobiology / by Elli Kohen, Joseph Hirschberg, Rene Santus.
        p.   cm.
    Includes bibliographical references and index.
    ISBN 0-12-417755-7
        1. Photobiology.   2. Photobiochemistry.   3. Phototherapy.
    I. Hirschberg, Joseph G.   II. Santus, Rene.   III. Title.
    QH515.K64   1995
    574.19'153--dc20
                                                        94-40230
                                    CIP

PRINTED IN THE UNITED STATES OF AMERICA
95   96   97   98   99   00   MM   9   8   7   6   5   4   3   2   1

*This book is dedicated
to the memory of
Ginette Tétard Hirschberg*

CONTENTS

PART I   INTRODUCING LIGHT

CHAPTER 1   The Nature of Light

CHAPTER 2 Pathways of Molecular Excitation and Deactivation

CHAPTER 3 Spectroscopy and Instrumentation

PART II LIGHT AND BIOLOGICAL SYSTEMS

CHAPTER 4 From Photophysics to Photochemistry: Determination of Primary Processes in Direct or Sensitized Photoreactions

CHAPTER 5   Photochemistry of Biological Molecules

CHAPTER 6 | Bioluminescence

CHAPTER 7 | Environmental Photobiology

CHAPTER 8 | Marine Photobiology

CHAPTER 9 Photosynthesis

SECTION I

CHAPTER 10 Photobioregulatory Mechanisms

SECTION I   *Photomorphogenesis*

CHAPTER 15 Photoimmunology

CHAPTER 16 **Photosensitive, Photoallergic, and Light-Aggravated (Photo-Koebner) Photodermatoses**

CHAPTER 17 **Phototherapy of Neonatal Bilirubinemia and Vitamin D Deficiency**

CHAPTER 18 **Photochemotherapy (PUVA Therapy) and UVB
Phototherapy**

CHAPTER 19   Photodynamic Phototherapy

PREFACE

The Sun is such a familiar companion that many people forget that sunlight is a prerequisite for life on Earth. Do we fully realize that conversion of the almost inexhaustible solar photon energy into free oxygen and fuel by the plant kingdom is the basic pathway by which we survive on our planet? Also, do we fully understand that conversion of light signals into electrochemical energy enables us to see? Paradoxically, whereas light is life, the endless creativity of man's brain in many areas of science has led us in the past decade to the recognition that certain rays of sunlight can become the fatal Trojan horse to living species.

The understanding of the mechanisms of life and death constitutes the science of photobiology. We have tried in this book to show the multiple impacts of photobiology, not only on a great number of biological and environmental processes, but also in biomedical research and medicine. The importance of the effect of light on living things is so great that photobiology has become a truly multidisciplinary field. People with many different kinds of scientific training find themselves pursuing research in photobiology or applying principles and using instruments relevant to photobiology. Among biologists this includes molecular biologists, geneticists, developmental biologists, immunologists, physiologists, toxicologists, nutritionists, radiation biologists, and environmental biologists. Among the chemists, biochemists, photochemists, physical chemists, and pharmacologists are likely to deal with photobiological processes. The same holds true for physicists studying atmospheric physics and biophysics, laser and optical physics, and radiation physics. Biomedical engineers play a significant role in updating and modernizing the instrumentation relevant to photobiology. Among physicians, dermatologists, ophthalmologists, oncologists, immunologists, pediatricians, psychiatrists, and endocrinologists are those most likely to have close interaction with photobiology. Mention should also be made of veterinary medicine and veterinary photopathology.

Since photobiology involves the interaction of light and molecules, in the first five chapters we present the physical and chemical bases of molecular photochemistry. These include concepts that are crucial in photobiology, such as quantum yield, quenching and chemical reactivity of the excited states resulting from absorption of light, emission and action spectra, and basic instrumentation. Thus we have principles that are sometimes difficult for the biologist, who often has no more than a distant awareness of molecular photobiology. Therefore, wherever possible, we have incorporated tables and figures to allow a smooth development and a clear understanding of the topic.

Chapters 6–11 include light–chemical energy transductions (bioluminescence and photosynthesis), light-controlled processes (photomorphogenesis, chronobiology, and photomovement), and vision. In Chapters 12–16 we present the biological effects of ultraviolet irradiation, photocarcinogenesis, and photoimmunology; the final three chapters address the diseases associated with light, and, paradoxically, the use of light as a tool in modern human health (e.g., phototherapy and photochemotherapy). We hope that our readers will be convinced that photobiology is a major scientific discipline, holding important potential for further progress in biology, biotechnology, and medicine.

Elli Kohen
René Santus
Joseph G. Hirschberg

ACKNOWLEDGMENTS

The authors are most appreciative of the valuable advice and tremendous help they have obtained from those named below, who were kind and caring enough to make available documentation, photographs, and other assistance at every step in the preparation of this book:

Dr. P. Bezille, Pathologie Médicale du Bétail et des Animaux de la Basse-Cour, Ecole Nationale Veterinaire de Lyon, France; Professor J. Brugère-Picoux, Ecole Nationale Vétérinaire d'Alfort, France; Professor B. Chance, Department of Biochemistry and Biophysics, School of Medicine, University of Pennsylvania, Philadelphia, Pennsylvania; Professor J. E. Cleaver, Laboratory of Radiobiology and Environment Health, University of San Francisco, San Francisco, California; Professor P. Cruz, Department of Dermatology, University of Texas Southwestern Medical Center, Dallas, Texas; V. A. De Leo, Department of Dermatology, Columbia Presbyterian Medical Center, New York, New York; Professor Thomas J. Dougherty, Division of Radiation Biology, Roswell Park Cancer Institute and Department of Radiation Medicine, Division of Radiation Biology and Photodynamic Therapy Center, Buffalo, New York; Professor L. Dubertret, Service de Dermatologie, Hôpital Saint Louis, Paris, France; Dr. G. Elgart, Department of Dermatology, School of Medicine, University of Miami, Miami, Florida; Professor R. M. Halder,

Director Vitiligo Center, Department of Dermatology, Howard University Hospital, Washington, DC; Dr. L. C. Harber, Rhodebeck Professor of Dermatology, College of Physicians and Surgeons of Columbia University, New York, New York; Professor W. Hastings, Department of Cellular and Developmental Biology, Harvard University, Cambridge, Massachusetts; Dr. A. J. Herron, Department of Pathology, Division of Comparative Pathology, University of Miami, School of Medicine, Miami, Florida; Dr. M. Jeanmougin, Policlinique de Dermatologie, Service du Professeur Dubertret, Hôpital Saint-Louis, Paris, France; Professor S. I. Katz, Dermatology Branch, NIH, Bethesda, Maryland; Dr. J. Kennedy, Queen's University, Kingston, Ontario, Canada; Dr. K. H. Kraemer, Laboratory of Molecular Carcinogenesis, National Cancer Institute, Bethesda, Maryland; Dr. M. Kripke, Departments of Cell Biology and Immunology, The University of Texas, MD Anderson Medical Center, Houston, Texas; Dr. P. Liebman, Department of Physiology, University of Pennsylvania, Philadelphia, Pennsylvania; Dr. N. Penneys, Department of Pathology, School of Medicine, University of Miami, Miami, Florida; Dr. Pottier, Department of Chemistry and Chemical Engineering, The Royal Military College of Canada, Kingston, Ontario, Canada; N. F. Rothfield, Division of Rheumatic Diseases, The School of Medicine of the University of Connecticut Health Center, Farmington, Connecticut; Professor L. Stryer, Department of Neurobiology, Stanford University, School of Medicine, Stanford, California; Professor P. S. Song, Department of Chemistry, University of Nebraska, Lincoln, Nebraska; Professor R. Steinman, The Department of Dermatology, Rockefeller University, New York, New York; Professor F. Urbach, Temple University Medical Practice, Fort Washington, Pennsylvania; Professor E. Vonderheid, Division of Dermatology, Hahneman University, Philadelphia, Pennsylvania.

The authors are also thankful to Professor M. Rougée, Drs. P. Morliere and M. Bazin, National Museum of Natural History, Paris, France for useful corrections and suggestions. They thankfully acknowledge the help of Mrs. J. Haigle, National Museum of Natural History, Paris and Mr. K. Ray, Medical Illustration, Department of Biochemistry and Biophysics, School of Medicine, in the preparation of drafts and drawings; the secretarial help of Miss J. Faurillon is also acknowledged.

The authors join in thanking INSERM (Institut National de La Sante et de la Recherche Medicale) and the JNICT (Junta Nacional de Investigacoes Cientifica e Tecnologiica) for the summer stay of one of them, R. Santus, at the Santa Maria Hospital of Lisbon, Portugal, where a part of this book was prepared.

INTRODUCING LIGHT

CHAPTER 1

The Nature of Light

Since the dawn of history man has been fascinated by the universe around him, most of which he senses by means of his eyes. He has therefore always considered light one of the most important phenomena.

1.1 REFLECTION

The ancient world was familiar with many optical principles and devices, among them the observation that light travels in straight lines, and the law of specular reflection for a ray of light, e.g., that the angles of incidence and reflection lie in a plane determined by the incident ray and a line perpendicular to the surface at the point where the incident ray intersects the reflecting surface, and that the angles of incidence and reflection are equal (see Figure 1.1, where i is the angle of incidence and r is the angle of reflection):

$$i = r. \tag{1.1}$$

Although refraction was also known in classical antiquity, the correct law

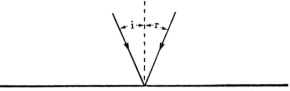

FIGURE 1.1   This shows the equality of the angles of incidence, *i*, and reflection, *r*, for specular (mirrorlike) reflection. Specular reflection occurs when the reflecting surface is smooth (variations in height are much less than a wavelength).

of refraction was not discovered until more than 2000 years later during the seventeenth century. Plane and spherical mirrors were also well-known to the ancients, and the latter as well as simple lenses were used to focus the sun's rays for starting fires.

The Greek philosopher Empedocles of Agrigentum (ca. 490–430 BC) was one of the first to put forth a theory of light, teaching essentially correct ideas of vision. Euclid, besides developing the famous axioms and theorems of geometry, was the author of a treatise *Optics* in which the fundamentals of geometrical optics were set forth, probably for the first time. Archimedes (ca. 287–212 BC), known for his famous Principle of Flotation, is said to have used concave mirrors to set the Roman ships afire during the Roman siege of Syracuse in 212 BC.

The actual nature of light was, however, unknown to the ancients and, for example, it was generally believed that its speed was infinite. After the close of the ancient period, roughly coinciding with the last western Roman emperor, Romulus Augustulus in 475, until the beginning of the Renaissance, which is often considered to have begun with the fall of the Eastern Roman Empire when Constantinople fell to the Ottoman Turks under Mehmet the Conqueror in 1453, there is a gap of nearly 1000 years during which very little occurred in the development of optical science. Spectacles, however, were first depicted in use by religious scribes in Europe during the thirteenth century, and were probably known in China long before.

1.2 REFRACTION

The revival of learning that occurred in Europe during the Renaissance did not neglect the study of light. Willebrod Snell (ca. 1620) and René Descartes (1637) independently discovered the correct law of refraction for a ray of light incident at the boundary of two transparent media (see

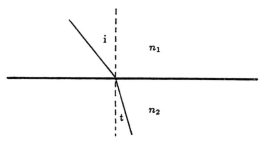

FIGURE 1.2 The angles of incidence, i, and refraction, t, for a ray as it passes through a plane separating two media with indices of refraction n_1 and n_2.

Figure 1.2):

$$n_1 \sin i = n_2 \sin t, \tag{1.2}$$

where n_1 and n_2 are quantities that depend on the properties of the two media, and are called their *indices of refraction*, and i and t are the angles measured from the perpendicular of a ray of light in each medium. As with reflection, the two rays lie in a plane determined by the incident ray and the line perpendicular to the surface at the point where the incident ray contacts the surface. With this deceptively simple law, together with the law of reflection, almost all of the principles of modern lens and optical mirror design can be derived.

1.3 FERMAT'S PRINCIPLE

Both the laws of refraction and reflection, together with that of recti-linear propagation, can, however, be derived from an even more funda-mental theorem, foreshadowed by Heron of Alexandria in antiquity (ca. 62 AD), and put forth by Pierre de Fermat about 1660. This is that *a ray of light follows the path of "least time."* In other words, if a ray of light is to pass from point A to Point B (see Figure 1.3), it will follow the path that will take the least time (in the case shown, the straight line). The laws of rectilinear propagation and of reflection follow immediately, and the law of refraction can also be shown, with a little calculation, to follow, be-cause the speed of light in a medium is inversely proportional to the medium's index of refraction.

Fermat's theorem implies something that had long been suspected but never proved: that the speed of light is not infinite, but that light propa-gates with a definite speed that can, in principle, be measured.

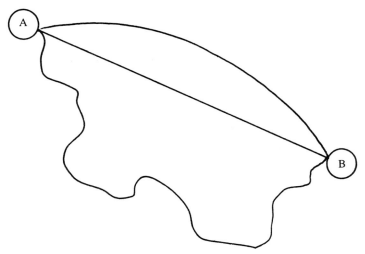

FIGURE 1.3   Three alternative paths between A and B; light "follows" the one that takes the shortest time. This is one of the versions of Fermat's principle.

1.4 THE SPEED OF LIGHT

Galileo (1564–1642) had apparently tried to measure the speed of light by uncovering a lantern on a mountain and having an assistant on a neighboring mountain uncover his as soon as he saw the light from the first lantern. The experiment failed because the speed of light was far to great to measure in that way.

Ole Römer, while developing a method to determine longitudes by using the rotations of the moons of the planet Jupiter, an idea originally suggested by Galileo, discovered that the speed of light was finite, and between 1672 and 1679 determined its value to within a few percent of its modern value of about 2.998×10^8 m/sec. At the present time the speed of light is among the most accurately known quantities, having been determined to better than one part in 10^{12}.

1.5 THE WAVE NATURE OF LIGHT

1.5.1 What Are Waves?

Waves are familiar to all of us. Perhaps the best known are the *gravity waves* that form on the surface of liquids such as seawater. These are of many forms. When they are rather low on the open water surface they are almost perfect sine shapes. When they grow large and approach the shore

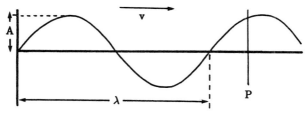

FIGURE 1.4 A sine-shaped wave is shown. A is the amplitude, λ is the wavelength, v represents the velocity of the wave, and P is a point (discussed in the text) where the frequency, f, of a moving wave may be measured.

they change form and the tops curl over, breaking as surf. It turns out that all shapes of waves can be treated as though they were combinations of simple sine waves, so for the time being, we will describe all waves as though they had the shape of the sine function, as shown in Figure 1.4.

1.5.2 Wave Characteristics

Waves are characterized by certain parameters that tell us a lot about them. These include their amplitude, their velocity, their wavelength, their frequency, and their period. In Figure 1.4, A is the amplitude and λ is the wavelength. Notice that the wavelength is the distance along the wave from a point to where the wave begins to repeat. When the wave moves with velocity, v, a number of wavelengths pass a given point, P, for example, in a given time. This number is the frequency, f. The period, T, is the reciprocal of the frequency, $1/f$, and is expressed in units of time, usually seconds. There is a relation between the velocity, the wavelength, and the frequency that can be expressed as follows:

The wavelength equals the velocity divided by the frequency

Algebraically it is written:

$$\lambda = v/f, \tag{1.3}$$

where λ is the wavelength, v the velocity, and f the frequency.

Let us look first at the units to see if they make sense. (This is a good idea whenever we deal with an equation, not only to help understand what is going on, but to check to see that the relation is correct.) Using SI units, the wavelength will be in meters, the velocity in meters divided by seconds, and the frequency in number of cycles, which has no units, divided by seconds, or 1/seconds. Putting these units together we will have a *units equation*. The units must turn out to be the same on both sides of the equation (apples = apples):

$$\text{meters} = (\text{meters/seconds}) \text{ divided by } 1/\text{seconds}.$$

Inverting the denominator, and multiplying:

$$\text{meters} = (\text{meters/seconds}) \times \text{seconds}.$$

The seconds cancel out, and we have meters = meters, which checks.

The wave equation can be written in several equivalent forms. Because they are often useful, they are collected here:

$$\lambda = v/f \qquad f = v/\lambda \qquad v = f\lambda$$
$$\lambda = vT \qquad T = \lambda/v \qquad v = \lambda/T. \qquad (1.4)$$

In the second set of three, the period, T, which is the reciprocal of the frequency, has been inserted.

> **Example:** The velocity of a water surface gravity wave is 2 m/sec. A small anchored boat is bobbing up and down once every 4 sec. What is the wavelength of the waves?
>
> **Answer:** v, the velocity, is given as 2 m/sec. The period, T, is the time between cycles of the boat's bobbing, or 4 sec. We use the equation relating velocity, period, and wavelength, which is solved for wavelength:
>
> $$\lambda = vT. \qquad (1.5)$$

Substituting values, we find that λ, the wavelength, equals the velocity, 2 m/sec, times the period, 4 sec, from which we obtain 8 m for λ.

The question of whether light consists of waves or bulletlike particles occupied the attention of scientists for almost the next two centuries. Christiaan Huygens (1629–1695), a Dutch astronomer, led those who believed that light consisted of waves, while Isaac Newton (1642–1726) championed the particle, or *corpuscular,* model. The corpuscular model turned out to be by far the more widely accepted model during the next century and a half. This was because light seemed to travel in strictly straight lines (except for an apparently unimportant small spreading, which occurred when the light passed through a very small hole—more about this in a moment), whereas the wave phenomena familiar at the time, e.g., sound, water waves, and the like, all were characterized by noticeable spreading. Bullets, of course, travel in straight lines, so the corpuscular model appealed to those guided by "common sense." Newton's overwhelming stature in science must also have contributed to the popularity of the corpuscular theory.

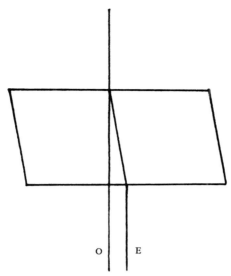

FIGURE 1.5 Here, a double-refracting crystal (calcite) is represented. The rays in the crystal (angles not to scale) exhibit the phenomenon of double refraction. The two rays, O for ordinary and E for extraordinary, are found to be plane-polarized at right angles to each other. To explain this, a transverse-wave model of light is implied.

There was, however, one phenomenon that stubbornly refused to fit into any sort of corpuscular model. This was *double refraction*. It was known that, if oriented in a certain way, some crystals, especially of calcite (also called iceland spar) produced two refracted beams when a single beam of light was directed at them, one, the *ordinary ray*, which followed the Snell–Descartes refraction rule, and another, the *extraordinary ray*, which did not [see Figure 1.5, where the ordinary (O) and the extraordinary (E) rays are marked; the extraordinary ray angle is somewhat exaggerated in the figure]. This was explained by Huygens using a wave model for light, but it could not be explained at all using a corpuscular theory.

Finally, in 1802, the British physician-*cum*-physicist Thomas Young (1773–1829) performed a definitive experiment that showed not only that light must have wave properties, but that allowed him to measure the wavelength. This experiment consisted of passing light from a small distant source through two neighboring slits and observing the resulting pattern on a screen (see Figure 1.6). On the screen were regularly spaced regions, or *fringes*, of greater and lesser light intensity, which could only be explained by the interference of waves. Where the waves from the two slits arrive in step (in phase), they add together, and where they are out

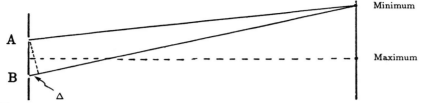

FIGURE 1.6   This shows Young's double-slit experiment (ca. 1801), which first provided a measurement of the wavelength of light. Light from a distant point source illuminates the two slits A and B. (The use of a distant point source ensures that the light coming from the left arrives at each of the two slits in phase.) After passing through the slits, the light falls on a screen. If the difference in the distance from the slits to the screen, Δ, is either zero or an integral multiple (multiplied by 1, 2, 3, etc.) of a wavelength, λ, a maximum of illumination occurs on the screen. This is because the two rays arrive at the screen in phase (see Figure 1.7). If, on the other hand, Δ is an integral multiple of $\lambda/2$, the rays will arrive out of phase (see Figure 1.7) and a minimum will result. The result is a series of light and dark spots on the screen. This experiment is often considered the first definitive proof of the wave model of light.

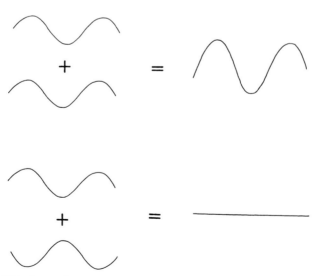

FIGURE 1.7   This shows the addition of two waves of equal amplitude, either in phase or out of phase. The upper pair shows the waves in phase, and when they add, the result, shown on the right, is a larger wave with its amplitude equal to the sum of the two individual amplitudes. The lower pair, on the other hand, is shown out of phase, and when they are combined they cancel, giving a zero sum.

of step (out of phase), they subtract and tend to cancel each other (see Figure 1.7). In Young's day it was difficult to see the fringes because the light was very weak, but today the experiment is easy to perform with a laser. Referring again to Figure 1.6, notice that when the path difference, Δ, from slit A and slit B is zero or an integral (by integral, we mean the integers 1, 2, 3, . . .) number n, of wavelengths, $n \times \lambda$, there are intensity maxima, because in these cases, the waves arrive on the screen exactly in step. When the path difference, on the other hand, is an integral multiple of a half-wavelength, $n \times \lambda/2$, the waves arrive out of step and they tend to cancel each other.

1.6 DIFFRACTION AND RESOLUTION LIMITS

The basis for a wave model of light was further reinforced by the theoretical work of the French physicist Augustin Fresnel (1788–1827). He was able to use waves to explain the fact that when light passes through a small aperture it tends to spread out at an angle, such that the smaller the aperture, the greater the spread. This is known as *diffraction* and was first described by the Italian physicist Francesco Grimaldi (1618–1663). At the time of Grimaldi's discovery the wavelength of light was unknown, so the significance of diffraction in supporting the wave theory was ignored. (Diffraction is an example of one of the fundamental laws of physics, the famous Heisenberg Uncertainty Principle.)

To simplify the situation, let us consider a single dimension (see Figure 1.8). Here a ray of light passes through a slit that has a breadth W. The actual pattern of the light after it passes through the slit is often complex, but one can consider the angular width of the resulting light pattern to be

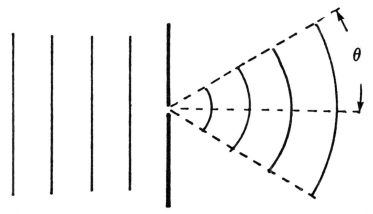

FIGURE 1.8 Here, waves from a distant source pass through a narrow (compared to the wavelength) aperture. After passing through, they spread out with half-angle θ.

between the places when the intensity has dropped to a certain fraction, usually taken to be about 80% of the maximum at the center. This is considered the condition where two equal neighboring images are just clearly distinguishable, and is called the *Rayleigh criterion* after Lord Rayleigh (1842–1919). It is commonly used to evaluate the resolving power of optical instruments such as the microscope, telescope, and spectrometer. In the case above, where we have a slit with width W, the spread in the angle, θ, of the light after passing through the slit, as measured between the 80% intensity points, is given by the simple relation

$$\theta = \lambda/W, \tag{1.6}$$

where λ is the wavelength of the light.

For a circular aperture, the situation is very much the same, but a correction factor has to be used, so the expression for θ becomes

$$\theta = 1.22\lambda/R, \tag{1.7}$$

where R is the diameter of the aperture.

For many optical instruments, such as microscopes, the apertures (usually the outer edges of the lenses) are circular, therefore the second form of the diffraction relationship is most commonly used. Let us consider a few examples:

(a) The human eye. The wavelengths of light for which the eye is sensitive extend from about 400 nm (1 nm $= 10^{-9}$ m) to about 750 nm. For convenience, take 550 nm (which produces the sensation of green light) as a typical value. The pupil of the eye varies in diameter depending on the brightness of the light from about 1 mm (1 mm $= 10^{-3}$ m) in very bright light to 7 mm in dim light. We can take 4 mm (or 4×10^6 nm) for an average value. Substituting 550 and 4×10^6 for λ and W in Eq. (1.7) we can calculate the angle of spread due to diffraction for light passing into the eye. Because diffraction tends to confuse an image, this value will give the angular limit of resolution, assuming a perfect eye:

$$\theta = \frac{1.22 \times 550}{4 \times 10^6} \tag{1.8}$$

or $\theta = 1.68 \times 10^{-4}$ rad. (A radian is a convenient measure of angle. It is equal to $360/2\pi$ or about $57.3°$, so that 1.68×10^{-4} rad corresponds to about 1/120th of a degree.) To see what this means, consider a pattern of white and black stripes (see Figure 1.9). Because the diffraction angle in radians is about 1.7 parts in 10,000, this means that if the stripes are 1.7 cm wide, and the pattern is at a distance, d, 10,000 cm away from the eye (100 m), the stripes could theoretically just be discerned. (Actually the human eye is not quite optically perfect; the best vision is about half this

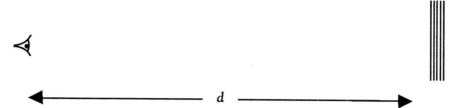

FIGURE 1.9 The observation of a series of equally spaced light and dark bands at a distance, d, from the eye.

good. A person with 20–20 vision would just be able to distinguish the pattern if it were about 50 m away.)

(b) A small telescope. Assuming that the telescope would be used visually, we can take the wavelength to be the same as for the human eye, 550 nm. A typical small telescope might have an objective lens with a diameter of 8 cm, or about 3 inches. This is 20 times the value of 4 mm we have taken for the eye in the first example. Thus the limiting angle, θ, would be 20 times smaller than for the unaided human eye, or 8.5×10^{-6} rad. This would mean that the pattern of 1.7-cm-wide stripes would be visible 20 times farther than in the example of the unaided eye, or about 2 km away.

(c) A large telescope. The Mount Palomar reflector has a mirror with a diameter of 200 inches, or 508 cm. This is 1270 times larger than the value we have taken for the eye, which makes the resulting angle so small that, assuming perfect imaging, the pattern would be discernible at a distance of 127 km! (Unfortunately, no earth-based telescope can do as well as this because the atmosphere is not homogeneous enough. This is one of the reasons for establishing an observatory in space.) Diffraction also limits the resolution of the microscope and the spectroscope.

1.7 THE QUANTUM THEORY AND PHOTONS

The idea that energy exists in small, discrete pieces, *quanta*, was first proposed in 1900 by Max Planck to explain the radiation from a *black body*. A black body is one that absorbs all the radiation incident on it. No perfect black body exists, but a good approximation is made by a hole drilled into a spherical or cylindrical shell. It turns out that the wavelength distribution of the radiation emitted by such a black body depends only on its temperature and not on its material, hence its usefulness. Figure 1.10 shows the spectrum (radiated energy plotted against wavelength) of the emission from a black body heated to various temperatures. The explanation of the shape of these curves remained a mystery at

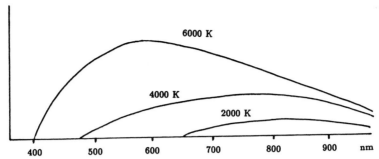

FIGURE 1.10 Shown on an arbitrary scale is the intensity, as a function of wavelength, of the radiation emitted by a black body (a black body is a perfect absorber of radiation) for three different temperatures. The temperatures are in degrees kelvin (1 K is equal to 1°C, and they are measured from absolute zero, or − 273.15°C; this means, for example, that the freezing point of water, at 0°C, is at 273.15 K).

the close of the nineteenth century. After several theoretical attempts by a number of other researchers, Planck finally succeeded in calculating the radiation distribution, but only by making the seemingly absurd assumption that the light must be emitted by myriads of radiators in the walls of the black body only in tiny discrete lumps, or quanta. He was forced to assume that the amount of energy in each such quantum depends only on the wavelength, and obeys the following relation:

$$E = hc/\lambda, \tag{1.9}$$

where h is a constant equal to 6.625×10^{-34} J · sec, c is the speed of light in a vacuum, λ is the wavelength of the radiation, and E is the energy in joules, J.

In the hands of Planck, the quantum was nothing more than a strange assumption that was necessary to derive the black body radiation curve, and had no general physical significance. It was Albert Einstein who, in 1905, brought the quantum into the mainstream of scientific thought, and he did it by suggesting that light was not to be considered just waves after all, but that it also had to be provided with particle properties. In other words, Newton's corpuscular theory of light, which had been apparently put to rest by Young, Fresnel, and all the wave theorists, had, in a new sense, to be resurrected. This is not to say that the wave theory is all wrong—light has wave properties to be sure—but light turns out to be more complex than that; it also has particle properties.

The phenomenon that suggested the particle picture of light to Einstein was the interaction of light with a solid surface, the photoelectric effect, which was first noticed by Heinrich Hertz (1857–1894) when he was in the process of discovering radio waves ca. 1887. The photoelectric

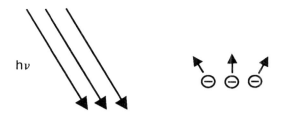

FIGURE 1.11   The photoelectric effect in schematic form. Radiation with energy equal to $h\nu$ is allowed to fall on a surface and electrons are emitted.

effect is when light strikes a solid in a vacuum and electrons are ejected from the surface (Figure 1.11). (There are other forms of the photoelectric effect involving gases and even liquids instead of solids, but we will not be concerned with them here.) Various metals and semiconductors are more or less efficient photoelectrically, the alkali metal, cesium, being especially so. The quantity and energy of the electrons emitted from the surface can be measured because they form an electric current. A surprising result, and the one that Einstein finally explained, is that the energy of the electrons ejected from the surface depends not at all on the *intensity* of the light falling on the metal surface, but only on its *wavelength*. This cannot be consistent with a pure wave model for light because the amplitude of the waves would certainly increase the energy of the particles (electrons) with which they interact. (Compare this with the effect of large and small ocean waves "exciting" pebbles on the beach.)

To escape from this dilemma, Einstein enlisted the help of Planck's quanta, whose energy, as we have seen, depends only on their wavelength. Einstein theorized that, when interacting with matter, light, which otherwise acts like waves, acts as though it consists of particles, which he called *photons*, each of which has the energy of one of Planck's oscillators, hc/λ. When the photons interact with the electrons in the surface, their energy is transferred to the electrons, which neatly explains the wavelength dependence of the photoelectrons' energy. (Einstein received the Nobel Prize in 1921, partially for this discovery.)

To describe light, we must therefore not only use a wave model but a particle model as well—waves during the transit of light from one place to another and particles when light is either emitted or absorbed. This dual nature of light has turned out to be not as unique as it was thought at first. The constituents of ordinary matter, that is, protons, neutrons, and electrons, ordinarily thought of as particles, are now known also to have both particle and wave properties. In the case of these matter waves, the wavelength is given by

$$\lambda = h/p, \tag{1.10}$$

where p is the momentum (mass multiplied by the velocity) of the particle. This expression was first suggested by Louis de Broglie (1892–1987) in 1924 and such waves are often called de Broglie waves. Notice that their wavelength varies inversely with the momentum, p, of the photon. According to Einstein's theory of relativity, the momentum of a photon is related to its energy, E, by the expression

$$E = pc, \tag{1.11}$$

where c is the speed of light. Combining Eqs. (1.10) and (1.11), we again have Eq. (1.9):

$$E = hc/\lambda. \tag{1.12}$$

1.8 THE PRODUCTION OF LIGHT

As we have pointed out, the photon nature of light is encountered whenever light is emitted or absorbed. To emit light, a source of energy must be present such that the photons can be provided with their necessary energy. In most cases, in fact in practically every case we shall be interested in, this energy is provided by electrons.

1.8.1 The Energy of Bound Electrons

When they are attached, or *bound*, to atoms and molecules, electrons can exist only with definite amounts of energy. The simplest picture of the atom that exhibits these levels of electron energy is the Bohr model, put forth in 1912 by Niels H. D. Bohr (1885–1962). He assumed that the atom was, as already (1911) had been suggested by Ernest Rutherford (1871–1937), formed resembling the solar system, with light, negative electrons revolving in circular orbits around a heavy, positive nucleus. Bohr also made the logical assumption that in place of the gravitational attraction that binds the planets to the sun, the atom is held together by the electrical attraction between the negative electrons and the positive nucleus.

Where Bohr's approach was revolutionary was in his next two assumptions. The first of these was that the angular momenta, p_ϕ, of the electrons in their orbits (angular momentum is equal to the time rate of change of the angle around the orbit times the mass of the electron times the radius of the orbit) were confined to integral multiples, l, of $h/2\pi$, or $\hbar$:

$$p_\phi = l\hbar. \tag{1.13}$$

Calculation from this assumption has the result that an electron circling a positive nucleus could exist only in orbits of definite energy according to the following law:

$$E = RZ^2/n^2, \tag{1.14}$$

where Z is the number of positive charges in the nucleus (one for hydrogen) and R is a constant called the Rydberg, which Bohr derived from fundamental quantities, given by

$$R = 2\pi^2 m e^4/ch^3, \tag{1.15}$$

where m is the mass of the electron, e is its electric charge, c is the speed of light, and h is Planck's constant. Thus the electrons orbit the nucleus in a manner similar to artificial satellites around the earth, but unlike such satellites, they are confined to orbits with definite amounts of energy (see Figure 1.12). It is a bit as though the NASA shuttle could be in an orbit at 100, 150, 200, and 300 km above the earth and nowhere in between.

The second of Bohr's revolutionary assumptions was that the rotating electrons do not radiate as would be expected according to classical physics (like the electrons vibrating back and forth in a radio broadcast antenna), but give rise to radiation only whey they "jump" from one orbit to another. If such a jump results in a decrease in electron energy, a photon is generally created whose energy equals the difference in the electron's energy. In other words, if the change in electron energy is ΔE, then the photon produced has a wavelength, λ, such that

$$\lambda = hc/\Delta E. \tag{1.16}$$

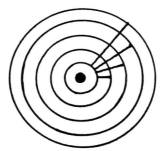

FIGURE 1.12   Shown (not to scale) are some of the Bohr orbits of the hydrogen atom.

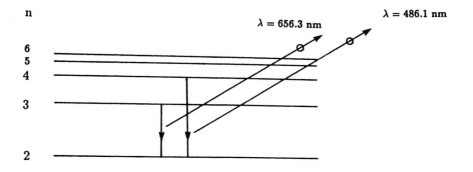

FIGURE 1.13   The first energy levels of the hydrogen atom are shown here, using an arbitrary energy scale. When the electrons fall from a higher to a lower level, radiation is emitted. Here *n* is the "principal quantum number" of the energy levels, and the spectral emissions at 486.1 nm (blue) and 656.3 nm (red) are shown. They are the brightest spectral "lines" of hydrogen in the visible region.

By measuring the wavelength of the light emitted by the atom, the spacing of its energy levels can therefore be determined. Collecting the results of such a study produces an energy level diagram, such as is shown for hydrogen in Figure 1.13. Here some of the jumps, or *transitions*, between energy levels are shown, together with the photons produced.

After de Broglie's discovery that electrons had wave properties, a new form of the theory of the atom was developed by E. Schrödinger (1887–1961), P. A. M. Dirac (1902–1984), and others; this new theory allowed more complete calculations than were possible using Bohr's model. This is now called *quantum mechanics* and it has provided an unparalleled expansion of our knowledge of atoms, molecules, and virtually all aspects of the natural world.

For atoms with more than one electron, the situation is more complex, but under ordinary conditions the electron transitions are performed only by the outer, or *valence*, electrons, usually numbering between one and four, so the appearance of the energy level diagram is not very different from that of hydrogen.

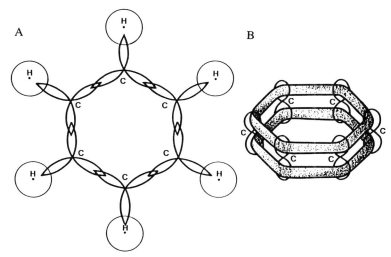

FIGURE 1.14   (A) Scheme of the σ electron orbitals of benzene; (B) scheme of the π electron orbitals of benzene. From Birks (1970). Copyright 1970, Wiley (Interscience); used with the permission of Wiley (Interscience).

1.8.2 Electron Spin

The appearance in the spectrum of sodium and other elements of closely spaced energy levels suggested that the electron acts like a tiny magnet and that its orientation in the atoms might be limited or quantized to certain directions, explaining the small differences in energy. Electrons are charged with negative electricity, and in 1925 the Dutch physicists S. Goudsmit and G. E. Uhlenbeck proposed that the magnetic behavior of electrons could be explained if they were spinning. A spinning charged body acts as though it is circled with an electric current, and an electric current has been known to produce a magnetic field since the time of Oerstead (1777–1851).

According to quantum mechanics, the angular momentum of the electron spin is $\hbar$ multiplied by $\frac{1}{2}$.

1.9 ELECTRONIC STRUCTURE OF ATOMS AND MOLECULES

Because carbon is essential for life, let us take it as an example for the qualitative description of molecular structure formation. The nucleus of the ^{12}C isotope contains 6 protons and 6 neutrons.

As shown in Section 1.8.1 and Figures 1.12 and 1.13, the principal energy levels of electrons circling around nuclei are characterized by a discrete number, the so-called *principal quantum number, n*, whose values

TABLE 1.1   The Electromagnetic Spectrum

Nature of radiation	λ (nm) (typical values)	Energy of einstein (J)	Origin of the radiation
Gamma rays	0.0010	1.2×10^{11}	Nuclear reactions
X-Rays	0.10	1.2×10^{9}	Transitions of inner atomic electrons
Solar ultraviolet	300	4×10^{5}	Transitions of outer atomic electrons
Visible	400	3×10^{5}	
	800	1.5×10^{5}	
Infrared	10^{3}	2×10^{5}	Molecular vibrations
Far-infrared	10^{4}	2×10^{4}	Molecular rotations
Radar	10^{8}	2	Oscillation of free electrons

can be 1, 2, 3, and so on. Each principal quantum number defining an atomic shell *can accommodate at most $2n^2$ electrons.* The electron's *orbital angular momentum quantum number, l,* must be taken into account *to accommodate all the electrons.* Thus, l can take the following values: 0, 1, 2, 3 . . . $(n - 1)$, to which correspond atomic *sub*shells: s $(l = 0)$, p $(l = 1)$, d $(l = 2)$. . . . The principal quantum number and the sublevels s, p, and d must be specified when describing the state of an electron. For example, an element such as carbon which has six electrons, $n = 1$, accommodates two electrons in an s subshell which is designated $1s$. Consequently, the *second* atomic shell, $n = 2$, must accommodate the remaining four electrons in the remaining two subshells s and p, noted as $2s$ and $2p$. These electrons are confined to specific regions of the second shell called *atomic orbitals* containing a maximum of two electrons each. The $1s$ level contains only one orbital with two electrons. These electrons are designated as $1s^2$. The four remaining electrons are therefore found in the $2s$ and $2p$ orbitals. According to the above nomenclature, these four electrons are defined as $2s^2$ and $2p^2$. The number of orbitals in a subshell is equal to $2l + 1$. As a result there are three p orbitals. The arrangement of electrons in atomic orbitals is referred to as the atom's *electron configuration.* The *Aufbau* (or building-up) principle of construction of electronic configuration for the carbon atom using the above rules is given as $1s^2\ 2s^2\ 2p^2$. The s orbitals have spherical symmetry whereas the three p orbitals are aligned along the x, y, and z axes of a cartesian coordinate system with nodes at the origin. The electronic configuration of the ground state of the C atom is $1s^2\ 2s^2\ 2p^2$. In forming compounds, a $2s$ electron is promoted to a $2p$ state, so that the electronic configuration of a C atom "prepared for binding" is $1s^2\ 2s^1\ 2p^3$. The four valence electrons $(2s^1\ 2p^3)$ have three possible configurations in forming molecules, corresponding to single bonds (CH_4, methane), double bonds (C_2H_4, ethylene; C_6H_6, benzene), or triple bonds (C_2H_2, acetylene). In all the cases, each carbon has a full octet of electrons.

In the tetragonal, or sp^3, hybridization, the four electron orbitals combine, leading to four equivalent hybrid orbitals directed toward the corners of a regular tetrahedron centered on the C nucleus. These four electron orbitals associate with those of other atoms to give a saturated molecule, such as methane (CH_4).

In the trigonal, or sp^2, hybridization, one of the original p orbitals (say p_z) is unchanged, and three equivalent hybrid orbitals are produced by mixing s, p_x, and p_y. These three sp^2 hybrid orbitals are in the xy plane and they are at 120° to each other. This configuration provides the hexagonal ring structure of benzene and the polycyclic aromatic hydrocarbons. The hybrid orbitals, which are symmetrical about their bonding axes and about the molecular plane, are known as σ electrons, and the bonds are called σ bonds. In benzene the σ sp^2 hybrid orbitals of C interact with each other and with the 1s orbitals of H to form the localized C—C and C—H σ bonds (Figure 1.14A).

The p_z atomic orbital of each C atom is unchanged by the sp^2 hybridization and contains the so-called π electrons. In benzene, the six p atomic orbitals of C interact to produce C—C π bonds. Unlike the σ electrons, the π electrons are delocalized. The six atomic orbitals interact to form six delocalized π molecular orbitals (Figure 1.14B). All aromatic molecules contain similar systems of delocalized π electrons. It is the reactivity of the excited states of these π electron systems that will be of interest for the photobiologist.

In the digonal, or sp, hybridization, the p_y and p_z orbitals of the C atom are unchanged, and two equivalent hybrid σ (or sp) orbitals are produced by mixing s and p_x. These σ orbitals are directed at 180° to each other along the x axis, which is the line formed by the intersection of the nodal planes of the two p orbitals. This configuration occurs in the linear molecule acetylene (H—C≡C—H). Beside the C—H and C—C σ bonds, the p_y and p_z atomic orbitals of each C atom are paired to form two C—C π bonds.

1.10 THE PLANCK EQUATION AND PHOTOCHEMICAL REACTIONS

The energy characteristic of a monochromatic beam of radiation is related to its wavelength, λ, or its frequency, ν, as expressed in Eq. (1.9). When dealing with photochemical reactions, the unit most commonly used is the mole, consisting of N molecules, where N is Avogadro's number, or 6.023×10^{23}. For radiation, a similar unit is used, called the *photon mole* (previously called the einstein) containing N photons. It follows that 1 photon mole has the radiant energy equal to $Nh\nu$.

$$1 \text{ photon mole} = 1.19629 \times 10^8 \text{ J}/\lambda, \qquad (1.17)$$

where λ is expressed in nanometers.

As described by Eqs. (1.17) and (1.9), the energy of the photon mole decreases enormously as we pass from the shortest wavelengths of the electromagnetic spectrum to the longest (as shown in Table 1.1), of which the near-ultraviolet and visible regions are the most important in photobiology. When these wavelengths are absorbed by biomolecules, the outermost electrons, e.g., those involved in atom binding, are those affected. As a result chemical changes can occur during light absorption.

Bibliography

Birks, J. B. (1970). "Photophysics of Aromatic Molecules," pp. 1–28. Wiley (Interscience), London.

Born, M., and Wolf, E. (1970). "Principles of Optics," p. xxi. Pergamon, Oxford.

Debus, A. G., ed. (1968). "World Who's Who in Science," p. 523. Marqis-Who's Who Inc., Chicago.

de Fermat, P. (1679) (post). "Varia Opera Mathematica."

CHAPTER 2

Pathways of Molecular Excitation and Deactivation

2.1 ELECTRONIC TRANSITIONS

The possible electron distributions defining the molecular orbitals and energy levels of molecules in their ground and excited states are given by the approximate solutions of a generalized Schrödinger equation. The latter takes into account electrostatic attractions and repulsions of protons and electrons, internuclear vibrations, and the rotational movement of molecules as well as magnetic interactions due to electron and nuclear spins and orbital motion. Molecular orbitals can contain no more than two electrons. A transition between two states of a molecule corresponds to the movement of one electron from one orbital to another. For instance, an electron in a π molecular orbital can be transfered to an excited π orbital (π^*) and the corresponding transition will be noted $\pi \rightarrow \pi^*$. Similarly, a σ electron can be promoted to a σ^* orbital and the transition will be noted $\sigma \rightarrow \sigma^*$. These transitions can be induced by absorption of radiation, provided the energy of the incident light quanta (given by the

Planck equation; see Chapter 1) is equal to the energy difference between two electronic levels.

A molecule will interact with the electromagnetic field and absorb (or create) a photon of frequency ν only if it possesses, at least transiently, an electric dipole oscillating at this frequency. The molecular transient polarization can be explained as follows. When a molecule such as benzene, with no permanent dipole moment, is placed in an external electric field, the field shifts the centers of positive and negative charges, thus polarizing the molecule and giving an induced electric dipole moment. The energy exchange (coupling) between the electric vector of the electromagnetic wave and this dipole will occur by a resonance phenomenon and will be called "resonance transfer."

In the case of absorption and emission spectra, this transient dipole is called the transition dipole moment. Its value can be determined from the quantum mechanical calculation of the change in the charge distribution in going from the initial state to the final state, and this calculation takes into account the wavefunctions of these two states. A high probability of electronic transition (e.g., characterized by an intense absorption band; see below) implies a large transition dipole moment.

Intuitively, Figures 1.14A and 1.14B suggest, and calculation confirms, that the energy required for a $\sigma \rightarrow \sigma^*$ transition (i.e., to excite an electron in bonding orbitals constituting the backbone of molecules) will be higher than that required for $\pi \rightarrow \pi^*$ transitions (i.e., excitation of delocalized electrons). As a result, $\sigma \rightarrow \sigma^*$ transitions need photons of wavelengths shorter than 200 nm, whereas $\pi \rightarrow \pi^*$ transitions can be observed with photons of nearer ultraviolet and visible radiations (in molecules with a large number of delocalized π electrons). This is the case for most biological molecules. As a result, photobiological reactions will most often be concerned with $\pi \rightarrow \pi^*$ transitions.

2.2 THE MULTIPLICITY OF THE ELECTRONIC STATES: ANOTHER IMPORTANT FACTOR OF THE PROBABILITY OF TRANSITION

According to the Pauli exclusion principle, in the ground state (unexcited) molecule, the electron spins are paired to give the ground electronic singlet state of the molecule. If a π electron is excited without change of spin, the resultant excited electronic state of the molecule is called a singlet state. If a π electron is excited and undergoes a spin reversal during the transition, the resultant excited electronic state of the molecule is a paramagnetic triplet state. The terms "singlet" and "triplet" represent the multiplicity of the electronic state. The latter is equal to $2S + 1$, where S is the resulting spin of the two electrons in the molecular orbital: $S = \frac{1}{2} \pm \frac{1}{2}$. The multiplicity describes the degree of degeneracy of the electronic state in the absence of a magnetic field. Application of a magnetic field to the molecule does not perturb a singlet state but it removes the degeneracy of a triplet state and splits it into three distinct Zeeman levels (this is why this state is called a triplet state).

Electric dipole transitions between electronic states of different multiplicity are forbidden. This *multiplicity selection rule* plays a key role in photophysical processes of aromatic molecules. The intensity of an absorption transition from the singlet ground state to the first excited triplet state is $\approx 10^{-8}$ that of the spin-allowed absorption transition to the excited singlet state. Because an electron has spin angular momentum, and because moving charges generate magnetic fields, an electron has a magnetic moment that arises from its spin. Similarly, an electron with orbital angular momentum is a circulating current, also generating a magnetic moment. The interaction of the spin and orbital magnetic moments is called *spin–orbit coupling*. The strength of the coupling, and its effect on the energy levels of the molecule, depend on the relative orientation of the two angular momenta. Small amounts of triplet wavefunctions are therefore mixed with the singlet wavefunctions, and vice versa, and a small, but finite, singlet–triplet transition probability is obtained.

All molecules of biological interest are in the singlet state in their ground state, with the notable exception of oxygen (O_2) (modern nomenclature favors dioxygen), which is a ground state triplet molecule. To all excited singlet states can be associated a triplet state. However, as in the case of atoms, the energy of the triplet state is lower than that of the corresponding singlet state. For the sake of clarity we will adopt the following nomenclature: singlet states of a molecule M will be noted ^1M_n and triplet states, ^3M_n. The state multiplicity is given by the superscript and the state (ground or excited) by the subscript. Thus for a ground state, $n = 0$.

2.3 MOLECULAR EXCITATION AND DEACTIVATION: THE JABLONSKI DIAGRAM

2.3.1 Absorption of Light by Molecules

As a result of light absorption by their singlet ground state ($^1M_0$), aromatic molecules (M) are promoted to singlet excited states. The excited singlet state (^1M_n) that is reached on light absorption depends on the incident quantum energy:

$$h\nu = E(^1M_n) - E(^1M_0), \tag{2.1}$$

where E is the energy of the corresponding state. As a result of this relationship, ν can only have discrete values (e.g., they are quantized).

The vibrational and rotational motions of a molecule are also quantized, for example, vibrational or rotational energy can be taken up or lost only in discrete quantum units. Thus the total energy (E) of a particular state of a molecule may be represented as the sum of the electronic excitation energy, E_e, the vibrational energy, E_v, and the rotational energy, E_r,

$$E = E_e + E_v + E_r. \tag{2.2}$$

Equation (2.1) may therefore be written as follows:

$$h\nu = \Delta E_e + \Delta E_v + \Delta E_r; \tag{2.3}$$

for example, the total change of energy on absorption of a quantum of light is the sum of the changes in the number of quanta of electronic, vibrational, and rotational levels. The size of the vibrational quanta is less than that of the quanta required to excite an electron electronically, and the size of the rotational quanta is still smaller. It is convenient to represent the various energy states available to the molecule by a simple energy level diagram, shown in Figure 2.1, known as the *Jablonski diagram*. Each electronic level is split into a series of vibrational levels, and each vibrational level is split into a series of closely spaced rotational levels.

2.3.2 The Extinction Coefficient: The Beer–Lambert Law

If a monochromatic beam of light of fluence rate I_0 crosses a specimen of thickness l cm, containing n molecules of the absorbing species per cm^3, the fluence rate I of the emergent beam

$$I = I_0 \exp{-\sigma n l} = I_0 \exp{-\mu l} \tag{2.4}$$

is a function of the molecular absorption cross section σ (cm^2) or of the absorption coefficient μ (cm^{-1}). These parameters are expressed in terms of the molar extinction coefficient ε, defined by

$$I = I_0 10^{-\varepsilon[M]l}, \tag{2.5}$$

where [M] is the molar concentration of the absorbing species and ε is expressed in units of $M^{-1} \cdot cm^{-1}$. Because $n = N[M] \times 10^{-3}$, where N (= 6.02×10^{23}) is Avogadro's number, comparison of Eqs. (2.4) and (2.5) gives

$$\sigma = 2.3\varepsilon/N = 3.8 \times 10^{-21}\varepsilon$$

With molecules such as porphyrins having a molecular mass of several hundred Daltons, the maximum values observed for ε are about $10^5 \, M^{-1}$ cm^{-1} at 400 nm. The effective cross section of such a molecule is about 10^{-16} cm^2; for instance, the molecules behave like opaque particles having

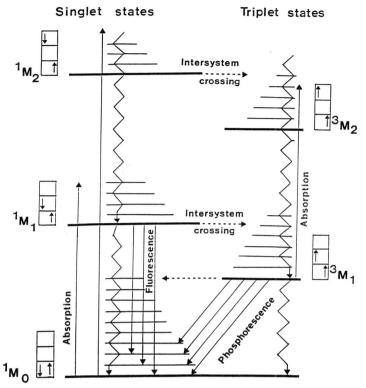

FIGURE 2.1 Jablonski diagram showing the possible transitions in aromatic molecules on excitation with light.

dimensions of about 0.1 nm (10^{-8} cm). By contrast, the aromatic amino acid tryptophan has a molar extinction coefficient of about 5800 M^{-1} cm^{-1} at 280 nm, hence with considerably less effective absorption cross section. As explained in a preceding section, the intensity of an absorption band (that is, its molar extinction coefficient) is directly related to the transition dipole moment of the transition.

The quantity $\log_{10}(I_0/I)$ is the absorbance (A) of the solution— formerly called optical density (O.D.)—and is proportional to the concentration of the absorbing species. If more than one absorbing species is present in the solution, the total A (A_T) is simply the sum of the individual optical densities (A_i).

The amount of light absorbed by the sample (I_{abs}) can be easily obtained from Eq. (2.5):

$$I_{abs} = I_0 - I = I_0(1 - 10^{-\varepsilon[M]l}) \quad \text{or} \quad I_{abs} = I_0(1 - 10^{-A}) \tag{2.6}$$

as a result, for weakly absorbing solutions, for example, $A \le 0.1$, the quantity 10^{-A} can be approximated by $1 - 2.3A$ and

$$I_{abs} \approx 2.3 I_0 A. \tag{2.7}$$

Absorption spectra simply consist of a plot of A against the wavelength of the absorbed light.

2.3.3 The Franck–Condon Principle

The bold horizontal lines of the Jablonski diagram (Figure 2.1) corresponds to the minimum of the potential energy diagram of the molecule. In a diatomic molecule, the potential energy diagram plots the electronic and vibrational energies of the molecule as a function of the nuclear separation r, and the wavefunctions of the vibrational modes approximate those of a harmonic oscillator (Figure 2.2). A similar diagram can be used for aromatic molecules.

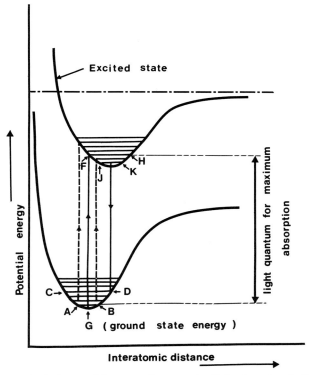

FIGURE 2.2 Hypothetical potential energy diagram of a diatomic molecule illustrating the Franck–Condon principle.

The Franck–Condon principle states that, because the time required for an electronic absorption is negligible ($\approx 10^{-15}$ sec; e.g., the period of vibration of the light waves) compared to that of nuclear motion, the most probable vibrational transition is one that involves no change in the nuclear coordinates. This transition, which is refered to as the Franck–Condon maximum, represents a vertical transition on the potential energy diagram (Figure 2.2).

The orbital occupied by the electron in the excited state is generally more extended than in the ground state. As a result, the energy corresponding to the equilibrium position in the excited state is at a greater interatomic distance than that in the ground state.

According to the Franck–Condon principle, the most probable transition (e.g., the most intense absorption band) is the one from the midpoint (G) of the zero-vibrational level (AB) of the ground state molecule (mean interatomic distance) to F on the potential energy diagram of the excited state at the third vibrational level. However, other transitions are possible, as shown by dashed lines in Figure 2.2 at both longer and shorter wavelengths, but with a smaller probability because of originating from ground state molecules in a less probable higher vibrational level at room temperature.

2.3.4 Deactivation from Singlet Excited States

Any molecule in an electronic excited state can dissipate the excitation energy by both radiative and radiationless processes. When excited to upper electronic excited states, a molecule undergoes a process called *internal conversion*, whereby, in 10^{-12}–10^{-13} sec, the molecule passes from a low vibrational level of the upper state to a high vibrational level of the lower excited electronic state (wavy lines in the Jablonski diagram). After internal conversion, the molecule rapidly loses its excess vibrational energy by collision with solvent molecules. As a result of all these processes, molecules raised to levels higher than the lowest vibrational level of the first excited state (e.g., level 0 of $^1M_1$) rapidly fall to the latter. Some molecules can undergo a photochemical reaction when excited to upper singlet states, but such a reaction (e.g., photodissociation) must take place rapidly to compete with the internal conversion and loss of vibrational energy by collisions. If one considers Figure 2.2, after excitation to a vibrational level such as FH, the molecule rapidly loses excess vibrational energy and falls to the lowest vibrational level JK, from where it can undergo a transition back to the ground state with the emission of radiation. The Franck–Condon principle tells us that the most intense emission band will be that corresponding to the vertical line drawn from the midpoint of JK to D. Thus, the important result of the application of the Franck–Condon principle to the fluorescence emission is that the fluorescence spectrum is shifted to the red with respect to the absorption spectrum. This is also called the Stokes Law, which states that the wavelength of the fluorescence is always longer than that of the exciting light (e.g., the so-called Stokes shift). Excited molecules can emit fluorescence by returning to any one of the vibrational levels of the ground state adjacent to CD. Because the molecules have absorbed photons to be raised in an electronically excited state, it is interesting to know how much light will be emitted in proportion of the absorbed energy. *This brings us to a very important quantity: the fluorescence quantum yield.*

The fluorescence quantum yield (Φ_F) is the ratio of the number of photons emitted to the number of photons absorbed. If all the molecules that have absorbed light return to the ground state by emitting a fluorescence, the quantum yield is unity. If a proportion of the excited molecules returns to the ground state by internal conversion (see Section 2.3.4.2), or undergoes a photochemical reaction either in the first or a higher excited state, the quantum yield of fluorescence will then be less than unity and could be almost zero.

At room temperature absorption occurs almost exclusively from the first vibrational level of the ground state, whereas the emission arises from the zero vibrational level of the first excited state; thus only one transition, namely, the 0–0 transition, can be common to both the absorption and emission spectra.

2.3.4.1 Fluorescence Excitation Spectrum

Because fluorescence is emitted from the lowest vibrational level of the first excited singlet state, the shape of the fluorescence emission spectrum of molecule M is independent of the excitation wavelength. The excitation spectrum must match the absorption spectrum but the emission transition will always be the $^1M_1 \rightarrow {}^1M_0$ transition. Because this emission generally originates from a π^* excited state, it will be said to be of $\pi^* \rightarrow \pi$ nature. In practice, if the fluorescence excitation spectrum is not the absorption spectrum of the fluorophore, this means that other processes, such as energy transfers or photochemical reactions (see below), take place. Conversely, if the fluorescence spectrum changes with the excitation wavelength, the presence of more than one fluorescent species should be suspected. It should be noted that the shape and position of the fluorescence also depends on other factors such as solvation or deprotonation in the excited state.

2.3.4.2 Deactivation of the First Singlet Excited State by Intersystem Crossing to the Triplet State

Population of the triplet state of a molecule M by direct light absorption is impractical because of the very weak probability of the $^1M_0 \rightarrow {}^3M_1$

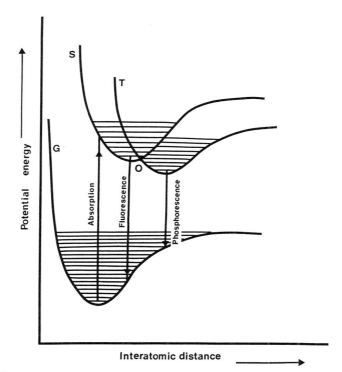

FIGURE 2.3 Potential energies of a hypothetical molecule in the ground (G), first singlet excited (S), and first triplet (T) states, illustrating the occurrence of intersystem crossing for populating the triplet state. Note that this scheme also illustrates the Stokes shift of the fluorescence and phosphorescence emissions relative to the absorption.

transitions (see Section 2.2). However, there is a much more efficient way of populating the triplet state via the $^1M_1$ state of a molecule (M), by an internal conversion process, which is called *intersystem crossing*, as indicated in the Jablonski diagram (Figure 2.1).

Let us consider (Figure 2.3) a potential energy diagram similar to that given in Figure 2.2 but containing the potential energy plot for the triplet (T) state in addition to the potential energy plots for ground (G) and first excited singlet (S) states.

The energy of the triplet state ($^3M_1$) is well below that of the $^1M_1$ state, and the curves for the $^3M_1$ and $^1M_1$ states cross at some point O. According to the Franck–Condon principle, on excitation *(hv)*, the molecule in an excited vibrational state of $^1M_1$ undergoes internal conversion toward the zero vibrational level of $^1M_1$, but when reaching point O, it crosses over to a vibrational level of the triplet state, thereby populating the first excited triplet state, $^3M_1$. As shown in the Jablonski diagram, the same process can take place from upper triplet and singlet excited states. As for upper excited singlet states, internal conversion from upper excited triplet states to the first excited triplet state $^3M_1$ is quite fast ($\sim 10^{-13}$ sec) and the lifetime of the triplet state molecule is governed by $^3M_1$ deactivation (see below).

2.3.5 The Decay of the $^1M_1$ and $^3M_1$ States

2.3.5.1 Effect of Multiplicity of States on Lifetimes

In dilute solutions, a molecule in the $^1M_1$ or $^3M_1$ state returns to the ground state independently of all other molecules. The decay is a monomolecular process whose kinetic behavior is quite similar to the decay of radioactive material.

Thus at time t the number of molecules emitting per second is proportional to the number n of excited molecules (singlets or triplets) present at that time; for example,

$$dn/dt = k_d n. \tag{2.8}$$

Integration leads to Eq. (2.9):

$$n = n_o \exp - k_d t, \tag{2.9}$$

where n_o is the number of excited molecules at time origin and k_d is the rate constant for the decay of these excited molecules (singlet or triplet). In the case of excited molecules that emit fluorescence with $\Phi_F = 1$, the rate of emission of light, R, is given by:

$$R = -dn/dt = k_F n_o \exp - k_F t = Q \exp - k_F t. \tag{2.10}$$

The intensity decays exponentially with a mean lifetime τ_F defined by

$$\tau_F = 1/k_F, \tag{2.11}$$

which is the time required for the fluorescence intensity to fall to $1/e$ of its initial value. The parameter k_F is the first-order rate constant for the process of fluorescence emission.

These transient conditions have a very practical application. During excitation of negligible duration with flashes of light absorbable by the fluorescent molecule, an initial concentration n_o of excited molecules is produced. The study of the fluorescence intensity response as a function of time leads to the molecular fluorescence lifetime. This very important parameter is sensitive to many environmental factors and molecular interactions; it is therefore a reporter of choice to understand and/or control subtle molecular perturbations.

When applying the same kinetic analysis to the triplet states, the triplet state lifetime (τ_T) is defined by

$$\tau_T = 1/k_T, \tag{2.12}$$

where k_T is now the first-order rate constant for the triplet decay.

A most important experimental observation is that $k_F \gg k_T$. This is because transitions between two states of the same multiplicity are allowed. This is the case for the fluorescence emission (two singlet states). On the other hand, the triplet $\rightarrow$ singlet transition responsible for the $^3M_1$ deactivation is "spin forbidden." As a consequence, it occurs at a much slower rate.

Typically, at low or room temperature, in rigid or fluid solutions the fluorescence lifetime of aromatic molecules is generally in the nanosecond time range, whereas the triplet lifetime can be in the microsecond to second time range. Interestingly, depending on experimental conditions, deactivation of the same triplet can be either radiative or nonradiative. The long-lived triplet emission is called phosphorescence. Phosphorescence is generally observed at low temperature in frozen solutions at 77 K (i.e., the temperature of liquid nitrogen) or at room temperature in very rigid solutions. Tryptophan provides an interesting example because its phosphorescence emission can be detected at 77 K, with a lifetime of about 7 sec, at room temperature (lifetime $\approx$ 0.1 sec) in plastics or boric acid glasses or in some proteins (e.g., keratin, alcohol dehydrogenase), where the Trp residue motions are blocked at room temperature in the tertiary or quaternary structure of the proteins.

In contrast to the fluorescence lifetime, the triplet state lifetime is thus very sensitive to viscosity. In fluid solutions, all $\pi^* \rightarrow \pi$ triplet deactivations are radiationless transitions. In some instances triplet states of aromatic molecules formed by a transition from a nonbinding lone-pair electron orbital of a *heteroatom* (noncarbon atom) such as oxygen to a π^* electronic orbital (the so-called $n \rightarrow \pi^*$ transitions) can be phosphorescent in fluid solution at room temperature. Benzophenone is the classical example.

As clearly suggested by the Jablonski diagram, the phosphorescence lies at longer wavelengths with respect to the fluorescence spectrum. For the sake of clarity the relative spectral positions of absorption, fluorescence, and phosphorescence spectra in aqueous media are presented for tryptophan in Figure 2.4. The width of electronic absorption bands in these solutions is the result of an unresolved vibrational structure due to simultaneous excitation of vibrational transitions during electronic excitation. However, some aromatic molecules such as anthracene, which in its ground or excited state does not interact with molecules of nonpolar solvents such as cyclohexane, can have well-resolved vibrational structure in their absorption spectra.

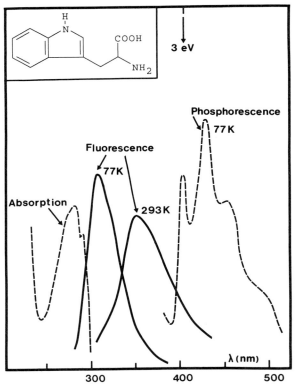

FIGURE 2.4   Absorption, fluorescence, and phosphorescence spectra of tryptophan, an aromatic amino acid frequently encountered in proteins. Note that the absorption spectrum extends beyond 290 nm, corresponding to the beginning of the solar UV spectrum at sea level. The onset of the phosphorescence spectrum at about 3 eV determines the energy of this state with respect to the ground state.

2.3.5.2   Delayed Fluorescence and Phosphorescence

The simplest way of populating the fluorescent $^1M_1$ state is, of course, light absorption by the ground state (e.g., 10^{-15} sec). However, other ways of populating the $^1M_1$ state exist.

1. The triplet state by the $^3M_1 \rightarrow {}^1M_1$ reverse intersystem crossing: Because the $^3M_1$ lifetime is much longer than that of $^1M_1$, a long-lived (delayed) fluorescence will be observed. It requires a small energy separation for the $^3T_1$ and $^1S_1$ levels, as in the case of eosin dye. This kind of delayed fluorescence is said to be of type E (for eosin).
2. Recombination of electrons and aromatic radical-cations produced by photoionization of chromophores: A good example is pro-

vided by aromatic hydrocarbons such as naphthalene (N) in frozen alcoholic solutions photoionized at 77 K by ultraviolet light via biphotonic processes involving the sequence

$$^1N_0 + h\nu_1 \longrightarrow {}^1N_1 \longrightarrow {}^3N_1 + h\nu_2 \longrightarrow {}^3N_n. \qquad (2.13)$$

If the energy of the second photon absorbed by the molecule in its $^3N_1$ state is sufficient to reach a very energetic ^3N_n state (see the Jablonski diagram, Figure 2.1), electron ejection competes with internal conversion to the $^3N_1$ state. The free electrons are temporarily trapped in defects of the glassy matrix, from which they are released by thermal activation and/or tunneling. Then they recombine with the parent radical-cation of the aromatic hydrocarbon. This recombination not only produces a delayed fluorescence but also a delayed phosphorescence. The aromatic amino acid tryptophan, either free or incorporated into proteins, gives rise to the same phenomena in various aqueous media at 77 K.

It is important to note that the reaction sequence (Eq. 2.13) is of great practical importance. Indeed, at room temperature in fluid solution, when, in most cases, no phosphorescence can be detected, the $^3M_1 \rightarrow {}^3M_n$ absorption spectrum allows the characterization and study of triplet states of photosensitizing molecules (M) of biological importance, as will be shown in Chapter 4. Under these conditions, $^1M_1 \rightarrow {}^1M_n$ absorptions are generally not observed because of the very short lifetime of the $^1M_1$ state.

2.3.6 Quantum Yields of Fluorescence and Triplet Formation

2.3.6.1 Fluorescence Quantum Yield

When an optically thin solution (e.g., the absorbance of the solution is small) of a fluorophore (F; molar extinction coefficient, ε) contained in a cuvette with optical pathway, l, is irradiated with suitable wavelengths (fluence rate, I_0), a constant fluorescent intensity is observed. Thus, an equilibrium has been reached between decay and production of the first singlet excited states of F; e.g., $^1F_1$.

A simple kinetic scheme can account for these photostationary conditions because the Jablonski diagram (Figure 2.1) provides us with all monomolecular pathways leading to $^1F_1$ deactivation.

process		rate
$^1F_0 \longrightarrow {}^1F_1$		$2303\, I_0\varepsilon[F]$
$^1F_1 \longrightarrow {}^1F_0$	(fluorescence)	$k_F[^1F_1]$
$^1F_1 \longrightarrow {}^3F_1$	(intersystem crossing)	$k_{ISC}[^1F_1]$
$^1F_1 \longrightarrow {}^1F_0$	(internal conversion)	$k_{IC}[^1F_1]$

For solutions having very low absorbances, it can be considered that the rate of light absorption ($2.303I_0\varepsilon[F]$) is constant throughout the solution and is therefore independent of the light path l. Furthermore, it can be seen that the rates of decay are in units of M sec^{-1}. It follows that the rate of formation must be expressed in the same units. Therefore, the fluence rate, I_0, conveniently expressed in units of mole sec^{-1} cm^{-2}, must be used in the calculations in compatible units; e.g., $M \cdot$ cm sec^{-1} = mole dm^{-3} sec^{-1} cm$^{-2} \times 10^3 \cdot$ cm^3. This change in units explains why the proportionality constant is 2303 and not 2.303 as obtained by expanding the exponential term of the Beer–Lambert law. Expressing I_0 in mole sec^{-1} m^{-2} (the S.I. unit) would lead to another value of the proportionality constant.

Under photostationary conditions,

$$0 = d[^1F_1]/dt = 2303I_0\varepsilon[F] - (k_F + k_{ISC} + k_{IC})[^1F_1]. \qquad (2.14)$$

If light is switched off, $^1F_1$ decays with a lifetime

$$\tau_F = (k_F + k_{IC} + k_{ISC})^{-1}. \qquad (2.14a)$$

According to the definition given in Section 2.3.4, the fluorescence quantum yield (Φ_F) can be expressed as

$$\Phi_F = k_F[^1F_1]/2303I_0\varepsilon[F] \qquad (2.15)$$

by virtue of Eq. (2.14), and Eq. (2.15) as

$$\Phi_F = k_F/(k_F + k_{IC} + k_{ISC}). \qquad (2.16)$$

The fluorescence quantum yield depends on the relative rates of the radiative process (k_F), the radiationless processes of intersystem crossing (k_{ISC}), and internal conversion (k_{IC}) [see Eq. (2.16)]. The rate of the radiative process does not vary with temperature and hence the variation of Φ_F reflects the variations of k_{ISC} and k_{IC}. These should increase as the temperature increases because at higher temperatures a larger proportion of the molecules is raised to upper vibrational levels of S_1 and the probability of passing through potential surface intersections is increased (see Figures 2.2 and 2.3). At low temperatures both rates tend to a limiting value corresponding to the probability of intersystem crossing or internal conversion from the lowest vibrational level of $^1F_1$. Thus a molecule with a fluorescence quantum yield close to unity at room temperature does not exhibit much change on lowering the temperature, whereas it may show a decrease at temperatures above room temperature. On the other hand, a weakly fluorescent compound may become strongly fluorescent at low temperature.

2.3.6.2 Quantum Yield of Triplet Formation Φ_T

By definition

$$\Phi_T = k_{ISC}[^1F_1]/2303I_0\varepsilon[F].\tag{2.17}$$

Combining Eqs. (2.15), (2.16), and (2.17) leads to

$$\Phi_T = k_{ISC}/(k_F + k_{IC} + k_{ISC}).\tag{2.18}$$

Thus if k_{IC} is negligible, which is true for many photosensitizing aromatic molecules, one obtains the important relationship:

$$\Phi_F + \Phi_T \approx 1.\tag{2.19}$$

2.3.7 The Quenching of Singlet and Triplet States: The Stern–Volmer Equation

It is frequently observed that addition of chemicals such as oxygen, inorganic anions or cations, and H_3O^+ or OH^- to fluorescent solutions decreases Φ_F. This quenching is due to competing bimolecular processes and is known as collisional quenching.

If one considers the sequence $^1F_0 + h\nu \rightarrow {}^1F_1$ followed by $^1F_1 + Q \rightarrow {}^1F_0$ (at rate k_Q), simple chemical kinetics tell us that the rate of quenching is $k_Q[Q][^1F_1]$. Because the concentration of the quencher [Q] is exceedingly large with respect to $^1F_1$, $k_Q[Q]$ can be incorporated as a constant term in Eq. (2.14). It follows that the new Φ_F in the presence of quencher (Φ_{FQ}) is

$$\Phi_{FQ} = k_F/(k_F + k_{IC} + k_{ISC} + k_Q[Q])^{-1}.$$

Similarly,

$$\tau_{FQ} = 1/(k_F + k_{IC} + k_{ISC} + k_Q[Q])^{-1}.$$

Using Eq. (2.14a) we have

$$\tau_F/\tau_{FQ} = 1 + \tau_F k_Q[Q].$$

Using Eqs. (2.14), (2.15), (2.16), and the above definition of Φ_{FQ} we also have

$$\tau_F/\tau_Q = \Phi F/\Phi_{FQ}.$$

Replacing τ_F with $(k_F + k_{IC} + k_{ISC})^{-1}$ we obtain

$$\Phi_F/\Phi_{FQ} = 1 + k_Q[Q] (k_F + k_{IC} + k_{ISC})^{-1},$$

which can be written as the Stern–Volmer equation:

$$\Phi_F/\Phi_{FQ} = 1 + K[Q] \qquad \text{with} \quad K = \tau_F k_Q.\tag{2.20}$$

Thus, by plotting the fluorescence intensity as a function of the quencher concentration, a straight line with slope K is obtained; K is called the Stern–Volmer constant.

It must be noticed that the same reasoning applies to the triplet state quenching ($^3F_1$) by various substrates, as will be discussed in Chapter 4. An equation equivalent to Eq. (2.20) is obtained, which includes the rate constants of formation and disappearance of the triplet state.

An increase in the fluorescent solute concentration commonly causes a decrease in Φ_F by collisional quenching accompanied by possible energy transfer processes (see below). The formation of an excited complex (or excimer) ($^1F_0 \ldots {}^1F_1$) during a collision can also be treated in the same way. Many aromatic hydrocarbons give rise to excimer formation having spectral characteristics different from those of the dilute fluorophore. For instance, pyrene monomers fluoresce at about 390 nm whereas the excimers fluoresce at about 475 nm.

Excited complexes are also formed with solvent molecules, leading, again, to a large Stokes shift of the fluorescent complexes. For instance, indole and the parent amino acid tryptophan (Figure 2.4) have, in nonpolar solvent or in low-temperature ices, a fluorescence at 320 nm, whereas in fluid aqueous media, the fluorescence maximum is shifted to 350 nm.

2.3.8 Quenching of First Excited Singlet and Triplet States by Energy Transfer Processes

Presented in this section are three quenching processes very useful to biologists, biophysicists, and photobiologists.

2.3.8.1 Energy Transfer between Singlet States: The Result of Energy Transfer Is Quenching of Donor Fluorescence and Electronic Excitation of the Acceptor Molecule

The trivial case of energy transfer involves the emission of fluorescence light by the donor molecule and the reabsorption of this fluorescence by the acceptor molecule. This is of little interest for the biologist because there is no distance requirement for such a process. A second type of transfer, which depends on the distance between the acceptor (A) and the donor (D) molecule, has been studied by Förster, who set out its theoretical conditions. In both cases of energy transfer, the absorption spectrum of the acceptor must overlap the emission spectrum of the donor. The Förster transfer can be schematized by Eq. (2.20):

$$^1D_1 + {}^1A_0 \longrightarrow {}^1D_0 + {}^1A_1. \tag{2.20}$$

The intermolecular transfer is a resonance process; e.g., it involves *isoenergetic* radiationless transitions in $^1D_1$ and $^1A_0$. The possible

$^1D_1 \rightarrow {}^1D_0$ transitions correspond to the donor fluorescence spectrum and occur into the various vibrational levels of $^1D_0$. The possible $^1A_0 \rightarrow {}^1A_1$ transitions correspond to the acceptor absorption spectrum and occur into the various vibrational levels of $^1A_1$. The energy transfer process is adiabatic, the difference in the electronic energy of $^1D_1$ and $^1A_1$ being partitioned as vibrational energy between the two final species $^1D_0$ and $^1A_1$. This energy transfer requires (1) a large overlap between the first absorption band of the acceptor molecule and the emission band of the donor molecule and (2) high fluorescence quantum yield of the donor molecule. Donor and acceptor may be the same molecules provided that there is appreciable overlap of absorption and fluorescence spectra. In this case one observes the so-called "concentration quenching" of fluorescence. The efficiency of the intermolecular interaction is expressed as the "critical separation distance" R_0 at which the probability of transfer is equal to that of the fluorescence decay. The critical distance, R_0, is given by the Förster equation:

$$(R_0)^6 = (20{,}723\, k^2 \Phi_F / 128\, \pi^5 n^4 N) \int f_D \varepsilon_A \, d\bar{\nu}/\bar{\nu}^4, \qquad (2.21)$$

in which N is Avogadro's number, n is the refractive index of the medium, k^2 is an orientation factor (approximately 2/3), Φ_F is the fluorescence yield, f_D is the fluorescence emission spectrum of the donor normalized to unity, ε_A is the absorption spectrum of the acceptor, and $\bar{\nu}$ is the wavenumber in cm^{-1}. For example, critical transfer distances of 5 nm for fluorescein and 8 nm for chlorophyll were determined by Förster. A mean molecular separation of 10 nm corresponds to a concentration of about $2 \times 10^{-3}\, M$. Practically, it can be assumed that if a solution has been sufficiently diluted, quenching by long-range energy transfer (i.e., Förster transfer) will be negligible. The efficiency of this nonradiative transfer process can, however, be considerably greater when the donor and acceptor are different molecular species having a favorable absorption/emission overlap and high donor fluorescence efficiency. An elegant proof of singlet-to-singlet energy transfer can be obtained from the comparison of the excitation spectra of the donor and acceptor taken separately from the spectrum of the mixture. If Förster-type energy transfer occurs in the mixture, then the excitation spectrum of the donor contributes to the excitation spectrum of the acceptor fluorescence in the mixture.

Very favorable conditions are provided by linking the acceptor and donor permanently. A classical example, found in all luminescence textbooks, was investigated by Weber and Teale, namely (see Figure 2.5), 1-dimethylaminonaphthalene-5-(N-benzyl)sulfonamide (I) and 1-dimethylaminonaphthalene-5-(N-phenyl)sulfonamide (II). In these fluoro-

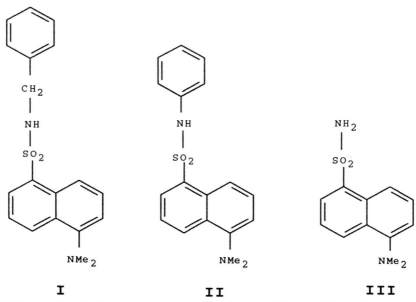

I   II   III

FIGURE 2.5 Chemical structure of 1-dimethylaminonaphthalene-5-(N-benzyl)sulfonamide (I), 1-dimethylaminonaphthalene-5-(N-phenyl)sulfonamide (II), and 1-dimethylamino-naphthalene-5-sulfonamide (Dansyl) (III).

phores, the π electron systems of naphthalene and benzene are not conjugated, and contribute separately to the absorption spectrum.

Both compounds emit only a fluorescence centered at 520 nm, sensitive to micropolarity and similar to that of 1-dimethylaminonaphthalene-5-sulfonamide (III) (Figure 2.5). This compound, also called "Dansyl," is a very popular reporter group that can be linked to many biological macromolecules, such as proteins, by free amino groups of lysine residues after suitable activation of the sulfonyl group. Thus, the Förster-type energy transfer is very useful to biologists to study proximity relationships in biological membranes or proteins because proteins can easily be labeled with fluorescent probes such as Dansyl. Modern cellular and molecular biology largely utilize fluorescent probes and energy transfer processes for *in situ* study of the molecular organization and interactions in the living cell, as will be shown later. Furthermore, residues such as tyrosine (the donor) and tryptophan (the acceptor) are very often used as intrinsic structural probes in proteins owing to the occurrence of Förster-type energy transfer with these residues.

2.3.8.2 Energy Transfer from Triplet to Triplet States of Aromatic Molecules

The energy transfer process

$$^3D_1 + {}^1A_0 \longrightarrow {}^1D_0 + {}^3A_1 \tag{2.22}$$

is also spin-allowed and was observed by Terenin and Ermolaev in frozen organic solutions at 77 K. When a solution of naphthalene (N) in ethanol was excited at this temperature with 366-nm radiation, no luminescence was observed because naphthalene does not absorb at this wavelength. However, if benzaldehyde (B) was added to the solution, a bright phosphorescence emission characteristic of naphthalene appeared. Thus the exciting light was absorbed by the benzaldehyde and the excitation energy was transferred to naphthalene, inducing the *sensitized phosphorescence*. It could not involve singlet-to-singlet transfer because the energy of the singlet state of benzaldehyde is located below the energy of the singlet state of naphthalene. The energy transfer can be explained by the following sequence:

$$^1B_0 + h\nu \quad (366 \text{ nm}) \longrightarrow {}^1B_1 \longrightarrow {}^3B_1$$
$$^3B_1 \longrightarrow {}^1B_0 + h\nu' \qquad \text{(phosphorescence)}$$
$$^3B_1 \longrightarrow {}^1B_0 \qquad \qquad \text{(internal conversion)}$$
$$^3B_1 + {}^1N_0 \longrightarrow {}^1B_0 + {}^3N_1. \quad \text{(energy transfer)}$$

In contrast to long-range singlet-to-singlet energy transfer, the triplet-to-triplet energy transfer requires orbital overlap between the acceptor and the donor. It is therefore a short-range transfer involving electron exchange interaction over distances (0.6–0.15 nm) of the order of the molecular diameter. Nevertheless, the triplet-to-triplet transfer can be observed in fluid solutions at room temperature because the triplet decay rate constant is much smaller than the mean diffusion rate constant of most usual solvents. Thus, the laser flash spectroscopy technique takes advantage of the triplet-to-triplet transfers to characterize triplet states and to determine their molar extinction coefficients. The latter is an essential parameter for the determination of triplet quantum yield. This will be studied in detail in another chapter.

2.3.8.3 Energy Transfer between Triplet and Singlet States

The most important example for the photobiologist is that of oxygen, a ground state triplet molecule, used as an acceptor for the transfer of energy of many triplet molecules ($^3D_1$). In an approach similar to that of Eqs. (2.20) and (2.22), we can write Eq. (2.23):

$$^3D_1 + {}^3O_2 \longrightarrow {}^1D_0 + {}^1O_2. \tag{2.23}$$

Because characteristic emission of the $^1O_2$ species occurs at 1270 nm, the energy of the transition $^3O_2 \rightarrow {}^1O_2$ is small. Therefore, this type of energy transfer is easy. It leads to the production of singlet oxygen, a very reactive species encountered in many biological systems, as will be shown in Chapter 4.

Bibliography

Birks, J. B. (1970). "Photophysics of Aromatic Molecules," pp. 29–544. Wiley (Interscience), London.

Parker, C. A. (1968). "Photoluminescence of Solutions, pp. 7–227. Elsevier, Amsterdam.

CHAPTER 3

Spectroscopy and Instrumentation

Because the photochemist and, hence, the photobiologist cannot ignore the concepts of the molecular spectroscopist and vice versa, investigators in both fields use very similar equipment such as light sources and spectrophotometers for absorption and luminescence measurements and time-resolved optical spectroscopies. Additionally, photobiologists may be involved in research at the level of the single living cell, and thus techniques of luminescence must be developed in conjunction with microscopic observation. This chapter will present an outline of the instrumentation generally used for these purposes.

3.1 GENERAL INSTRUMENTATION OF PHOTOCHEMICAL LABORATORIES

3.1.1 "Conventional" Continuous Light Sources

The main sources (Figure 3.1) used by the photochemist as ultraviolet (UV) light sources are the super-high-pressure mercury lamp and the xenon lamp, which present a mixed spectrum containing both line and

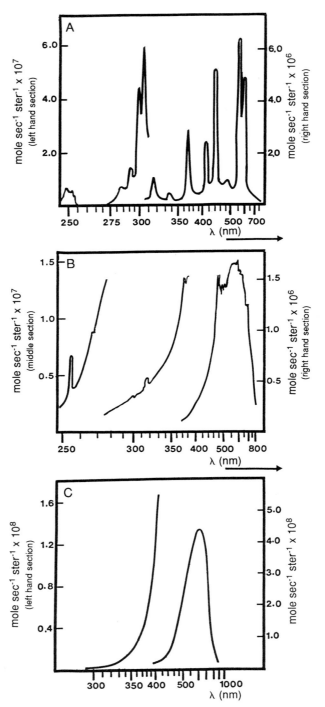

FIGURE 3.1 Typical radiant intensity and shape of the emission spectra of super-high-pressure mercury (A), xenon (B), and tungsten (C) lamps. Adapted from Parker (1968). Copyright 1968, Elsevier. Used with the permission of Elsevier Science Publishers.

continuous emissions. For irradiation in the visible range, tungsten fila- ment or tungsten–halogen lamps may be suitable and give a true contin- uous spectrum.

3.1.1.1 Mercury Lamps

3.1.1.1.1 Low-Pressure Mercury Lamps

The simplest low-pressure lamp consists of a quartz or fused silica tube 10–50 cm long and about 1 cm in diameter containing a droplet of mercury and a few millimeters pressure of inert gas to facilitate starting. These lamps provide only tens of watts and are "cold" lamps. The vapor pressure of the mercury is close to its room temperature value ($\approx 10^{-3}$ mm). Most of the light is emitted at 253.7 and 185.0 nm.

Filters are used to isolate the 253.7 nm line, which is the only useful line. The lamps may be purchased in a coiled form that can be arranged to surround a cylindrical cuvette so as to obtain the greatest light flux through the sample.

3.1.1.1.2 Super-High-Pressure Mercury Lamps

These lamps operate at high temperature and at pressures of hundreds of atmospheres. Tem- perature and pressure broadening of the spectral lines is important and a strong continuum is also present (Figure 3.1A). Spectral emission below 275 nm is almost absent from these lamps. The most useful types consist of a small quartz bulb with two electrodes, having a small arc gap. The arc is thus compressed into a small volume and very high intrinsic brightness is obtained. With their small size and extreme brightness these lamps are particularly valuable for use in conjunction with monochro- mators or filters to isolate the mercury "lines" at wavelengths above 297 nm.

Moderately expensive, these lamps are particularly useful to the pho- tochemists and photobiologists working on the photochemical effects of UV light on biomolecules or organisms. Because intense emissions can be isolated at 313, 334, 365, and 404 nm, they provide good standards for actinometry (see below). They also provide a series of very intense lines in the whole visible range and the light fluence is very reproducible from one lamp to another. In short, these lamps are an indispensable tool for the molecular photobiologist.

3.1.1.2 Xenon Arc Lamps

These lamps can be used by both the molecular spectroscopist wish- ing, for example, to determine a fluorescence excitation spectrum, and the photobiologist for the determination of action spectra. Xenon arc lamps delivering 50 W to 2.5 kW are commonly used. The larger sizes of lamps, together with their associated control equipment, are expen- sive and also produce an inconvenient amount of heat. The source is small and of high intrinsic brightness and the photon irradiance (or fluence rate) falls off in the shorter wavelengths of the UV region (see Figure 3.1B).

3.1.1.3 Tungsten Lamps

These lamps have the great advantage of being inexpensive. The emission from a 100-W filament lamp is shown in Figure 3.1C and comparison with the xenon arc lamp shows that it can compete with it only in the longer wavelengths of the visible spectrum, where its intrinsic brightness is still considerably less. It has the advantages of a completely continuous spectrum and a very high stability. For long wavelength excitation, particularly if high fluence rates are not required, the tungsten lamp is thus suitable. For instance, many photosensitizers show very broad absorption bands in the visible range. Thus the integrated absorbed light provided by a tungsten lamp can be large enough to compare favorably with a mercury arc.

The tungsten–halogen lamps have more output in the near-UV region. However, they still cannot compete with the mercury and xenon lamps.

3.1.2 Monochromators and Optical Filters

Luminescence spectroscopies and photochemistry cannot be correctly performed without monochromatic excitation light. Thus, a fluorometer that cannot provide excitation spectra is practically useless for research work or analytical biochemistry using fluorescence techniques. Accordingly, a photochemist or a photobiologist needs monochromatic light for quantitative photochemistry. Furthermore, the use of monochromatic excitation could reduce unwanted photochemistry of photochemical products with substantial absorbance outside the absorption range of the initial product.

Two types of device can be used to render a continuous spectrum more or less monochromatic. The most elaborate equipment is the monochromator, which can provide light of very narrow bandwidth ($\approx 1-2$ nm); the smaller the bandwidth, the smaller the light output. This resolution may, however, be required by fluorescence or phosphorescence spectroscopy and also for action spectra determination.

An inexpensive substitute for a monochromator can be a filter, which will provide a rather large range of bandwidths (from 5 to 200–300 nm) and a higher fluence rate than a monochromator. Filters may be used in conjunction with a monochromator to increase its efficiency, but their main utilization is in preparative photochemistry, which does not require highly monochromatic light, and in photosensitized reactions in which the substrate does not directly absorb light.

3.1.2.1 Monochromators

The essential components of a monochromator are (1) an entrance slit, (2) a collimating device (to produce parallel light), (3) a wavelength selection or dispersing system, (4) a focusing lens or mirror, and (5) an exit slit.

Modern instruments favor diffraction gratings to produce monochromatic light. Compared with prisms, gratings offer better resolution with linear dispersion and therefore constant bandwidths. Mechanical drive requirements are also simplified (Figure 3.2A).

The main characteristic of a grating monochromator is the linear dispersion (LD) expressed in millimeters per nanometer. The bandwidth ($\Delta\lambda$) is defined as the band of wavelengths corresponding to half-peak intensity (Figure 3.2B). It can be shown that for entrance and exit slits of equal width, the amount of light passed by the monochromator from a continuous source is proportional to the square of the slit width, e.g., the square of the bandwidth. Thus if twice the bandwidth can be tolerated for a particular application, the fluence rate can be increased four times by doubling the slit width.

On the other hand, the intense lines of specific wavelengths are superposed on a weak continuous background, but the fluence rate of a line varies as the first power of the bandwidth whereas the fluence rate of the continuous spectra varies as the square of the bandwidth. This property must be kept in mind when, for the purpose of increasing the rate of a photochemical reaction, the slit widths of a monochromator illuminated with mercury or xenon lamps are broadened.

3.1.2.2 Filters

A large variety of glass and gelatin filters are commercially available, as well as many homemade solution filters. The choice of substances suitable for bandpass filters is limited because these require a compound that has a large wavelength separation between the first and second absorption bands, with low extinction coefficients in the intermediate region. Some dyestuffs are suitable for wavelengths in the visible region and solutions of transition metal salts are also widely used. Interference filters having a specified narrow bandpass may also be obtained from commercial sources. Generally, the peak transmission is considerably lower than that obtainable with broad bandpass filters of other types. Furthermore, if wide coverage of wavelengths is required, the total cost of a set of interference filters may well approach that of a small grating monochromator giving equivalent performance.

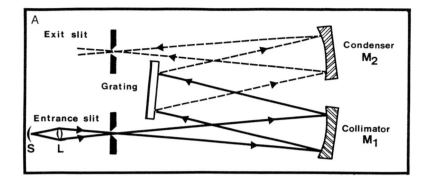

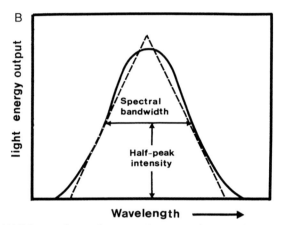

FIGURE 3.2   (A) Scheme of a grating monochromator showing the arrangement of the light source (S), the condensing Lens (L), and the mirror optics (M) producing a monochromatic light at the exit slit. (B) Energy distribution as a function of the wavelength at the exit slit, illustrating the significance of the bandwidth ($\Delta\lambda$) of the monochromator. Typically $\Delta\lambda$ can be about 1 nm for a standard absorption spectrophotometer and up to 10–15 nm for the excitation or emission monochromator of a fluorometer.

When mercury lamps are used, a high fluence rate of nearly mono-chromatic light can be obained by isolating the desired mercury emission lines. For instance, isolation of the 365-nm emission line can be readily done with a Wood-type glass filter (isolating the 275- to 400-nm UV region) placed behind a 2-cm light-path cuvette containing a $2\,M$ aqueous solution of potassium nitrate. Chemical filters are generally preferred to interference filters because they are more stable to light, especially at high fluences.

3.1.3 Actinometry

Determination of luminescence and photochemical quantum yields requires knowledge of the number of photons (preferably monochromatic) absorbed during a particular experiment. The easiest way of measuring actinic light in absolute units is to follow a photochemical change. As a matter of fact, chemical methods have important advantages over physical methods employing calibrated thermopiles or photoelectric devices. The main advantages are as follows: (1) The results are readily reproducible in different laboratories; they provide a simple laboratory reference standard for absolute light measurements. (2) The actinometer is an integrating system recording the total dose of light (also called fluence or radiant exposure) received during irradiation. (3) By choice of suitable shape and size of the containing vessel, the actinometer can integrate any beam. (4) With a quantum yield independent of wavelength, the actinometer may act as a quantum counter for a polychromatic light. The ferrioxalate actinometer of Parker and Hatchard largely fulfills these conditions for wavelengths between 250 and 500 nm. It consists of a solution of potassium ferrioxalate, $K_3F_e(C_2O_4)_3 \cdot 3H_2O$, in 0.1 N sulfuric acid. In acid solution the trioxalato-ferric ions are dissociated into monoxalato and dioxalato complexes. On irradiation, the following reactions occur:

$$[Fe(C_2O_4)]^+ + (h\nu) \longrightarrow Fe^{2+} + (C_2O_4)^-$$
$$(C_2O_4)^- + [Fe(C_2O_4)]^+ \longrightarrow 2\,CO_2 + Fe^{2+} + (C_2O_4)^{2-}.$$

After photolysis the ferrous ion formed is converted to its 1,10-phenanthroline complex, absorbing at 510 nm. The quantum yield of ferrous ion formation has been accurately determined over the wavelength range 250–500 nm. The minimum detectable amount of light is about 2×10^{-10} einstein/ml. The actinometer can be used to measure the very low fluences isolated by a monochromator or very high fluences such as those obtained with a laser.

The quantum yields (Φ) used for the ferrioxalate actinometer at 22°C are 0.15, 0.86, 1.1, 1.13, 1.23 on irradiation with 334, 365, 405, 436, 509, and 546 nm, respectively. A 1-cm length of 0.006 M potassium ferrioxalate solution absorbs 99% or more of the light of wavelengths up to 390 nm. It can be used conveniently for wavelengths up to about 430 nm (absorption about 50%/cm). For longer wavelengths, a 0.15 M solution is usually more convenient.

Thus, *under conditions whereby all incident light is absorbed,* let us consider a volume V (dm³) of ferrioxalate at concentration C. The area of the light beam is generally smaller than the area of the optical cell. By definition,

Φ = moles of ferrioxalate decomposed/moles of photons absorbed.

The number of moles of ferrioxalate decomposed per second is $V\Delta C/\Delta t$, where, $\Delta C/\Delta t$ is the slope of the plot of C as function of the irradiation time, e.g., the rate of destruction of the ferrioxalate. Hence, the absorbed light (I_{abs}) in units of mole sec^{-1} is

$$I_{abs} = (V\Delta C/\Delta t)/\Phi. \qquad (3.1)$$

Thus the incident light fluence rate is simply related to the rate of ferrioxalate disappearance. This chemical actinometer is also very useful for calibrating in the UV–visible range other photometers, such as thermopiles, photodiodes, and photomultipliers.

3.1.4 Coherent Light Sources: Lasers

Lasers have rapidly become the light source of choice for applications in many fields. The word laser is an acronym (light amplification by stimulated emission of radiation); in 1960 T. Maiman invented the ruby laser, and now a bewildering array of different types exist, including gas lasers, chemical lasers, free electron lasers, diode lasers, and X-ray lasers. All of the various types of lasers share the use of stimulated emission, predicted by Einstein early in the twentieth century. Stimulated emission is the production of light by an atom or molecule when irradiated with light of the same wavelength.

In order for the total amount of radiation to increase in the lasing region, there must be a population inversion of the energy levels of the lasing material. Such an inversion occurs when an upper state (see the energy level diagram of hydrogen in Figure 1.13) is populated more than a lower state to which it can decay. If now the lasing atom or molecule is irradiated with light of the same wavelength as that corresponding to the energy difference between the two levels, transitions will be stimulated, adding to the quantity of light present in the medium. If there are enough electrons available in the upper state, a sort of chain reaction occurs, and the total amount of radiation may greatly increase, leading to the tremendous intensities often associated with lasers. In addition, the stimulated photons occur with the same phase as the exciting photons, and this is what causes the unique coherence observed. In most lasers, the intensity of the radiation in the lasing region, often a tube, is enhanced by placing mirrors at either end. This is not really necessary, however, and lasers have been prepared without mirrors. In fact, X-ray lasers usually have no mirrors because, in the X-ray region, high-reflectance mirrors are not available, except under special circumstances.

The properties of lasers include the following characteristics:

1. Very high spatial and temporal coherence. Spatial coherence is measured by allowing light from different regions of the laser beam perpendicular to the beam propagation direction to interfere. For example, light from the center of the beam and light from a distance from the center are brought together on a screen. If the light is coherent, and a thin plate is rotated in one of the beams to change the optical path slightly, the intensity of the light will fluctuate as the difference in optical paths of the two beams passes through several wavelengths (see Figure 1.7). The measure of the amount of coherence is the *visibility* of the interference fringes formed. The more the intensity varies as the path difference is changed, the greater the coherence.

Temporal coherence is measured in the same way, except that the two sample beams are extracted at different positions along the beam, and therefore correspond to a difference in time. For instance, it is not unusual for a laser to show coherence for two points a meter apart along the beam. Taking the speed of light as 3×10^8 m/sec, this gives a coherence time of 3.33×10^{-9} sec. This seems unremarkable until we remember that visible light (red) has a frequency of 5×10^{14} Hz, so during the coherence time the light stayed in phase for 1.667 million vibrations; thus the remarkable character of laser light can be seen.

This property allows the use of techniques involving interference, perhaps the best known of which is the hologram. Coherence may, however, cause special problems in using laser light for imaging, because the interference patterns due to the light often intrude on the desired images. Because of their coherent properties, laser beams can be used for communications, and are rapidly replacing electricity in many long- and short-distance applications.

2. Lasers can produce uniquely narrow beams of light with very high intrinsic brightness. This property depends on the region in which the lasing medium operates. In most lasers, a resonant cavity is provided with mirrors at either end. In this case, an extremely sharp beam is formed, often limited only by diffraction (see Section 1.6). Surgery is now routinely performed using narrow laser beams, and they are used for cutting metals and even rock. Laser beams have also been propagated for long distances.

3. Lasers can also be made to provide extremely short pulses, reaching even into the femtosecond range (a femtosecond is 10^{-15} sec). Short pulses have been used to measure fluorescence lifetimes, which are of the order of 10^{-8} to 10^{-10} sec, of various fluorophores. The investigator is sometimes enabled, by analyzing the lifetime observations, to disentangle a complicated mix of several fluorescing agents in living tissues, because most fluorophores exhibit different lifetimes.

4. Gas lasers exhibit finer spectral lines than other sources, because the stimulated emission process tends to narrow the line produced. In order to produce such narrow spectral lines, single-mode emission must be used, which entails careful alignment of the resonant cavity and the use of a small aperture to screen out all but the central mode.

5. Employing dyes and spectral selection in the resonant cavity, lasers can be made to operate at almost any wavelength from the near-UV to infrared. This property has made lasers popular for exciting fluorescence, because the exciting wavelength can be chosen and varied to provide excitation spectral information. Dye lasers can also be rapidly pulsed, making them useful for fluorescence lifetime measurements, as mentioned above.

Because of the unique properties of lasers, as outlined above, they have been very widely used in photochemistry and allied disciplines, despite their initial cost, which may be higher compared to other light sources. Dye lasers can provide monochromatic light at the various wavelengths especially necessary for luminescence spectroscopy, as mentioned previously, in most spectral regions with greater intensity than other sources.

3.2 CONVENTIONAL OPTICAL SPECTROSCOPY EQUIPMENT FOR THE PHOTOCHEMIST AND THE PHOTOBIOLOGIST

3.2.1 Absorption Spectrophotometry

As detailed in Chapter 2, because electrons in a molecule have unique ground state energy and because the levels to which electrons may jump are also unique, it follows that there will be a finite and predictable set of transitions possible for the electrons of a given molecule. Each of the transitions, or jumps, requires the absorption of a quantum of energy and there will be a direct relationship between the wavelength of the radiation and the particular transition that it stimulates. This relationship is known as the specific absorption, and a plot of those points along the wavelength scale at which a given substance shows absorption is called the absorption spectrum (see Figures 3.3A and 2.4 for a concrete example). The absorption spectrum is a kind of an "identity card" of a molecule. Traditionally the preferred technique has been double-beam geometry in the sample handling area with a time-sharing system that alternates between the unknown and a reference. A wavelength scanning mechanism is usually provided as well (Figure 3.3B).

Electronic developments, particularly the application of microprocessors allowing baseline storage and substraction, have made it possible to

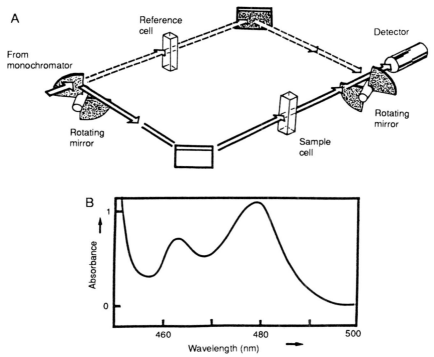

FIGURE 3.3 (A) Schematic diagram of a dual-beam spectrophotometer. Light choppers allow the alternate irradiation of the reference and sample curvettes and the alternate measure of the energy in the reference and sample beams by the detector. (B) Representation of the absorption spectrum of a hypothetic molecule (M). The energy (2100 cm^{-1}) of the first absorption maximum at 480 nm corresponds to the energy of the $^1M_0 \rightarrow {}^1M_1$ electronic transition.

achieve comparable results with a single-beam configuration. Optical and mechanical construction are greatly simplified in modern spectrophotometers. A good absorption spectrophotometer must cover at least the whole UV–visible range (e.g., 200–800 nm). Because it is not possible to provide suitable emission from a single light source to cover the whole region, two light sources are normally supplied. Ultraviolet light is generally obtained from a deuterium arc with a quartz or fused silica envelope, used in the 190- to 350-nm range. A tungsten (or tungsten–halogen) lamp is provided for the longer wavelengths.

Two types of detectors can be encountered: photomultipliers and the rapidly developing silicon diode. The latter, with its operational amplifier, compares favorably to photomultipliers because they are mechanically robust. Linked to digital readout and a microprocessor, the variation of the fluence of the analyzing light beam passing through the sample cuvette can be converted into optical densities (O.D.). By application of the Beer–Lambert law, O.D. $= \varepsilon[C]l$, the molar extinction coefficient, ε (M^{-1} cm^{-1}), at each wavelength of the absorption spectrum or the concentration [C] of the substance that is in solution in a parallelepipedic cell of 1-cm light path (see Figure 3.3A) can be determined.

3.2.2 Spectrofluorometry and Spectrophosphorometry

3.2.2.1 Spectrofluorometry

3.2.2.1.1 General Description
During the past decade, fluorescence spectroscopy has become an essential technique for the biologist. Not only are fluorescence microscopy and immunofluorescence used as routine analytical methods in medicine, but also many enzyme-linked immunosorbent assay tests have been developed using fluorescent markers. Thus, fluorescence is not only an essential research tool but also an everyday technique. As a consequence, it is of utmost importance for the student or the researcher to understand the basic principles of fluorometry.

A modern spectrofluorometer must be a versatile instrument. It generally contains a source of light to excite fluorescence in the sample, a sample holder, and a detector to record the fluorescence. A monochromator is used to select the required wavelengths from excitation and emission beams. The spectral distribution of the fluorescence light emitted—the fluorescence *emission* spectrum—or the measure of the variation of fluorescence intensity with wavelength of the exciting light —the fluorescence *excitation* spectrum—can thus be obtained.

A diagram of a typical spectrofluorometer is shown in Figure 3.4. It consists of a source giving a continuum of visible and ultraviolet light, a monochromator to select the required wavelength for excitation, a sample holder, and a second monochromator fitted with a photomultiplier or a silicon diode array to analyze the fluorescent light. In most spectrofluorometers the excitation beam is at right angles to the emission beam, which minimizes stray light and fluorescence of the optical cell.

3.2.2.1.2 Determination of Fluorescence Quantum Yields
For photochemists or photobiologists the best method for measuring a fluorescence quantum yield is to compare the total fluorescence emission of the sample to that of a known fluorophore. In dilute solutions, the fluorescence intensity (F) will be proportional to the rate (R) of light emission. By definition, this rate is equal to the rate of light absorption multiplied by the fluorescence quantum yield (Φ_F).

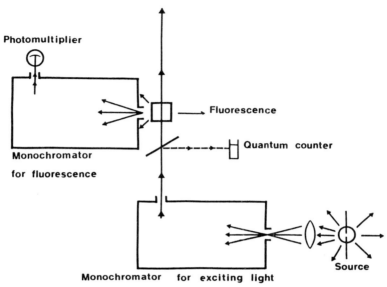

FIGURE 3.4 Scheme of the essential parts of a fluorometer. Reproduced from Parker (1868). Copyright 1968, Elsevier. Used with the permission of Elsevier Science Publishers.

Applying Eq. (2.7) solutions for which the (O.D.) < 0.1, one gets $R = 2303 I_0 (\text{O.D.}) \Phi_F$, provided the fluence rate is expressed in mole sec^{-1} cm^{-2}. Because R corresponds to the total emission spectrum, the integrated area under the corrected fluorescence spectrum is proportional to R.

Thus, if the fluorescence emission spectra of two solutions are measured with the same instrumental geometry and optics in the same solvent and at the same fluence rate of exciting light, the ratio of the two *observed* fluorescence intensities is given by:

$$F_2/F_1 = R_2/R_1 = \text{area } 2/\text{area } 1 = I_0 \varepsilon_2 c_2 l \Phi_2 / I_0 \varepsilon_1 c_1 l \Phi_1$$

$$= \Phi_2 (\text{O.D.})_2 / \Phi_1 (\text{O.D.})_1. \tag{3.2}$$

If the absolute fluorescence quantum yield (Φ_1) of one of the substances is known, that of the other can be calculated. The most popular fluorescence standard for molecules emitting in the 400- to 500-nm region is quinine bisulfate in 0.1 N sulfuric acid, for which $\Phi_F = 0.51$. Its corrected fluorescence spectrum is known, allowing in turn the determination of the instrumental response (emission monochromator and photomultipler) in the studied spectral range. As a result, unknown fluorescence spectra can then be corrected. Another fluorescence standard useful in the 330- to 400-nm region is the aromatic amino acid trypto-phan, whose Φ_F is 0.14 in water at 20°C. Some precautions must be taken in determining fluorescence quantum yields. One must avoid errors due to inner filter effects (e.g., the optical density should be lower than 0.1 so that no appreciable incident light fluence attenuation occurs through the optical cell), "autofluorescence" of solvent, oxygen quenching, and pho-tochemistry.

3.2.2.1.3 Determination of Fluorescence Excitation Spectra   Equa-tion (3.2) shows that the fluorescence intensity changes by varying the wavelength of the exciting light. For a solution of a single solute, the fluorescence intensity is proportional to $I_0 \varepsilon \Phi_F$. Thus if the fluence rate of exciting light is kept constant as the wavelength is varied, the fluores-cence intensity will be proportional to $\varepsilon \Phi_F$. Plots of $\varepsilon \Phi_F$ against wave-length of the exciting light are called fluorescence excitation spectra. For most aromatic molecules in solution, the fluorescence quantum yield is approximately independent of the wavelength of the exciting light. Thus the excitation spectrum of a dilute solution of a single solute will be a replica of the absorption spectrum of the compound. The absorption spectra of fluorescent compounds can thus be obtained at concentrations far lower than would be needed to measure the absorption spectrum directly by absorption spectrophotometry. The absorption spectrum of one component of a mixture of absorbing compounds can be obtained by tuning to the wavelength of the appropriate fluorescence emission band, provided that the fluorescence spectra are not totally overlapping.

In practice, the fluence rate of the exciting light is variable (see Figure 3.1). Its variation during scanning of the excitation wavelength can be monitored using "quantum counters." The independence of fluorescence spectra of many solutions with the wavelength of excitation can be used in the design of a detector having a response directly proportional to the fluence of the incident beam, and independent of its wavelength. For this purpose the concentration of the fluorescence solute must be sufficient to absorb all of the incident light over the complete spectral range re-quired, in a thickness that is small compared with the optical depth of the cell (see Figure 3.5). The resulting fluorescence penetrates an approxi-mately constant depth of liquid before reaching the photodetector. The solution thus acts as a filter for the exciting light and at the same time ensures that substantially the same band of fluorescence reaches the detector.

In the modern spectrofluorometer such quantum counters are perma-nently added to the fluorometer to continuously monitor the fluence rate of the exciting light beam. Corrected excitation spectra can thus be di-rectly recorded and the spectral response of the emission monochromator can be determined.

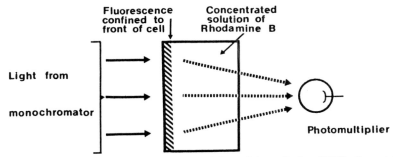

FIGURE 3.5 Principle of a quantum counter. Adapted from Parker (1968). Copyright © 1968, Elsevier. Used with the permission of Elsevier Science Publishers.

3.2.2.2 Spectrophosphorometry

Because fluorescence and phosphorescence differ by several orders of magnitude in their lifetimes, to distinguish between these emissions it is necessary to interrupt the beam of exciting light periodically and to record the phosphorescence (or delayed fluorescence as well) when the fluorescence has completely decayed.

The first phosphoroscope was built by E. Becquerel at the Museum National d'Histoire Naturelle in Paris. It consisted of two circular disks mounted on the same axis. The disk had holes cut around the circumference, the holes in the first disk being offset from those in the second. The sample was placed between the disks and when the latter were rotated the sample was irradiated intermittently by the beam of exciting light passing through the holes in the first disk, and was viewed during the periods of darkness through the holes in the second disk. Instead of using disks, the sample can be placed inside a rotating cylindrical can having a slot cut in the circumference so that, as the can rotates about its axis, the sample is alternately irradiated with the exciting light passing through the slot, and viewed through the same slot. Similar arrangements are employed in some commercial spectrophosphorometers. With both of these arrangements only the long-lived emission can be measured. To measure the total luminescence (e.g., fluorescence plus phosphorescence) the rotating disks or the can must be removed.

3.2.3 Photoacoustic Spectroscopy

Photoacoustic spectroscopy (PAS) is a technique recording the energy emitted as heat after absorption of monochromatic light by a sample. In contrast to absorption spectroscopy, which deals with the optical constant of the sample, the thermal properties of the sample play the major role in PAS. The theory of PAS is based on Boyle-Mariotte's law, which

A

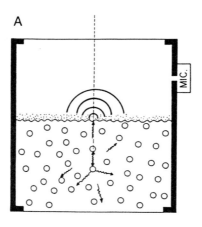

B

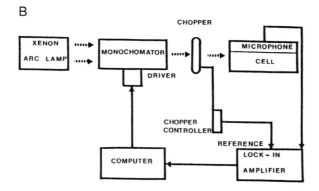

C

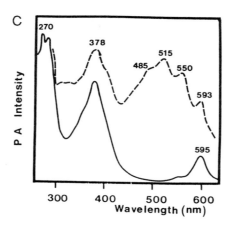

states that if the temperature is constant the product of pressure (*P*) and volume (*V*) of a gas is constant: that is, $PV = nRT$, where *n* is the number of gas molecules, *R* is the universal constant for ideal gas, and *T* is temperature expressed in degrees Kelvin. Hence, $P = nRT/V$.

This gas law states that any factor affecting the temperature of a gas contained in a closed vessel will affect pressure. This simple principle is the basis of photoacoustic spectroscopy. The sample is placed in a hermetically closed photoacoustic cell. A monochromatic beam, chopped at an audiofrequency, impinges on the sample through a transparent quartz window to excite the molecules. The deactivation of the excited state can be either radiative (e.g., fluorescence or phosphorescence) or nonradiative. The latter is measured by PAS. Thus, the sample is subjected alternatively to light excitation and thermal deactivation. This cycle induces a periodic heating and cooling of the gas in contact with the sample (Figure 3.6A). The resulting pressure oscillation, as predicted by the ideal gas law, produces sound waves that are sensed by a highly sensitive microphone that generates corresponding electrical signals at the wavelengths being absorbed. The electrical signals generated are amplified by a phase-sensitive amplifier and sent to a computer for analysis. A photoacoustic spectrum is thereby obtained, which is equivalent to an absorption spectrum. A schematic diagram of a standard PA spectrometer is shown in Figure 3.6b.

The PA signal is proportional to three parameters, e.g., the radiant power absorbed, the radiationless internal conversion quantum yield, and the thermal transfer efficiency. It is important to note that the thermal transfer efficiency depends on the frequency of the modulated light beam. Indeed, the important parameter of PAS is the thermal diffusion length, μ_s, which is proportional to $1/\omega$, where ω is the angular frequency of light modulation. The coefficient μ_s is equivalent to *k* (cm), the optical absorption length, that is, the reciprocal of the absorption coefficient μ (cm^{-1}) of the Beer–Lambert law [see Eq. (2.4)]. The optical absorption length represents the depth at which the incident light fluence has decreased by a factor of $1/e$. As a result, *by simply changing the frequency of modulation, useful information from different depths of the sample can be obtained.*

FIGURE 3.6 (A) Principle of photoacoustic effect. Molecules (circles) have absorbed light. Part of the energy is reemitted as heat; arrows show the directions of the heat diffusion. Part of the heat reaches the air–solution interface and causes, through microthermal changes in the boundary layer (points), pressure changes in the cell, which are detected by a microphone (MIC). (B) Scheme of a typical photoacoustic spectrometer equipped with a gas microphone cell. (C) Photoacoustic spectra of wild-type *Rhodospirillum rubrum* (---) and its carotenoid-negative mutant (—). The absorptions at 485, 515, and 550 nm are characteristic of the spirilloxanthin carotenoid. These spectra were obtained with light modulation at 85 Hz.

As an example, Figure 3.6C shows the PAS spectrum obtained from carotenoids in photosynthetic bacteria. One can note the excellent resolution of the carotenoid bands (450–550 nm). The PAS technique has also been applied to intact retina, eye lenses, and pharmaceutical powders. This shows the versatility of PAS, which makes it possible to obtain comparable spectra for a compound in opaque, translucent, and transparent samples. It is a sensitive and nondestructive technique and can be applied to *in vivo* materials without tedious and time-consuming preparations.

3.2.4 An Outline of Raman Spectroscopy

Raman spectroscopy is quite different from absorption spectroscopy in that it involves light *scattered* by a sample rather than light *absorbed* or *emitted*. Suppose a photon collides with a molecule. Whatever the energy of the photon, the photon–molecule collision may scatter the photon, meaning that the direction of motion of the photon is changed. Although most of the scattered photons undergo no change in frequency and energy (Rayleigh scattering), a small fraction of the scattered photons exchange energy with the molecule during the collision. The resulting increase or decrease in energy of the scattered photons is the Raman effect, discovered by C. V. Raman in 1928.

If v_0 and v_{scat} are the frequencies of the incident photon and the Raman-scattered photon, respectively, and E_a and E_b are the energies of the molecule before and after it scatters the photon, conservation of energy gives

$$h v_0 + E_0 = h v_{scat} + E_b \quad \text{or} \quad \Delta E = E_b - E_a = h(v_0 - v_{scat}).$$

The energy difference ΔE is the difference between two stationary state energies of the molecule, so the Raman shifts, $v_0 - v_{scat}$, give molecular energy level differences.

In Raman spectroscopy, the sample (gas, liquid, or solid) is irradiated with monochromatic radiation of any convenient frequency v_0. Unlike absorption spectroscopy, v_0 need not have a relationship with the difference between energy levels of the sample molecules. Usually v_0 lies in the visible or near-UV region. The Raman effect lines are extremely weak (only about 0.001% of the incident radiation is scattered, and only about 1% of the scattered radiation is Raman scattered). Very intense laser beams are used as the exciting radiation. Scattered light at right angles to the laser beam is focused on the entrance slit of a spectrometer, light intensity is recorded versus wavenumber expressed in cm^{-1} ($\bar{v}$), giving the Raman spectrum.

For a polyatomic molecule, the Raman spectrum shows several bands, each corresponding to a vibrational transition involving two adjacent

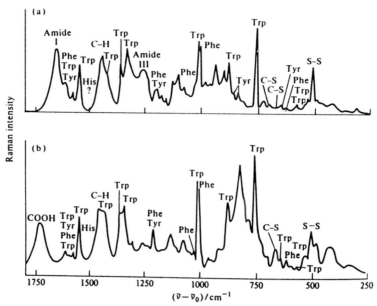

FIGURE 3.7 (a) Vibrational Raman spectrum of lysozyme, a proteolytic enzyme, in water. (b) Superposition of the mixture of amino acids constituting the enzyme. Reproduced from Long (1977). Copyright 1977, McGraw-Hill. Used with the permission of McGraw-Hill Book Company.

molecular vibrational levels similar to those shown in the Jablonski diagram (Figure 2.1) for ground state or electronically excited molecules. A normal mode is Raman active if it changes the molecular polarizability (see Chapter 2). Because this requirement is different from that for infrared (IR) activity, the Raman spectrum often allows studies of the frequencies of IR-inactive bands.

An important advantage of vibrational Raman spectroscopy over IR spectroscopy arises from the fact that liquid water shows only weak vibrational Raman scattering in the Raman shift range 300–3000 cm^{-1}, but has strong, broad IR absorptions in this range. Thus, vibrational Raman spectra of substances in aqueous solution can readily be studied without solvent interference. Vibrational Raman spectra of biological macromolecules in aqueous solutions can provide information on conformation and hydrogen bonding. An example of the application of this technique to lysozyme is shown in Figure 3.7, which presents the vibrational Raman spectrum of an aqueous solution of lysozyme and, for comparison, a superposition of the Raman spectra of the constituent amino acids. The differences are indications of the effects of conformation, environment, and specific interactions (such as S—S linking) in the enzyme molecule.

In resonance Raman spectroscopy, the exciting frequency ν_0 is chosen to coincide with an electronic absorption frequency of the molecule. This dramatically increases the intensities of the Raman-scattered radiation for those vibrational modes that are localized in the portion of the molecule that is responsible for the electronic absorption at ν_0. The two important advantages of resonance Raman spectroscopy in the study of biological molecules are (1) the increased scattering intensity, which allows study of solutions at the high dilutions (10^{-3} to 10^{-6}) characteristic of biopolymers in organisms, and (2) the selectivity of the intensity enhancement. Thus only the vibrations in one region of the molecule are observed, simplifying the spectrum and allowing study of the bonding in that region.

This technique is now applied to the single living cell in conjunction with microspectrofluorometry. It should be helpful to the photobiologist or the biologist who desires to look at particular reactive sites of biological macromolecules in the microenvironment of the living cells. However, great care must be taken to ensure that photochemical reactions induced by the intense laser beam do not interfere with detection of Raman emission.

3.3 TIME-RESOLVED SPECTROSCOPIES

3.3.1 Determination of Fluorescence Lifetime

There are two kinds of equipment for recording the times of fluorescence decay: pulsed source fluorometers and phase fluorometers.

3.3.1.1 Pulsed Source Fluorometers

Pulsed source fluorometers provide the most direct and simplest method of determining fluorescence lifetimes. The sample is irradiated with a light flash of duration less than that of the lifetime. The decay of the fluorescence is then recorded. Because of the fast rate of decay of fluorescence, the photomultiplier is pulsed to give high gain for a period of several lifetimes and the voltage output corresponding to the emission is stored in oscilloscope. Then, the trace is plotted on an x–y recorder (Figure 3.8).

In the past, primarily hydrogen-filled flash lamps have been used in pulse fluorometers. They are now replaced by lasers delivering light pulses of duration ranging from the femtosecond to the picosecond time scale. [It should be noted that the determination of phosphorescence lifetimes in the millisecond or second time scale is, of course, much simpler, because a mechanical shutter placed in front of the exciting light

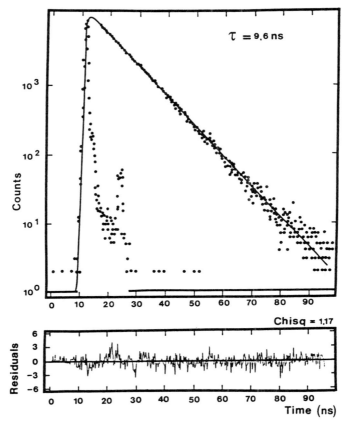

FIGURE 3.8 Decay of the fluorescence of the H_2-tetraphenylporphine in chloroform. The decay is fitted by a single exponential of lifetime $\tau = 9.6$ nsec. The normalized residuals of the fluorescence decay are also shown. The dots extending to about 30 nsec show the light pulse profile of the exciting hydrogen flash lamps. (Courtesy of M. Rougeé, Museum National d'Histoire Naturelle, Paris).

beam produced by a continuous source is used to stop the irradiation of the sample. Then, phosphorescence is simply recorded with an oscilloscope and photographed or directly plotted on a $x-y$ recorder. Generally, the phosphorescence of aromatic solutes is measured in frozen solutions at 77 K or, in a few cases, in deaerated fluid solution to avoid quenching of the phosphorescent triplet state by oxygen (see Chapter 2).]

3.3.1.2 Phase Fluorometers

In phase fluorometers (see Figure 3.9) the fluorescence is excited with a light beam modulated at high frequency. The phase of the fluorescence is then compared with that of the exciting light. For fluorescence decaying exponentially with lifetime τ, the phase Θ of the fluorescence relative to

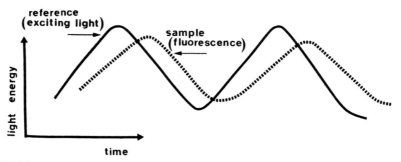

FIGURE 3.9 Principle of the phase shift and demodulation between the exciting light and the fluorescence signal in phase fluorometers. Because of the fluorescence lifetime, the light emission is delayed with respect to the excitation. The phase meter is "zeroed" with a scattering solution replacing the fluorescent sample.

that of the exciting beam is given by

$$\omega\tau = \tan \Theta,$$

where $\omega = 2\pi N$ is the angular frequency and N is the frequency of the modulation. The signals from the source and the fluorescent sample are fed into a phase meter and ratio digital voltmeter. A known delay is introduced between the two signals so as to "zero" the indicating circuit. For example, a circuit that effectively subtracts the two signals gives a minimum output when the two signals are in phase.

An alternative method of measuring Θ is to determine the degree of modulation of the two signals. Clearly, the longer the lifetime of the fluorescence the lower will be its degree of modulation. The degree of modulation in the exciting beam m_e and in the fluorescence beam (m_f) are related to Θ by

$$m_f / m_e = \cos \Theta.$$

The degree of modulation, m_f or m_e, is the ratio of the alternating current (AC; i_a) to the direct current (DC; i_d) components in the fluorescence and the exciting beam, respectively:

$$m = i_a / i_d.$$

For exponential decay, values of Θ obtained by these two methods are in good agreement. However, it is much more difficult to use this method for polyexponential decays as compared to flash fluorometers. Light sources are generally hydrogen lamps fed with a high-frequency modulated current source. In most modern equipment, fast computerized analysis makes it possible to obtain lifetimes and time-resolved fluorescence spectra in real time. This is particularly important when investigating by fluorescence spectroscopy the time-dependent interactions of macromolecules with various substrates or the mechanisms of their alteration by light.

3.3.2 Time-Resolved Absorption Spectrophotometry: Laser Flash Spectroscopy

Time-resolved absorption spectroscopy can be used by the molecular photobiologist for several purposes. The first excited triplet state ($^3M_1$) of aromatic molecules is a central intermediate in many photochemical reactions, as will be seen in Chapter 4. Owing to the relatively long lifetime of the triplet molecule in fluid solutions, $^3M_1 \rightarrow {}^3M_n$ transitions can be observed from this metastable state (the lifetime of triplet states is typically several microseconds at room temperature) by fast kinetics absorption spectroscopy.

Time-resolved absorption spectrophotometry can also be very useful for the study of short-lived radical species. A radical is a paramagnetic species with a single unpaired electron. Photochemically, it can result from the loss or gain of one electron by an excited chromophore reacting with a ground state molecule (called a substrate), or it can be produced by the photochemical cleavage of chemical bonds of molecules in their singlet or triplet excited states. Because of the presence of an unpaired electron, radicals are very reactive and thus generally short-lived.

Two conditions must be fulfilled to allow the observation of these transient species by time-resolved absorption spectrophotometry. First, a high transient concentration must be produced to obtain a detectable transient absorption (optical densities $\geq 10^{-3}$). As a result, a very intense exciting light must be used. Second, the duration of the excitation must be short with respect to the transient half-time (time for which the transient concentration is reduced by half that obtained at the time origin). Time-resolved laser flash spectroscopy meets these two requirements.

3.3.2.1 General Description of Laser Flash Photolysis Equipment

As shown in Figure 3.10, the laser pulse excites the sample contained in an optical quartz cell (QC) on one side, whereas the analyzing beam (fluence rate: I_0) crosses the 1-cm light-path cell in a direction perpendicular to the excitation beam. Because the transient concentration (thus the transient O.D.) is very small, the absorbed light (e.g., the product $I_0 \times$ O.D.) must be the highest possible to produce significant photomultiplier current and good signal/noise ratio, implying that the analyzing light fluence rate must be the highest possible. Generally, an intense xenon arc (F) triggered by the laser control unit is used as the analyzing light beam for short-lived transients (half-life < 100 μsec), whereas a tungsten–

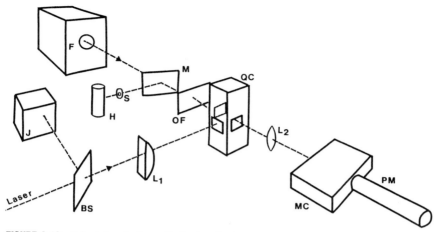

FIGURE 3.10   Principle of a laser flash photolysis equipment. In addition to the discussion given in the text, note the quartz beam splitter (BS) that allows laser output measurement with a joulemeter (J); L_1 and L_2 are lenses, S is a shutter, M is a revolving mirror, OF are optical filters, and PM is a photomultiplier.

halogen continuous source (H) is preferred for long-lived transients. A monochromator (MC) is placed in the analyzing light beam to record transient absorptions at various wavelengths. Transient absorptions are normalized to a constant laser pulse dose and are displayed on an oscilloscope and/or digitalized for computerized kinetic analysis and to obtain time-resolved transient absorption spectra. An example is shown in Figure 3.11 for the $^3Trp_1 \rightarrow \,^3Trp_n$ transient absorption of the Trp-215 residue in human serum albumin.

In Chapter 4 some applications of laser flash photolysis to the study of primary mechanisms of photochemical reactions of biological interest and the determination of triplet state parameters (molar extinction coefficients, triplet formation quantum yield) will be presented.

3.4 MICROSPECTROSCOPIC METHODS FOR THE STUDY OF LIVING CELLS

Microspectrofluorometry has been used in conjunction with fluorescence micrography for metabolic control analysis in a variety of normal and malignant cells in culture, including mouse and human fibroblasts, mouse sarcoma, hepatoma, human glia and glioma, genetically deficient human fibroblasts (cystic fibrosis, Gaucher disease), as well as human fibroblasts. These studies point to the role of mitochondria as the "cell's

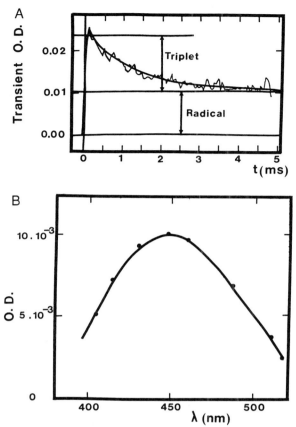

FIGURE 3.11 (A) Transient absorption at 460 nm of the triplet state of the single Trp-215 residue of human serum albumin on excitation with 265-nm laser pulses of deaerated albumin solutions at 20°C. Note the residual absorption of the longer lived radical produced by the photolysis of the Trp residue after the triplet molecules have decayed. (B) Corresponding triplet–triplet absorption of Trp-215.

policeman" with regard to metabolic control. Cytotoxic agents active on mitochondrial structure and function (i.e., anthralin, azelaic acid) produce an unleashing of extramitochondrial pathways characterized by large and out-of-control NAD(P)H transients elicited by microinjected subsrates. An interesting aspect has been the demonstration of an active nuclear energy metabolism, by NAD(P)H fluorescence excited at 365 nm, which may help to link cell bioenergetics to gene expression in eukaryotes by the use of DNA probes. The *metabolic control analysis* of cell bioenergetics has been extended to the pathways involved in the cell's handling of cytotoxic agents. Cooperative interactions of the endoplasmic

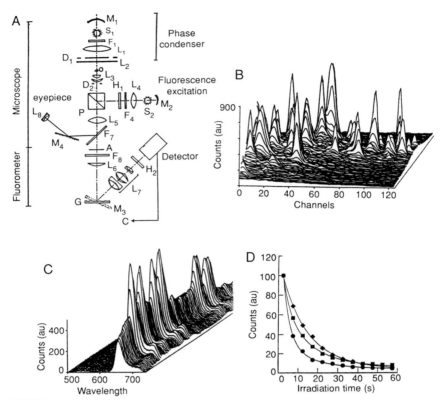

FIGURE 3.12 (A) Scheme of a microspectrofluorometer showing the essential parts of the equipment, including an inverted microscope to which is adapted a custom-made fluorometer with gratings (G), mirrors (M), and detection (generally a SIT, an ISIT, or a cooled CCD camera) with the associated computerized imaging. L, Lenses; S, light sources; D, diaphragm; P, illuminator block; F, optical filters; A, bidimensional slit; O, the object plane with the studied cell. (B) Topographic image of a human skin fibroblast incubated for 17 hr with 2 μg/ml TPPQ in serum-supplemented cultured medium. TPPQ is an acronym for a porphyrine derivative (tetraphenylporphine) covalently linked to the lysosomotropic chloroquinoline ring. The TPPQ fluorescence, which appears as sharp peaks, originates from lysosomes and was excited with 435-nm radiation (fluence rate 0.2 W/cm^2) and read at $\lambda > 645$ nm. (C) Spectrotopographic image corresponding to an area of about 2 μm $\times$ 42 μm. Each spectrum was recorded from an area of 0.8 μm^2. The fluorescence maximum is at 652 nm. The conditions are the same as in B, but the fluence rate was 0.6 W/cm^2. (D) Illustration of a photochemical reaction in real time in a living cell. This represents the photobleaching of Photofrin 2, a photosensitizer used in the photochemotherapy of cancers (see Chapter 19). The photobleaching was carried out in skin fibroblasts incubated as in B and C but with 10 μg/ml Photofrin 2. Irradiation was at 435 nm (fluence rate: 0.7 W/cm^2) and observation was over an area of 1.75 μm $\times$ 2 μm at three wavelengths corresponding to the fluorescence spectrum of Photofrin 2: •, 635 nm; ◆, 665 nm; ■, 695 nm. The bandwidth was 10 nm. The differences in the photobleaching rates as a function of the observation wavelength are due to the uneven distribution of the photosensitizer molecules, which is reflected by slightly different fluorescence spectra. Several populations of molecules in the small area thus contribute to the photobleaching kinetics.

reticulum, Golgi apparatus, lysosomes, and nuclear membrane, which are paralleled by the formation of giant lysosomes (myelinosomes), are being unraveled. New fluorogenic probes are being standardized to study lysosomal deficiency associated with gene defects. The microspectrofluorometric approach is suitable to study *in situ* within the living cells the mechanisms of action of photosensitizer drugs used in photodynamic therapy. Noninvasive fluorescence equipment at the developmental stage offers possibilities for diagnostics and therapeutic evaluations in dermatology (see Figure 3.12).

Bibliography

Hélène, C., Charlier, M., Montenay-Garestier, T., and Laustriat, G., eds. (1982). "Trends in Photobiology." Plenum, New York.

Long, D. A. (1977). "Raman Spectroscopy." McGraw-Hill, New York.

Moreno, G., Pottier, R. H., and Truscott, T. G., eds. (1988). "Photosensitization; Molecular, Cellular and Medical Aspects," NATO ASI Ser., Vol. 15. Springer-Verlag, Berlin.

Parker, C. A. (1968). "Photoluminescence of Solutions," pp. 37–227. Elsevier, Amsterdam.

LIGHT AND BIOLOGICAL SYSTEMS

From Photophysics to Photochemistry: Determination of Primary Processes in Direct or Sensitized Photoreactions

4.1 INTRODUCTION

The involvement of sunlight in promoting chemical reactions, such as the production of chlorophyll by plants, is common knowledge. Indeed, plants use sunlight to perform photochemical reactions that ultimately provide all the food supplies of the animal world. The goal of the photobiologist is to understand the quantitative aspects of this holistic relationship between sunlight (or in a broader sense, electromagnetic radiations) and man.

The first quantitative law of photochemistry was formulated by Grotthus (1817), that is, *"only light absorbed by a system can cause chemical change."* The second law, known as the Stark–Einstein law (1912), states that *"one quantum of light is absorbed per molecule of absorbing and reacting substance that disappears."* *Stricto sensus, this law applies only to the primary photochemical process, e.g., the production of the excited species by light absorption,* because some of the excited molecules may return to their ground state by other processes, such as fluorescence, intersystem crossing, etc. Furthermore, the final products are the result of many unstable intermediates following the excitation. Einstein was the first to emphasize the importance of the quantum yield in a photochemical reaction. In Chapter 2, the quantum yield Φ of a photochemical reaction was defined as

Φ = moles decomposed or produced/moles of absorbed photons.

Owing to the frequent complexity of photochemical reactions, Φ can vary from a million (the explosive $Cl_2 + H_2$) to a small fraction (production of thymine dimers in DNA or photooxidation of tryptophane (Trp) residues in wool keratin). It must be understood that the task of the photobiologist implies recognition of the primary photochemical processes and the elucidation of the subsequent reactions. In the forthcoming sections of this chapter it will be emphasized that laser flash spectroscopy is the essential tool of the molecular photobiologist (1) for characterizing and studying singlet and triplet-state properties of important biological molecules and photosensitizers and (2) for investigating primary photochemical processes of biological chromophores. Then it will be shown how mechanistic aspects and product analyses of direct or sensitized photochemical reactions with biological relevance can also be studied by much less expensive steady state irradiation techniques.

4.2 STUDY OF EXCITED STATE FORMATION AND FATE BY LASER FLASH SPECTROSCOPY

4.2.1 Laser Flash Spectroscopy as a Tool to Unravel Macromolecular Dynamics: Probing Hemoglobin Biochemistry with Light

As early as 1937, Felix Haurovitz observed that the structure of deoxyhemoglobin was different from that of oxyhemoglobin in that crystals of the former disintegrated on oxygenation. X-Ray crystallography by Nobel Prize winner M. F. Perutz showed that oxyhemoglobin and deoxyhemoglobin differ markedly in quaternary structure. Hemoglobin absorbs light in the visible and near-ultraviolet region because of the allowed $\pi \rightarrow \pi^*$ transitions of the porphyrin ring. The most intense transition is at 430 nm, corresponding to the so-called Soret band. The photodissociation of carbonmonoxyhemoglobin was first observed by Haldane and Lorraine-Smith in 1896. In the 1970s, it was shown by subnanosecond laser flash photolysis that oxyhemoglobin and carbonmonoxyhemoglobin or myoglobin were photolysed with a quantum yield of unity. On light excitation, the deligation of the O_2 or CO molecule is accompanied by a fast geminate recombination process whose time

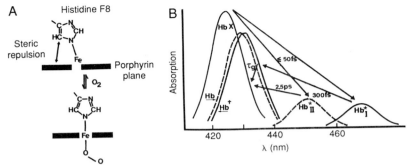

FIGURE 4.1 (A) Movement of the ferrous ion and proximal histidine on oxygen binding to hemoglobin. Reproduced from Stryer (1981). Copyright 1981, W. H. Freeman and Company. Used with the permission of W. H. Freeman and Company. (B) Transient absorption spectra of hemoglobin and its photoproducts after femtosecond laser pulse spectroscopy with 580-nm light. All the pathways are fully described in the text; τ_g reflects the rate for the geminate recombination between the deoxylike species and the ligand, e.g., the return of the system to ground state. Reproduced from Petrich et al. (1987). Copyright 1987, American Chemical Society. Used with the permission of the American Chemical Society.

scale spreads from picoseconds for O_2 to nanoseconds for CO. Thus if one wants to study the structural and ligand dynamics of these hemoproteins, it should be within this time scale. Biochemists have shown that ligand binding to hemoglobin is a cooperative process (e.g., the fixation of one O_2 molecule to the deoxyhemoglobin increases the affinity of the macromolecule for the other O_2 molecules). Perutz proposed that the displacement of the ion toward the heme plane on O_2 or CO binding (Figure 4.1A) triggers tertiary structural changes in the protein and induces the cooperativity by going from the tense (T) form to a relaxed (R) form of the quaternary structure. In the ligated state, the quaternary structure of hemoglobin is in the R form with few bonds between the subunits, the heme pocket being wide open to the entrance of the O_2 molecule. In the deligated T state, with stronger bonds between the subunits, the heme pocket is narrow and impedes the O_2 entry. Thus, by studying the early events after photodissociation, it is, in principle, possible to verify some of the hypotheses raised above.

We noted that the time scale for O_2 rebinding after photodissociation is the picosecond, thus femtosecond laser flash spectroscopy has to be used to explore fully the picosecond time domain. The change from the nanosecond to the picosecond and femtosecond time scale implies modification of the laser flash spectroscopy technology presented in Chapter 3. Thus the analyzing light beam is a spectrally broad probe pulse. This pulse is conventionally obtained by continuum generation in a water cell from the exciting laser pulse. Therefore excitation and analysis arise from the same laser pulse that is split after amplification.

Figure 4.1 shows the transient absorptions and the lifetime of the various species obtained with the femtosecond laser flash photolysis of hemoglobin with 580-nm light. The excited state unligated species Hb_I^* and Hb_{II}^* are formed in less than 50 fsec from the photoexcited HbX, where X can be O_2, CO, or NO, all strong ligands of the central Fe^{2+} ion. Hb_I^* is a species in which the heme is already partially domed (i.e., the Fe^{2+} ion is out of the heme plane). Within 300 fsec, Hb_I^* decays to the ground state unligated species Hb†. Hb_{II}^* is populated via a competition channel with a yield strongly dependent on the ligand. Thus Hb_{II}^* is efficiently formed when X is O_2 or NO and could have a planar heme, because it recombines very rapidly (within 2.5 psec) with O_2 and NO. This ultrafast recombination of a significant fraction of the dissociated ligand with this highly reactive species is in part at the origin of the low yield ($\Phi = 0.008$) of photodissociated O_2 heme compounds as measured on a microsecond time scale.

After a few picoseconds, the deoxylike species, Hb†, is the only remaining species. *This species represents the first identified intermediate state on the physiological pathway between the ligated conformation and the deoxy structure.* In hemoglobin, the spectrum of the Hb† species is shifted to the red (3 to 4 nm) with respect to relaxed unligated Hb. The 3- to 4-nm red shift occurs in the presence of the Fe^{2+}—N (His) bonding but not in protoheme, suggesting that the spectral changes in the 431-nm region may be attributed to the doming of the porphyrin ring following the iron displacement from the heme plane and reinforcement of the interaction of the Fe ion with the His-58 residue. The slow relaxation of the species (nanosecond to microsecond) is thus attributed to the constraint in the hemoglobin tetramer that results from an ultrafast increased interaction between the (at least partially) domed heme and its proximal environment. *The time scale of the relaxation corresponds to the time required for the change in quaternary structure of the hemoglobin molecule.* This information about the structural dynamics of the protein environment of the heme is obtained with ultrafast spectroscopy of vibrational rather than electronic transitions. The response of the protein to doming of the heme can be characterized by the respective position of the His-58 residue and the heme plane. In particular, the proximal histidine tilts with respect to the heme plane and accordingly the $\bar{\nu}_{Fe-His}$ stretching mode (230 cm^{-1}) appears 25 psec after the photolysis and ligand dissociation. The time required for the appearance of this stretching mode represents the time that is necessary for the transmission of the movement of the Fe^{2+} to other subunits by the proximal histidine.

After the ultrafast photodissociation, ligands may rebind to the heme. The time scale (τ_g in Figure 4.1) of this geminate recombination depends on the ligand. Recombination of CO takes place on the time scale of nanoseconds and slower, whereas recombination of NO and O_2 occurs mainly on the picosecond time scale.

The hemoglobin example illustrates how, by means of ultrafast kinetic absorption and Raman spectroscopies, one can elucidate subtle changes in macromolecular structures that determine a key biological activity. Examples of the power of such techniques applied to vision or photosynthesis will be provided in other chapters.

4.2.2 Detection, Formation Quantum Yield, and Molar Extinction Coefficient of the First Excited Triplet State in Fluid Solutions

Because the triplet state is a key intermediate in many photobiological processes, we will present a rather detailed description of its spectroscopic properties as studied by fast kinetics optical spectroscopy. The $^3A_1 \rightarrow {}^3A_q$ transient absorption of anthracene (A) shown in Figure 4.2A is a good model for presenting the basic methodology of the determination of the triplet state parameters that govern its chemical reactivity.

4.2.2.1 Identification of Triplet–Triplet Transient Absorption

Three criteria must be simultaneously fulfilled for attributing the transient absorption obtained after pulsed laser excitation of anthracene in its $^1A_0$ state to $^3A_1 \rightarrow {}^3A_q$ transitions (Figure 4.2A):

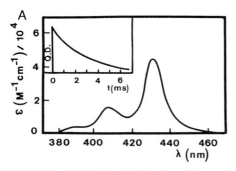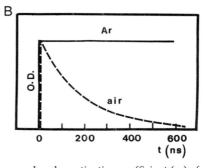

FIGURE 4.2   (A) Triplet–triplet absorption spectrum and molar extinction coefficient (ϵ_T) of anthracene in argon-saturated benzene at room temperature. The exponential decay of the triplet transient is presented in the insert. The triplet concentration is proportional to the optical density (O.D.) monitored at 435 nm after 353-nm laser pulse excitation. The lifetime of the $^3A_1$ excited state (τ_T = 2.9 msec) is readily calculated from these data. (B) Quenching of the $^3A_1$ triplet by oxygen in benzene. At this time scale, the long-lived triplet state (2.9 msec) looks like a very stable species under argon (Ar) saturation, but its lifetime is reduced to 200 nsec in an air-saturated solution. The lifetime shortening can be explained in terms of the collisional quenching reaction of $^3A_1$ by molecular oxygen, applying pseudo-first-order kinetics, with $k_T(O_2)$ given in Table 4.1. Pseudo-first-order kinetics is applicable because $[{}^3O_2]$ is largely in excess of $[{}^3A_1]$.

TABLE 4.1   Oxygen Concentrations: Triplet and Singlet State Bimolecular Reaction Rates[a]

Compound	Solvent	$[O_2]$ (μM)	$k_T(O_2)$ $(/10^9\ M^{-1}\ sec^{-1})$	$k_F(O_2)$ $(/10^{10}\ M^{-1}\ sec^{-1})$
Anthracene	Benzene	1900	2.8	2.6
Naphthalene	Ethanol	2100	2	2.1
Indole	Cyclohexane	2300	16	3.1
Indole	Water	270	7	1.2
Trptophan residue in corticotropin	Phosphate buffer	180	5.1	0.8
Protoporphyrin dimethylester	Benzene	1900	2.7	1.8

[a] Oxygen concentrations under air saturation; reaction rate constants are for the quenching of fluorescence $[k_F(O_2)]$ and for the triplet state $[k_T(O_2)]$ of aromatic molecules and heterocycles by oxygen in a variety of solvents at room temperature.

1. *This transient has an exponential decay* from which the lifetime of the $^3A_1$ state (τ_T) is obtained (Figure 4.2B). The reciprocal of τ_T is k_T, the rate constant for the $^3A_1 \rightarrow {}^1A_0$ transition (see Chapter 2). This lifetime must be wavelength independent through the $^3A_1 \rightarrow {}^3A_q$ absorption band. Any deviation from this rule suggests the presence of an underlying transient absorption, as was illustrated for tryptophan in Figure 3.11.

2. *This transient must be quenched by oxygen* (Figure 4.2B). Table 4.1 gives oxygen concentration at air saturation and the bimolecular reaction rate of triplet states of several aromatic molecules (M), including some molecules of biological interest in various solvents:

$$^3M_1 + {}^3O_2 \longrightarrow ({}^3M_1{}^3O_2) \longrightarrow {}^1M_0$$

It can be seen that the k_T (O_2) constant is much lower than the rate constant for the fluorescence quenching by oxygen $[k_F\ (O_2)]$, which is also given in Table 4.1. The reason for this discrepancy arises from several factors, which will be discussed in Section 4.4, which deals with singlet oxygen.

3. *This transient absorption must be quenched by a ground state acceptor molecule with concomitant formation of the acceptor triplet state due to triplet–triplet energy transfer.* The best acceptor in deaerated nonpolar solvent is β-carotene, which has a triplet energy of 75 kJ mol^{-1}. Thus, when β-carotene (C) is submitted to direct laser photolysis, no triplet carotene is observed (Figure 4.3) because of its low intersystem crossing quantum yield (see Chapter 2). On the other hand, in deaerated solutions containing anthracene (A) and β-carotene, one observes after the laser flash the sensitized production of triplet β-carotene due to the triplet–triplet energy transfer (Figure 4.3). The $^3C_1$ deactivation leads to ground state

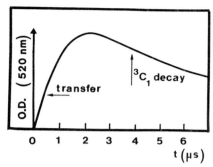

FIGURE 4.3 Transfer of triplet anthracene ($^3A_1$) energy to β-carotene at room temperature in deaerated benzene as solvent. The $^3A_1$ is formed during the 10-nsec, 353-nm laser pulse. The optical density of the β-carotene triplet ($^3C_1$) is measured at 520 nm, where $^3C_1$ maximally absorbs and $^3A_1$ does not (see Figure 4.2a). The β-carotene concentration ($^1C_0$) (500 μM) must be chosen so that it is always larger in excess of [$^3A_1$]. Note the time scale (about 2 μsec) for the triplet transfer at [$^1C_0$] = 500 μM.

β-carotene; e.g., $^3A_1 + {}^1C_0 \rightarrow {}^3C_1 + {}^1A_0$ and $^3C_1 \rightarrow {}^1C_0$, where $^1A_0$ is the ground state of the triplet donor. The β-carotene triplet–triplet transient absorption is maximal at 520 nm and decays with a lifetime of 7.7 μsec in benzene at room temperature. Another useful probe of triplet formation evidenced by triplet–triplet transfer is *retinol*, which is soluble in short aliphatic alcohols such as methanol and ethanol. Like β-carotene, retinol has a low intersystem quantum yield (0.03) and a low-lying first excited triplet state at 155 kJ mol^{-1}. In ethanol, the transient triplet state of retinol has an absorption maximum at 405 nm.

4.2.2.2 Determination of Molar Extinction Coefficients and Triplet Formation Quantum Yield

The quantum yield of triplet formation, or intersystem crossing quantum yield (Φ_T), of an aromatic molecule (M) can be obtained by direct measurement of the initial transient absorbance (or initial optical density: (O.D.$_T$) and molar extinction coefficients [$\epsilon_{(TA)}$] of the $^3M_1 \rightarrow {}^3M_q$ absorption and of the light absorbed by the sample (I_{abs}) to raise molecules from the $^1M_0$ to the $^1M_1$ state:

$$\Phi_T = O.D._T / \epsilon_{(TA)} I_{abs} \qquad (4.1)$$

The laser flash excitation energy is controlled so that less than 5% of the chromophore molecules are raised from the $^1M_0$ state to the $^3M_1$ state in order to avoid significant ground state depopulation. The incident light fluence can be measured by chemical actinometry or with a calibrated joulemeter and transformed into appropriate units to be used in Eq. (4.1) [see also Chapter 2, Eq. (2.14)]. The molar excitation coefficient $\epsilon_{(TA)}$ is obtained from a separate experiment performed with a laser flash of high fluence to produce full depopulation of the ground state. Alternatively, $\epsilon_{(TA)}$ can be determined from a triplet–triplet transfer. Thus, assuming total transfer, the ratio between the optical density of the donor triplet at the transient maximum $(O.D._T)_D$ and that of the acceptor $(O.D._T)_A$ is equal to the ratio of the corresponding coefficients $(\epsilon_T)_D/(\epsilon_T)_A$. The molecule studied (M) can be either the donor or the acceptor.

A comparison method can also be used for the determination of Φ_T. It is based on relative actinometry by comparing the transient triplet absorption of the chromophore (M) $(O.D._T)_M$ to that of a reference compound (R) for which the molar extinction coefficients $(\epsilon_T)_R$ and the quantum yields of triplet formation $(\Phi_T)_R$ are known. Thus, under conditions where $(O.D._T)_R$ is proportional to the incident laser light energy (I), the light absorbed (I_{abs}) by ground state R is

$$I_{abs} = [I(O.D._T)_R]/(\Phi_T)_R(\epsilon_T)_R \qquad (4.2)$$

Hence, if the optical density of solutions containing ground state R and M are identical, on excitation with the same number of incident photons one has

$$(\Phi_T)_M = (\Phi_T)_R(O.D._T)_M(\epsilon_T)_R/(O.D._T)_R(\epsilon_T)_M \qquad (4.3)$$

Again, this calculation is valid only under conditions wherein ground state depopulation is $<5\%$ for the two solutions. Two reference compounds are generally used, depending on the absorption range of M. Thus for excitation wavelengths shorter than 300 nm, solutions of

TABLE 4.2   Molar Extinction Coefficients and Triplet Formation Quantum Yields[a]

Compound	Solvent	$\epsilon_T(\lambda_{max})$ $(M^{-1}\,cm^{-1})$	Φ_T
Anthracene	Cyclohexane	64,700 (422.5)	0.71 ± 0.1
β-Carotene	Benzene	200,000 (500)	≈ 0.001
Chlorophyll a	Ether	40,000 (430)	0.61
Chlorophyll b	Ether	21,000 (450)	0.88
Tryptophan	Aqueous, pH 7	5000 (450)	0.18
Eosin	Aqueous, pH 9	28,000 (520)	0.71
Riboflavin	Water	5000 (670)	0.7
Retinol	Ethanol	80,000 (405)	≈ 0.01
Naphthalene	Cyclohexane	24,500 (414)	0.75
Uroporphyrin octamethyl ester	Benzene	32,000 (440)	0.7
8-Methoxypsoralen	Acetonitrile	10,000 (440)	0.27
Nalidixic acid	Aqueous, pH 9	≤ 9000 (620)	>0.6

[a] Molar extinction coefficients (ϵ_T) are at maximum transient absorption $(\lambda_{max}$, in parentheses, in nanometers); the triplet formation quantum yields (Φ_T) are for various molecules at room temperature in fluid solution.

naphthalene in cyclohexane are preferred, whereas anthracene is used as a standard for longer ultraviolet wavelengths. Table 4.2 provides the molar extinction coefficient at the absorption maximum and the quantum yield of formation of some molecules in their first excited triplet state.

4.3 LASER FLASH PHOTOLYSIS: A TOOL FOR THE ELUCIDATION OF EXCITED STATES INVOLVED IN PHOTOCHEMICAL PROCESSES

Whereas the preceding section dealt with reversible photochemical processes, in this section it will be explained how laser flash photolysis can help in elucidating very complicated direct or photosensitized chemical reactions. Thus, mechanistic studies of photochemical reactions at either the singlet or triplet excited state will be exemplified.

4.3.1 Photoionization of Tryptophan in Aqueous Medium

On pulsed laser excitation of deaerated aqueous tryptophan solutions with wavelengths shorter than 305 nm, a broad, transient absorption appears in the wavelength range 310–750 nm (Figure 4.4A). This transient absorption decays according to complex kinetics, suggesting the presence of several species (Figure 4.4B). The initial transient with maximum at 750 nm ($t = 0$ in Figure 4.4A) is suppressed by saturating the Trp solution with N_2O prior to laser excitation, and its lifetime is shortened

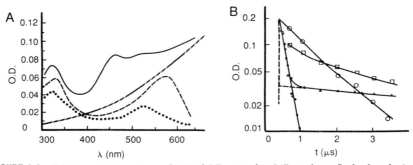

FIGURE 4.4 (A) Transient spectrum obtained 0.7 μsec after 265-nm laser flash photolysis of tryptophan in aqueous solution at room temperature. (—), Argon saturation, pH 7; (---), oxygen saturation, pH 3, corresponds to the cationic Trp^+ transient; (·······), oxygen saturation, pH 7, corresponds to the neutral $Trp\cdot$ transient. The absorption spectrum of hydrated electrons (e_{aq}) obtained by pulse radiolysis of water is also represented (-- --). (B) Semilogarithmic plot of the transient optical density (O.D.) decays corresponding to (a). (○), 650 nm, argon saturation, decay of e_{aq}; (●), 650 nm, air saturation, reaction of e_{aq} with O_2; (□), 510 nm, argon saturation, decay of $^3Trp_1$ (with some contribution of e_{aq} at short delay and $Trp\cdot$ after 3 μsec); (▲), 510 nm, air saturation, decay of $Trp\cdot$ (with some contribution of e_{aq} and $^3Trp_1$ at short delay).

under air saturation (Figure 4.4B). All the above properties characterize the formation of hydrated electrons (e_{aq}) because the following scavenging reactions,

$$N_2O + e_{aq} \longrightarrow N_2 + OH\cdot + OH^-$$

$$O_2 + e_{aq} \longrightarrow O_2^-$$

occur at diffusion-controlled rates; e.g., the bimolecular reaction rate constant for these reactions is $\approx 2.10^{10}\ M^{-1}\ \text{sec}^{-1}$. Because N_2O is much more soluble than O_2 in water (by an order of magnitude), hydrated electrons react with N_2O within the time response of the laser flash photolysis equipment (for instance, about 30 nsec with laser pulses of 10-nsec duration).

The electron formation can therefore be attributed to the photo*ionization* (that is to say, a photo*oxidation*) of the $^1Trp_1$ singlet excited state with a quantum yield (Φ_i) of ≈ 0.08:

$$^1Trp_0 + h\nu \longrightarrow Trp^{\ddagger} + e_{aq} \tag{4.4}$$

The transient absorptions that remain after the e_{aq} decay in deaerated solutions can now be identified. Thus the transient absorption peaking at 450 nm decays exponentially with a lifetime of 10 μsec and is quenched by oxygen at rate $2 \times 10^9\ M^{-1}\ \text{sec}^{-1}$ (Figure 4.4B). It is attributed to the $^3Trp_1$ transient absorption.

A glance at Eq. (4.4) may suggest that the long-lived transient remaining after 100 μsec is the tryptophan radical cation. *It must be both a cation, because tryptophan has been photoionized, and a radical, because tryptophan with all electrons paired in the ground state has lost one electron by photoionization.* Actually, this radical-cation, with absorption maxima at 330 and 600 nm, rapidly deprotonates ($pK_a = 4.3$), leading to the neutral radical

$$Trp^{\ddagger} + H_2O \longrightarrow H_3O^+ + Trp\cdot$$

which absorbs at 320 and 520 nm with molar extinction coefficients of 3500 and 1750 $M^{-1}\ cm^{-1}$, respectively. This neutral tryptophyl radical decays over several milliseconds by bimolecular reaction, with rate constant $\approx 10^8\ M^{-1}\ \text{sec}^{-1}$, according to the following equation:

$$Trp\cdot + Trp\cdot \longrightarrow \text{products}$$

It should be noted that the general rule, $\Phi_F + \Phi_T \approx 1$, valid for most aromatic molecules in organic solvents (see Chapter 2), does not hold at all for Trp in aqueous solutions ($\Phi_F + \Phi_T \approx 0.4$). Photoionization does not explain this discrepancy because $\Phi_i \approx 0.08$. Strong interactions with solvent molecules responsible for the large fluorescence Stokes shift

(10,000 cm^{-1}) of Trp in aqueous solutions may explain the high quantum yield of internal conversion ($\Phi_{ic} \approx 0.5$).

As will be shown in Chapter 5, Trp is a key target of solar UV radiation because it is the only aromatic amino acid of proteins that can absorb this radiation. Basically, in proteins the primary photochemical process is still photoionization of tryptophan. However, due to the protein structure the photoionization quantum yield varies from protein to protein. In proteins containing disulfide bonds, the electron photoejected from Trp can either be solvated (e.g., forming e_{aq}) or may directly react with cystine residues forming the Cys S—S$^{\cdot}$—Cys radical, which strongly absorbs in the visible ($\lambda_{max} \approx 410$ nm, $\epsilon = 9000\ M^{-1}$ cm^{-1}). The photoreduction of the disulfide bond leads to Cys—S—S—Cys bond splitting and protein photoinactivation. The e_{aq} produced by Trp residue photoionization can also react with protonated residues such as His, leading to their destructive reduction. *The reactions in which chemical energy is transferred from Trp to other residues in the protein are examples of photosensitized reactions induced by the first excited singlet state.*

4.3.2 An Insight into the Reactivity of the Triplet State of Aromatic Molecules: A Clue to the Understanding of Photosensitized Reactions

The rather long lifetime of the first triplet state with respect to that of the first excited singlet state has an important consequence; e.g., the probability for collision of triplet molecules with reactive molecules (substrates) is enhanced in the same proportion.

A pedagogic study of the reactivity of the triplet state with another molecule (called a substrate) can best be illustrated by laser flash photolysis of deaerated aqueous solutions of nalidixic acid in the presence or in the absence of a substrate of biological interest. Nalidixic acid (1,4-dihydro-1-ethyl-7-methyl-4-oxo-1,8-naphthyridine-3-carboxylic acid) is commonly used all over the world as an antibacterial agent of the urinary tract. Its main side effect is the marked photosensitivity of the skin of patients, who experience bullous eruptions after exposure to ultraviolet light. Because we have already characterized the main transient species that can be obtained with tryptophan, we choose this important biological substrate as an illustrative example.

Figure 4.5A shows the transient absorption obtained on 353-nm laser flash photolysis of neutral or slightly basic aqueous nalidixic acid (NA) solutions. Under physiological conditions (pH 7.4), the predominating species are the anionic form (NA^{-}) because the pK_a of the carboxylic group is 6.02. This anion absorbs light in the ultraviolet. The absorption maximum is at 336 nm with a molar extinction coefficient of 18,300 M^{-1}

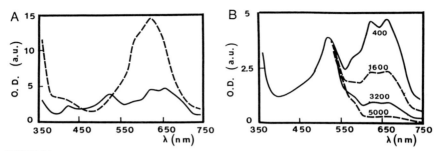

FIGURE 4.5 (A) Room temperature transient absorption spectrum of a deaerated solution of 400 μM nalidixic acid anion (NA^-) on 353-nm laser flash photolysis in the presence of 1 mM tryptophan. (----), Immediately after the laser flash (i.e., transient triplet spectrum); (—), 5-μsec delay, the $Trp\cdot$ is identified by its absorption at 510 nm and the maximums at 640 and 660 nm are due to the $NA^{2-} + Trp^{\ddot +}$ radical-dianion. (B) Same as (a) but under air saturation. While the $Trp\cdot$ transient is unaffected by oxygen, the NA^{2-} rapidly vanishes, as shown by the transient spectra recorded from 400 to 5000 μsec.

cm^{-1}. Immediately after the laser flash the oxygen-sensitive transient absorption with $\lambda_{max} = 620$ nm is attributed to the $^3(NA^-)_1$ spectrum. The $^3(NA^-)_1$ transient exponentially decays with a lifetime $\tau_0 = 100$ μsec. In the absence of Trp, the $^3(NA^-)_1$ transient lifetime is reduced to 500 nsec in oxygen-saturated medium.

The addition of Trp to deaerated NA^- solution leads to a shortening of the $^3(NA^-)_1$ lifetime (quenching of the triplet molecule by Trp) and the appearance of new transient absorbances (Figure 4.5A).

In the presence of 1 mM Trp, 5 μsec after the laser flash, formation of the long-lived Trp radical is observed (λ_{max} at 520 nm) overlapping another long lived-transient. These observations are consistent with the following reaction sequence,

$$^1(NA^-)_0 \xrightarrow{h\nu\ (353\ nm)} {}^3(NA^-)_1$$

$$^3(NA^-)_1 \longrightarrow {}^1(NA^-)_0 \qquad \text{rate constant: } k_T(k_T = 1/\tau_0)$$

$$^3(NA^-)_1 + {}^1Trp_0 \longrightarrow NA^{2-} + Trp^{\ddot +} \qquad \text{rate constant: } k_q(Trp) \quad (4.5)$$

where NA^{2-} is the semireduced NA^- species (e.g., a radical-dianion) and $Trp^{\ddot +}$ is the semioxidized substrate. From the above, we know that this radical-cation rapidly deprotonates at neutral pH:

$$Trp^{\ddot +} \longrightarrow Trp\cdot + H_3O^+$$

Hence, in the presence of 1 mM Trp, the decay of $^3(NA^-)_1$ can be represented by

$$d[^3(NA^-)_1]/dt = k_T[^3(NA^-)_1] + k_q(Trp)[^3(NA^-)_1][Trp] \qquad (4.6)$$

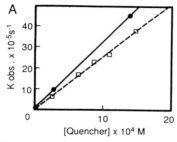
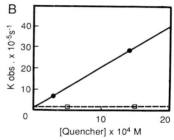

FIGURE 4.6 Stern–Volmer plots for the quenching of (A) $^3(NA^-)_1$ and (B) NA^{2-} by Trp (□) (deaerated solutions) or oxygen (●). The ordinate (K_{obs}) is simply $(\tau_0/\tau) - 1$ in Eq. (4.6).

The triplet concentration at time origin, $[^3(NA^-)_1]_0$, can be obtained with the Beer–Lambert law and data in Table 4.2. Under usual experimental conditions, the laser pulse energy is chosen so that no more than 5% of $^1(NA^-)_0$ is transformed into $^3(NA^-)_1$. Thus, [Trp] is always largely in excess of $[^3(NA^-)_1]$ and pseudo-first-order kinetics apply. Therefore, the time dependence of $[^3(NA^-)_1]$ is represented by:

$$[^3(NA^-)_1] = [^3(NA^-)_1]_0 \exp - \{k_T + k_q(Trp)[Trp]\}t$$

Thus, in the presence of Trp, the $^3(NA^-)_1$ lifetime is

$$1/\tau = 1/\tau_0 + k_q(Trp)[Trp]$$

which can be re-written as

$$(\tau_0/\tau) - 1 = \tau_0 k_q(Trp)[Trp] \qquad (4.7)$$

This is strictly equivalent to the Stern–Volmer law for quenching of the first excited singlet state that we established in Chapter 2. Data in Figure 4.6 make it possible to determine a bimolecular rate constant $k_q(Trp)$ equal to $1.4 \times 10^9 \ M^{-1} \ sec^{-1}$. The relative magnitude of $k_q(S)$ for various substrates (S) would characterize their oxidizability by $^3(NA^-)_1$.

This example is of quite general interest for many photosensitized reactions encountered in photobiology. *The photosensitizer (here NA) absorbs the light that is not absorbed by the substrate (Trp).* The substrate can be either semioxidized (good electron or hydrogen donors such as Trp, vitamin C, and guanine) or very rarely semireduced (electrophilic compounds such as nitromidazoles). As a result, the photosensitizer is either semireduced or semioxidized. The semireduced NA^{2-} transient species absorb at about 650 nm (Figures 4.5A and 4.5B).

In deaerated aqueous solutions, recombination processes such as

$$2NA^{2\cdot-} \longrightarrow \text{products}$$

$$NA^{2\cdot-} + Trp\cdot + H_3O^+ \longrightarrow NA^- + Trp \quad \text{(or products)}$$

$$Trp\cdot + Trp\cdot \longrightarrow \text{products}$$

may take place. However, photobiology is mainly concerned with *in vivo* situations, wherein oxygen is a determining factor. Importantly, *a photosensitized reaction occurring in the presence of oxygen is called a photodynamic reaction*. If, as above (see Eq. (4.5)), the substrate first reacts with the excited state (generally the triplet state) and then the semioxidized substrate reacts with O_2, the photosensitized reaction (a photosensitization) is called a *type I photodynamic reaction*. In the above example, the fate of the radical species depends on the presence of oxygen:

$$Trp\cdot + O_2 \longrightarrow \text{products} \quad \text{(very slow reaction: see Figure 4.6)}$$

accompanied by

$$NA^{2\cdot-} + O_2 \longrightarrow NA^- + O_2^{\cdot-} \quad \text{(very fast reaction: see Figure 4.6).}$$

Overall, the photosensitizer is repaired whereas the substrate is fully oxidized. As mentioned in this chapter, triplet states are deactivated by oxygen with a rate constant approaching $2 \times 10^9 \, M^{-1} \, sec^{-1}$. As a result, in aerated solutions ($[O_2] = 2 \times 10^{-4} \, M$), an additional deactivation pathway $k_q[O_2][^3(NA^-)_1]$ must be added on the right-hand side of Eq. (4.6). The overall triplet lifetime becomes

$$1/\tau = 1/\tau_0 + k_q(Trp)[Trp] + k_q(O_2)[O_2] \qquad (4.7)$$

Thus, in the presence of oxygen, the extent of a type I photodynamic oxidation of Trp will be governed by the following ratio:

$$k_q(Trp)[Trp]/k_q(O_2)[O_2] \qquad (4.8)$$

We will now study in detail the critical role of the deactivation of triplet photosensitizers by oxygen (Figure 4.2b) in type II photodynamic reactions. This will of course apply to triplet nalidixic acid.

4.4 TYPE II PHOTODYNAMIC REACTIONS: FORMATION AND REACTIVITY OF SINGLET OXYGEN ($^1O_2$)

4.4.1 An Introduction to the Mechanism of $^1O_2$ Formation

In 1939, Kautsky proposed that singlet molecular oxygen ($^1O_2$) was the intermediate species in the dye-sensitized photooxidation of leucomalachite green absorbed on silica gel. It took many years before this proposal was accepted. Interest in $^1O_2$ was renewed in the 1960s and 1970s with the discovery of the importance of photosensitized reactions in medicine and biology.

Singlet oxygen can be produced by collision of ground state triplet oxygen with both the first singlet and triplet excited states of aromatic molecules (M) via the following processes:

$$^1M_1 + {}^3O_2 \longrightarrow ({}^1M_1 \cdots {}^3O_2) \longrightarrow {}^1O_2 + {}^3M_1 \qquad (a)$$

$$^3M_1 + {}^3O_2 \longrightarrow ({}^3M_1 \cdots {}^3O_2) \longrightarrow {}^1O_2 + {}^1M_0 \qquad (b)$$

Most biological photosensitizers have an $^1M_1$ lifetime in the nanosecond time scale and are often used in *aqueous environments* where the oxygen concentration under air saturation in smaller than 1 mM. Under these conditions, $^1O_2$ formation by process (a) can be neglected. However, Table 4.1 shows that for hydrophobic photosensitizers with a long $^1M_1$ lifetime in nonpolar solvents (a good example is pyrene, for which $\tau_F \sim$ 400 nsec), process (a) is no longer negligible.

Thus the most common way of producing $^1O_2$ is the quenching of the photosensitizer triplet state by triplet ground state oxygen. As already mentioned in Chapter 2, process (b) is formally equivalent to a triplet–singlet energy transfer. Table 4.1 data show that the bimolecular reaction rate constant for the quenching of triplet state of $^3O_2$ is much lower than that for quenching of the fluorescent state. It has been suggested by Stevens that quenching of triplet states ($^3M_1$) by $^3O_2$ yields $^1O_2$ with a rate constant that is limited by the spin statistical factor of 1/9, which results from the number of possible spin states that can be formed by combining individual spins in the ($^3M_1 \cdots {}^3O_2$) collisional complex.

$$\text{Probability} = 1/9: {}^3M_1 + {}^3O_2 \longrightarrow {}^1({}^3M_1 \cdots {}^3O_2) \longrightarrow {}^1M_0 + {}^1O_2$$

$$\text{Probability} = 3/9: {}^3M_1 + {}^3O_2 \longrightarrow {}^3({}^3M_1 \cdots {}^3O_2) \longrightarrow {}^1M_0 + {}^3O_2$$

$$\text{Probability} = 5/9: {}^3M_1 + {}^3O_2 \longrightarrow {}^5({}^3M_1 \cdots {}^3O_2) \longrightarrow {}^1M_0 + {}^3O_2$$

However, observed quenching rates in water are rather close to 4/9 of the diffusion rate constant of oxygen, which indicates that other factors, such as viscosity and polarity, can intervene in these processes. Also, quenching of the triplet state of tryptophan by oxygen in proteins can be very inefficient because of hindered oxygen penetration within the protein core.

4.4.2 Decay of $^1O_2$ in the Absence of a Chemical Reaction

The return of excited $^1O_2$ to ground state $^3O_2$ is a spin-forbidden transition. As a result, it can be expected that $^1O_2$ is a relatively long-lived species. As illustrated in Table 4.3, the lifetime of $^1O_2$, $\tau(^1O_2)$, is highly

TABLE 4.3   Lifetime of $^1O_2$ in Solution

Solvent	Lifetime (μsec)
Water	4
D_2O	60
Methanol	10
Ethanol	20
Benzene	24
Acetone	42
Chloroform	250
Dioxane	32
Acetonitrile	60

solvent dependent, being as low as ~ 4 μsec in water and as high as 700 μsec in carbon tetrachloride.

The most convenient and accurate method for measuring $\tau(^1O_2)$ is to record the decay kinetics of the $^1O_2 \rightarrow {}^3O_2$ spin-forbidden transition, which is responsible for chemiluminescence at 1270 nm. Indirect methods based on the bleaching of reactive substrates by $^1O_2$ or on energy transfer to β-carotene triplet have also been used, as outlined below. Kearns explained the nearly three orders of magnitude difference in solvent quenching rate constants by the transfer of the electronic energy of singlet oxygen into the vibrational modes of the solvent molecule. The solvent effects on the $^1O_2$ lifetime are very often used as a test for $^1O_2$ involvement in photosensitized reactions in aqueous media, because replacing H_2O by D_2O enhances $\tau(^1O_2)$ 15-fold. The photooxidation rate of a molecule reacting with $^1O_2$ is therefore increased in the same proportion in going from water to D_2O.

4.4.3 Mechanistic and Kinetic Study of the Chemical Reactivity of $^1O_2$ in Photosensitized Reactions Involving Biological Substrates

Many dye-sensitized photooxygenations of various acceptors (A) involve singlet oxygen, because they give products and relative reactivities identical to those obtained following chemical production of singlet oxygen, such as the reaction of hypochlorites with hydrogen peroxide. The following simple mechanism for photooxygenation by a type II mechanism with a photosensitizer (P) is well established:

$$^1P_0 + h\nu \longrightarrow {}^1P_0 \longrightarrow {}^3P_1 \qquad (\text{yield } \Phi_T)$$

$$^3P_0 + {}^3O_2 \longrightarrow {}^1O_2 + {}^1P_0$$

$$^1O_2 \longrightarrow {}^3O_2 \qquad (\text{rate constant } k_1)$$

$$^1O_2 + A \longrightarrow AO_2 \quad \text{or} \quad \text{other products} \qquad (\text{rate constant } k_r; \\ \text{chemical quenching})$$

$$^1O_2 + A \longrightarrow {}^3O_2 + A \qquad (\text{rate constant } k_q; \text{physical quenching})$$

It follows that under photostationary conditions, a calculation similar to that carried out for the establishment of the Stern–Volmer law for fluorescence quenching (see Chapter 2) leads to

$$\Phi(AO_2) = \Phi_T k_r[A]/k_1 + k_r[A] + k_q[A] \qquad (4.9)$$

or

$$1/\Phi(AO_2) = (1/\Phi_T)(\alpha + \beta/[A]) \qquad (4.10)$$

where $\alpha = (k_r + k_q)/k_r$ and $\beta = k_1/k_r$. From the slope and intercept of the linear plots of $1/\Phi(AO_2)$ versus $[A]^{-1}$, values of β can be obtained provided Φ_T is known, and thus relative reactivities under identical solvent conditions can be determined. In methanol, $\tau(^1O_2)$ is $\sim 10~\mu$sec, $k_1 \approx 1 \times 10^5$ sec^{-1}, and k_r can reach values such as $10^9~M^{-1}$ sec^{-1}, as shown in Table 4.4; consequently, compounds have β values as low as $10^{-4}~M$. Table 4.4 also gives k_r for a number of biological substrates and other molecules or ions of particular interest.

In general, the degradation of substrates during the photosensitized reaction can be followed by spectrophotometry (DPBF), fluorescence (Trp residues in proteins), or high-performance liquid chromatography (HPLC) (for most compounds). The k_q value represents the so-called physical quenching rate constant of $^1O_2$ by the substrate A; e.g., in the collisional complex ($^1O_2 \cdots A$), $^3O_2$ is restored by various processes that will be examined below. The combined bimolecular rate constant for the $^1O_2$ quenching ($k_q + k_r$) can easily be obtained by time-resolved spectroscopy of the $^1O_2 \rightarrow {}^3O_2$ transition emission at 1270 nm. Basically a Stern–Volmer relationship similar to Eq. (4.6) is obtained by recording the variation of $\tau(^1O_2)$ as a function of the substrate concentration [A]. Thus combination of continuous irradiation and $\tau(^1O_2)$ determination allows [see Eq. (4.10)] the determination of the triplet formation quantum yield of the photosensitizer [Φ_T in Eq. (4.10)].

TABLE 4.4   Bimolecular Reaction Rate Constants of the Chemical (k_r) and Chemical Plus Physical Quenching ($k_r + k_q$) of $^1O_2$ by Selected Compounds

Substrate (solvent)	k_r $(M^{-1} sec^{-1})$	$k_r + k_q$ $(M^{-1} sec^{-1})$
Trp (methanol)	4×10^6	4×10^7
His (methanol)	7×10^6	5×10^7
Met (methanol)	5×10^6	3×10^7
CysH (D_2O, pD = 7)	—	8.9×10^6
Vitamin E (methanol)	4.6×10^7	6.2×10^8
Vitamin C (D_2O)	—	1.6×10^8
Cholesterol (pyridine)	—	6.6×10^4
Linolenic acid (C_6D_6)	—	1×10^5
Methyl arachidonate (pyridine)	2.2×10^5	—
Oleic acid (C_6D_6)	—	5.3×10^3
Guanosine (D_2O)	—	5×10^6
NADH (acetonitrile–methanol)	—	7.4×10^7
1,3-Diphenylisobenzofurane (DPBF) (methanol)	9.8×10^8	1.2×10^9
β-Carotene (methanol)	9.1×10^4	1.5×10^{10}
N_3^- (water)	—	7.9×10^8
1,4-Diazabicyclo[2.2.2]octane (DABCO) (water)	—	2.4×10^7
Bilirubin (CH_3Cl)	4×10^8	2.1×10^9
Chlorophyll a (CCl_4)	—	7.0×10^8

4.4.3.1 Main Mechanisms of the Singlet Oxygen-Mediated Reactions

The two principal reaction pathways of singlet oxygen are cycloaddition and "ene" reaction. They are shown in Figure 4.7. The formation of endoperoxides by cycloaddition requires a conjugated double bond, whereas the formation of hydroperoxides requires an allylic hydrogen. When the double bonds are surrounded by electron-donating groups, the activation energy for both reactions is nearly zero and the rate constant then approaches its maximum value of about 10^9 M^{-1} sec^{-1}. The formation of dioxetanes in singlet oxygen reactions has a much higher activation energy and is observed chiefly when other paths are blocked. Dioxetanes are produced via the process of 1,2-addition and are thermally decomposed into carbonyl compounds.

Among the most important biological substrates for singlet oxygen (see Table 4.4), the lipid-soluble vitamin E and the hydrophilic vitamin C are efficient $^1O_2$ scavengers. This explains their use as "photoprotectors" in some sun-tanning preparations (also see Chapter 5). The amino acids Trp, His, Met, and Cys readily react (rate constant $\approx 10^7$ M^{-1} sec^{-1}) with $^1O_2$. Whereas His and Met are specific type II photodynamic substrates, Trp and CysH are equally good type I photodynamic substrates, as shown above. The photosensitized degradation of these residues in pro-

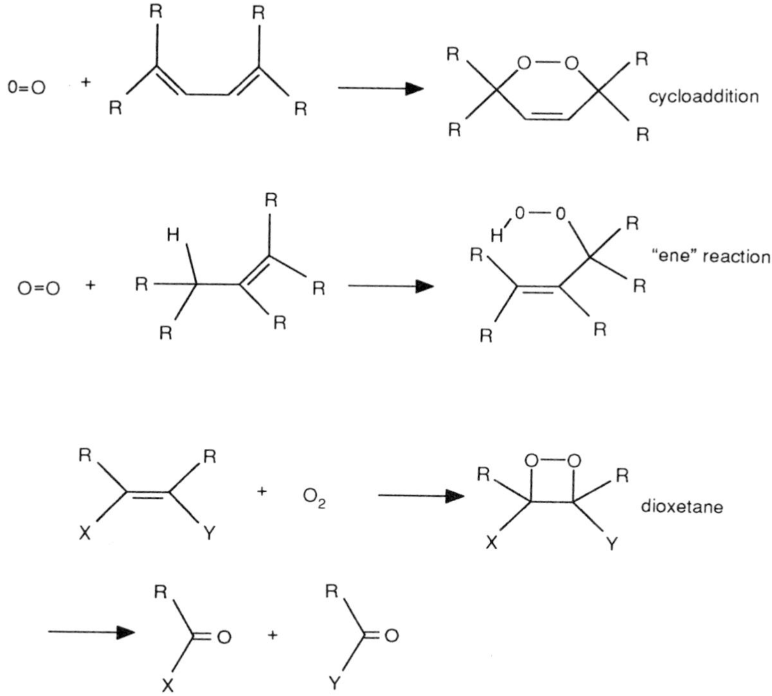

FIGURE 4.7 Characteristic reactions of singlet oxygen.

teins has important consequences, such as the inactivation of enzymes. The singlet oxygen involvement in the photosensitized inactivation can be easily proved using the D_2O/N_3^- test: D_2O enhances the photoinactivation rate whereas addition of N_3^- leads to inhibition of the photodynamic inactivation (Figure 4.8A).

Because only a few residues are susceptible to photodynamic reactions, these reactions can be used as a tool for the determination of crucial residues in the active site of enzymes. For instance, the methylene blue- and eosin-photosensitized degradation of Trp-108 in lysozyme is responsible for most of the loss of lysozyme activity, suggesting the participation of this key residue in the proteolytic activity of this enzyme, although lysozyme contains six Trp residues, of which only the fluorescent Trp-108 is destroyed (Figure 4.8B).

Among nucleic acid bases, only xanthine and to a lesser extent guanine are significant $^1O_2$ substrates, whereas fatty acids with unsaturated bonds moderately react with $^1O_2$ (the greater the number of double bonds, the faster the rate constant). Both endoperoxides and hydroperoxides of fatty acids are formed. Like unsaturated fatty acids, they may have double

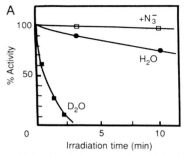
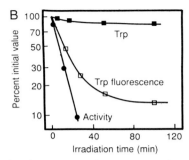

FIGURE 4.8   (A) Effect of D$_2$O and 0.01 M N$_3^-$ on the photosensitized inactivation of trypsin by 10 μM methylene blue in air-saturated aqueous solutions irradiated at room temperature with red light from a tungsten–halogen lamp. (B) Photodynamic inactivation of lysozyme by 10 μM methylene blue in pH 8 phosphate buffer. Note the excellent correlation between the loss in Trp fluorescence and enzyme activity. Meanwhile, only 1 out of 6 Trp residues is destroyed.

bonds and allylic hydrogens. The comparatively low reactivity of fatty acids with $^1O_2$ is compensated by the dark reactions of the lipid peroxidation radical chain, which will be examined in more detail in Chapter 5.

Cholesterol (see Figure 4.9) is a very important $^1O_2$ substrate because it gives rise by the "ene reaction" to the 5α-OOH-cholesterol, a specific product that is often used as proof for $^1O_2$ involvement in photosensitized reactions in biological membranes. The analytical application of the $^1O_2$ quenching by the last three substrates given in Table 4.4 (β-carotene, N$_3^-$, and DABCO) will be examined below.

4.4.3.2 Mechanisms of Physical Quenching of Singlet Oxygen

Apart from its physical quenching by solvent molecules, two other mechanisms have been postulated for the quenching of $^1O_2$.

4.4.3.2.1 Charge Transfer Quenching of $^1O_2$
Different compounds, including amines, sulfides, and bilirubin, have been suggested to quench singlet oxygen by a charge transfer (CT) mechanism. The suggested

FIGURE 4.9   Site of singlet oxygen reaction on cholesterol.

mechanism is

$$^1O_2 + A \longrightarrow \overset{\delta^- \quad \delta^+}{[O_2 \cdots A]} \longrightarrow {}^3O_2 + A$$

However, many of these compounds (some of them are listed in Table 4.4) not only physically quench but also react with $^1O_2$. It seems that the ratio of quenching to reaction depends on the polarity of the solvent. Thus, amines with low ionization potential are better quenchers. This results in the reactivity sequence tertiary > secondary > primary amines. A small spin–orbit coupling between the singlet and triplet states in the CT intermediate would allow a spin flip to occur and hence a facile intersystem crossing from singlet to triplet oxygen.

DABCO is only moderately effective as a physical $^1O_2$ quencher ($k_q \sim 2.4 \times 10^7\ M^{-1}\ sec^{-1}$; see Table 4.4). However, like the related N_3^- ions, it is a stable water-soluble compound and it does not absorb light of wavelengths longer than 300 nm. Therefore both compounds are widely used as a diagnostic test for the participation of $^1O_2$ in photodynamic reactions, as illustrated in Figure 4.8A.

4.4.3.2.2 Quenching by Energy Transfer Various compounds have been reported as effective quenchers of singlet oxygen in solution via an energy transfer mechanism. One of the most effective of these is β-carotene (C), which exhibits diffusion-controlled quenching (see Table 4.4).

$$^1O_2 + {}^1C_0 \longrightarrow {}^3O_2 + {}^3C_1$$

In fact, most natural polyenes, including not only β-carotene but also related compounds, the so-called carotenoids, are very efficient $^1O_2$ quenchers. Actually, to be effective $^1O_2$ quenchers, carotenoids must have at least 8 to 9 conjugated double bounds (Figure 4.10).

The destruction of biopolymers in living organisms by photosensitized oxidations was recognized long before the discovery of $^1O_2$. It also became rapidly clear that carotenoids were present in large quantities in living systems irradiated with solar light and containing a photosensitizer and oxygen. The total production of carotenoids in nature has been conservatively estimated at about 10^8 tons per year. Most of this huge output is in the form of four major carotenoids (fucoxanthin, lutein, violaxanthin, and neoxanthin) containing at least 9 conjugated double bonds, a prerequisite for high quenching activity. The probable mechanisms by which carotenoid pigments protect components of the living organism against photosensitized oxygenation can now be understood by taking into account all the information contained in this chapter; these mechanisms are as follows:

1. β-Carotene is an effective quencher of all natural photosensitizers with triplet state energy greater than 75 kJ mol^{-1}. Thus, chlorophyll a and

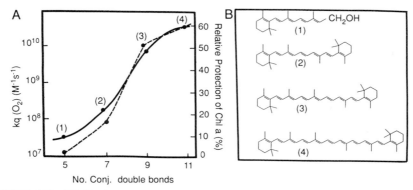

FIGURE 4.10 (A) Singlet oxygen quenching rate $[k_q(O_2); (\text{—})]$ and protective action (---) against photobleaching of chlorophyll a (Chl a) as a function of the number of conjugated double bonds of polyenes. (B) Structure of polyenes: (1), retinol; (2 and 3), two synthetic β-carotene analogs; (4), β-carotene. Reproduced from Foote *et al.* (1970). Copyright 1970, American Chemical Society. Used with the permission of the American Chemical Society.

b triplets (their energy is about 130 kJ mol^{-1}) and porphyrin triplets can be effectively quenched by β-carotene. However, triplet–triplet energy transfer requires overlap of the donor and acceptor molecules. The strong hydrophobicity of β-carotene and all other major carotenoids implies that only triplet states of hydrophobic photosensitizers (chlorophylls are good examples!) can be effectively quenched by carotenoids.

2. Carotenoids with more than 8 to 9 double bonds quench $^1O_2$ formed in hydrophobic structures. Thus, chlorophyll a, which readily reacts with $^1O_2$ (see Table 4.4), is rapidly bleached by self-photosensitization (Figure 4.10). In nature, the ability of these carotenoids to quench the chlorophyll photobleaching by $^1O_2$ may well be part of their protective function.

3. In addition to the deactivation of photosensitizing triplets, the polyene chain of β-carotene also reacts with oxygenated radicals or peroxides. β-Carotene can therefore scavenge other reactive oxygen species formed, for instance during lipid peroxidation, and that can be very harmful to living systems (see Chapter 5).

4.4.4 Transformation of a Type II Photodynamic Reaction into Type I

In contrast with photosensitizers related to ketones, pyronins, acridines, etc., which are good type I photosensitizers, chorophylls and porphyrins are almost exclusively type II photosensitizers. Their triplet state cannot directly oxidize substrates in competition with $^1O_2$ formation. The low electron affinity of the porphyrin triplet state is coupled to its electron-donating properties. As a result, electron affinic nitroheterocyclic compounds such as 2-nitroimidazole derivatives (some of them are

TABLE 4.5 Biomolecular Reaction Rate Constants[a]

Quencher	k_q ($/10^9$ M^{-1} sec^{-1})
O_2	1
Metronidazole	0.85
Misonidazole	0.67

[a] Bimolecular reaction rate constants (k_q) at room temperature in aqueous medium for the quenching of the uroporphyrin triplet state by oxygen and nitroimidazoles. Metronidazole is (2'-hydroxyethyl)-1-methyl-2-nitro-5-imidazole and misonidazole is α-(methoxymethyl)-2-nitro-1-H-imidazole-1-ethanol.

powerful antibiotics) can oxidize the triplet state of chlorophyll or porphyrins (P) according to a reaction scheme that can be easily deduced from laser flash photolysis experiments. This scheme is similar to those described in this chapter regarding the reaction of the nalidixic acid triplet with Trp (Section 4.3.2).

$$^1P_0 \xrightarrow{h\nu(530\,nm)} {}^3P_1$$

$$^3P_1 + {}^1(NI)_0 \longrightarrow P^{\ddot{+}} + (NI)^{\ddot{-}}$$

Table 4.5 shows that nitromidazoles (NI) are as effective triplet quenchers as ground state oxygen. Although the semireduced NI$^{\ddot{-}}$ species is repaired by oxygen according to the reaction

$$(NI)^{\ddot{-}} + O_2 \longrightarrow {}^1(NI)_0 + O^{2\ddot{-}}$$

the long-lived porphyrin radical cation $P^{\ddot{+}}$ (half-time of several hundred milliseconds) can in turn oxidize at neutral pH a variety of type I substrates, such as Trp or CysH, and other nonphotodynamically active molecules, such as Tyr without $^1O_2$ involvement. With Trp as a substrate, one has the following reaction:

$$P^{\ddot{+}} + Trp \longrightarrow Trp\cdot + {}^1P_0$$

followed by

$$O_2 + Trp\cdot \longrightarrow products$$

4.5 OXYGEN-INDEPENDENT PHOTOSENSITIZED REACTIONS

We have already encountered an example of photosensitized reactions not involving oxygen. Thus, on excitation of Trp in its $^1Trp_1$ state, basic residues such as His or disulfide bridges could be reduced by the hy-

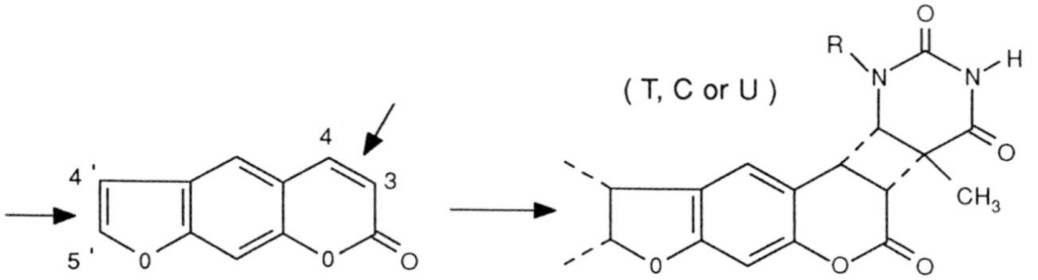

FIGURE 4.11   Sites of photochemical reactivity of the psoralen ring with pyrimidine bases (T, C, U), explaining the formation of monoadducts and diadducts.

drated electrons produced by Trp photoionization. Another example of oxygen-independent reaction is provided by the psoralens. The psoralens, a class of molecules absorbing in the near-UV, are used in the phototherapy of some proliferative diseases such as psoriasis. Psoralens intercalate into nucleic acids and can form either monoadducts or diadducts with pyrimidine bases of the nucleic acids (uracil, thymine, cytosine) on irradiation with near-UV light. In practice, on irradiation with wavelengths longer than 320 nm, adduct formation between a double bond of the pyrimidine and any one of the 3,4 or 4′,5′ double bonds of the furocoumarin ring occurs, whereas diadduct formation involves both bonds. It should be noted that on irradiation with wavelengths greater than 370 nm monoadducts formed with 4′,5′ bonds of the furocoumarins can be further converted into diadducts because these 4′,5′ monoadducts absorb light above 320 nm, whereas monoadducts formed with the 3,4 double bond do not. On photobinding, a cyclobutane ring (Figure 4.11) is created between the reacting bonds. Formation of a cyclobutane ring can be prevented by adding a bulky substituent on the 4′,5′ or 3,4 double bond of the psoralen ring. The therapeutic benefit of such a substitution will be examined in Chapter 18.

Bibliography

Bensasson, R. V., Land, E. J., and Truscott, T. G. (1983). "Flash Photolysis and Pulse Radiolysis: Contributions to the Chemistry of Biology and Medicine, Pergamon, Oxford.

Foote, C. S., Chang, Y. C., and Denny, R. W. (1970). Chemistry of Singlet Oxygen. X. Carotenoid quenching parallels biological protection. *J. Am. Chem. Soc.* **92**, 5216–5218.

Petrich, J. W., Martin, J. L., Houde, D., Poyart, C., and Orszag, A. (1987). Time-resolved Raman spectroscopy with subpicosecond resolution: Vibrational cooling and delocalization of strain energy in photodissociated (carbonmoxy)hemoglobin. *Biochemistry* **26**, 7914–7923.

Stryer, L. (1981). "Biochemistry." Freeman, San Francisco.

Wasserman, H. H., and Murray, R. W. (1979). Singlet oxygen. *Org. Chem. (N.Y.)* **40**.

CHAPTER 5

Photochemistry of Biological Molecules

In the preceding chapter we examined the primary photochemical events following the excitation of chromophores. These primary processes determine the sequence of secondary dark reactions and structural rearrangements giving rise to stable products whose chemical structure can be established through the use of the technical armamentarium of the chemist. In this chapter we give an *outline* of some important photoproducts of the biological molecules (amino acids, nucleic acid bases, and lipids) constituting the basic units of living cells. As far as direct photochemistry is concerned, we limit the discussion to the photochemical changes of these molecules on excitation in their first excited singlet state. Furthermore, we must keep in mind that the bandwidth of the optical absorption corresponding to this excited state must somewhat overlap the solar radiation spectrum reaching the ground (e.g., wavelengths >290 nm), to be of biological significance. Also, because most of the photoreactions essential to life occur in the presence of oxygen, we will only consider photoproducts that can form by direct photolysis under aerobic conditions or during photosensitized reactions.

5.1 PHOTOCHEMISTRY OF NUCLEIC ACIDS

Because of the central importance of DNA in cell life, we will first study the photochemistry of DNA components. However, carbohydrates, which make up 40% of the nucleic acids and do not absorb light above 230 nm, will not be considered.

5.1.1 Photochemical Reactivity of the Pyrimidine

The most photosensitive nucleic acid bases, at least *in vitro*, are the pyrimidines. They undergo four main photochemical reactions, which are described in Sections 5.1.1.1–5.1.1.4.

5.1.1.1 Hydration Products

Pyrimidine hydration products are formed through *addition* of a water molecule to the 5,6 double bond. On hydration, the optical absorption of the pyrimidines centered at 260 nm is lost, whereas a new band appears at about 240 nm. The highest quantum yield is obtained with uracil ($\Phi \approx 0.004$). These hydrates (Figure 5.1) are formed from a singlet state, because triplet quenchers and photosensitizers do not form the hydrates. They are readily decomposed to the parent compound by heat, alkaline, or acidic treatment. The cytosine photohydrate formation has been ob-

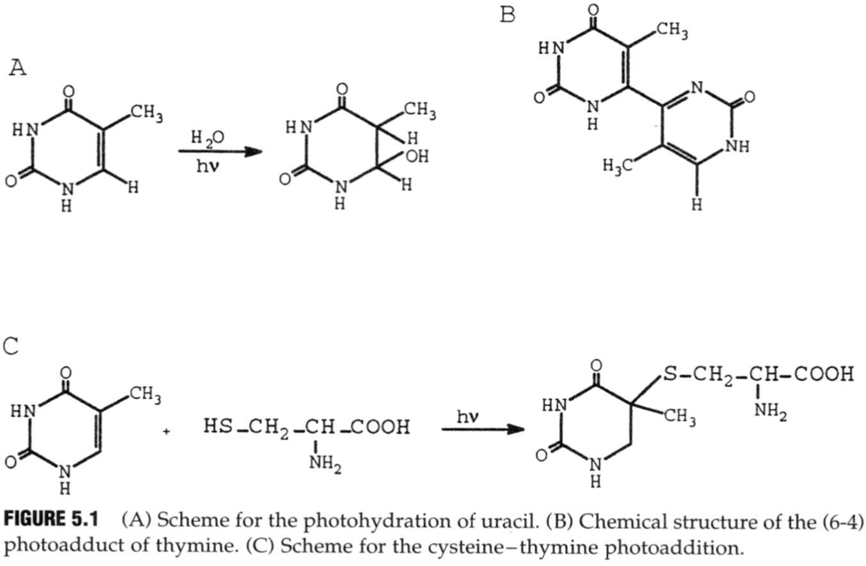

FIGURE 5.1   (A) Scheme for the photohydration of uracil. (B) Chemical structure of the (6-4) photoadduct of thymine. (C) Scheme for the cysteine–thymine photoaddition.

served in irradiated denatured DNA. The mutagenecity of these photolesions, although shown *in vitro*, remains to be demonstrated *in vivo*.

5.1.1.2 Cyclobutane-Type Dimers

If an aqueous solution of thymine is frozen and then UV-irradiated, the major product formed is a dimer. Aggregates formed on freezing favor bimolecular photochemical reaction between stacked thymines. The dimer thus produced is formed at high yield ($\Phi \approx 0.2$). In the thymine dimer, two thymine molecules are linked to each other between their respective C-5 and C-6 atoms, forming a cyclobutane ring (four-carbon ring) between the two thymines, which lose their characteristic absorption band at about 260 nm. There are six possible isomers of the thymine dimer, and these have been isolated from irradiated thymine oligomers. The cis–syn thymine dimer (Figure 5.2A) is formed between adjacent thymines in the same strand of DNA. Some of these isomers are stable to acid hydrolysis, whereas others are not. Because acid hydrolysis is the usual method for liberating photoproducts from irradiated DNA, the labile photoproducts are destroyed.

Because of their absorption spectrum (Figure 5.2B), there is a wavelength dependency both for the formation and for the monomerization of the cyclobutane-type thymine dimer. At the longer wavelengths (near 290 nm) the formation of the dimer is favored; at the shorter wavelengths (near 240 nm) monomerization is favored, because the dimer absorbs radiation only below 260 nm. The quantum yields for the formation and splitting of the dimer are different.

Five other cyclobutane-type dimers of the natural pyrimidines are also known. These are the dimers of uracil and cytosine, and the cyclobutane-type adducts of uracil–thymine, cytosine–thymine, and uracil–cytosine. The isolation of cytosine dimers is not possible because cytosine deaminates readily when its 5,6 double bond is saturated. Cytosine dimers are therefore converted to uracil dimers. If a cytosine dimer in the DNA of a cell should deaminate to form a uracil dimer, and if this dimer is then split *in situ* by a photoreactivating enzyme, a mutation could result because the uracil residues would be paired with adenine instead of guanine.

Thymine dimers can be formed by triplet–triplet energy transfer. This transfer requires that the triplet state of the photosensitizer be higher in energy than the triplet state of thymine. An example is the formation of thymine dimers by light of wavelengths >300 nm when DNA is irradiated in the presence of acetophenone. The advantage of the use of triplet state photosensitization to perform the dimerization is that it can be done with much longer UV wavelengths, which the nucleobases do not absorb.

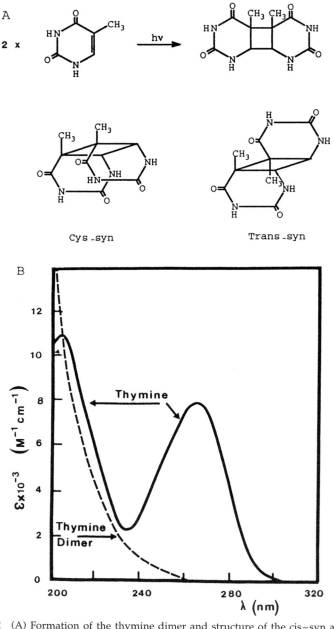

FIGURE 5.2   (A) Formation of the thymine dimer and structure of the cis–syn and trans–syn isomers. (B) Optical absorption spectrum of both thymine and the thymine dimer in water.

5.1.1.3 Pyrimidine–Pyrimidine (6-4) Adducts

When a solution of DNA is UV-irradiated, or when a frozen solution of thymine is UV-irradiated and thawed, a new peak appears at 320 nm in the absorption spectrum. This is due to the photochemical formation of (6-4) adducts (Figure 5.1). Only the (6-4) adducts of TC, CC, and TT are observed in UV-irradiated DNA lesions. In contrast to cyclobutane-type pyrimidine dimers, these lesions are not reversible on reirradiation at short wavelengths, but are labile in hot alkali. The yield of (6-4) adducts is about 1/100th that of the pyrimidine dimers. The (6-4) adducts play a role in UV radiation mutagenesis at specific sites in DNA.

5.1.1.4 Heteroadducts of the Pyrimidines

The pyrimidines also react photochemically with many other compounds. For example, the amino acid cysteine photochemically adds to the 5,6 double bond of the pyrimidines and forms a stable product (Figure 5.1), which may be an important mechanism for the formation of DNA–protein cross-links. Cysteine, arginine, lysine, tyrosine, tryptophan, and cystine also form photoadducts with thymine, whereas glycine, serine, cysteine, cystine, methionine, lysine, arginine, histidine, tryptophan, phenylalanine, and tyrosine can form adducts with uracil. The pyrimidines can also form photoadducts with the purines (e.g., adenine–thymine), but their biological importance has yet to be determined.

5.1.1.5 DNA–Protein Cross-links

In vitro experiments demonstrated that DNA and proteins could be cross-linked by UV irradiation. In the nucleus, these cross-linking reactions also take place, as shown by the progressive decrease in the amount of DNA that can be extracted free of proteins from bacteria and mammalian cells after increasing doses of UV radiation. The DNA that becomes nonextractable after UV irradiation is found in the precipitate containing the denatured proteins. Digestion of this precipitate with trypsin leads to free DNA, showing that the DNA has been cross-linked to proteins during UV irradiation. The chemical mechanisms by which DNA and proteins are cross-linked *in vivo* are still unclear. Cysteine, lysine, phenylalanine, tryptophan, and tyrosine react photochemically with DNA.

Although the photochemical reactions of pyrimidine discussed herein are possible under direct light absorption, the pyrimidine bases are rather insensitive to photosensitized reactions. As a result, a compilation of the photosensitized degradation of pyrimidines is not of prime importance for the understanding of DNA photochemistry and will not be further considered.

5.1.2 Photochemistry of Purines

By contrast, purines are rather insensitive to *direct* photochemical degradation after light absorption in their first excited singlet state, although they are somewhat sensitive to photosensitized reactions. However, only xanthine and to a lesser extent guanine are susceptible to photodynamic reactions. The extremely low reactivity of guanine nucleosides with singlet oxygen ($^1O_2$) formed in type II photodynamic reactions is illustrated in Table 4.4 (Chapter 4), which compares the reactivity of many biological substrates toward $^1O_2$.

As shown by the rose bengal-photooxidized reactions of DNA and nucleotides, only the guanine moiety appears to have a detectable reactivity toward singlet oxygen, DNA being a poorer substrate than dGMP. Evidence that guanine is the nucleobase most sensitive to photodynamic action has been provided by liquid chromatographic analysis of the acidic DNA hydrolysate and by analysis of the template activity of DNA polymerase. The latter was inhibited when photosensitized DNA fragments were used as the substrates. Termination site synthesis occurs one base prior to the guanine residue, suggesting that these purine nucleobases have been specifically damaged by the photodynamic action.

The use of the 3′,5′-di-*O*-acetylated guanosine to improve the HPLC separation of the polar oxidation products has allowed characterization of the specific type I and II photooxidation products. The two main $^1O_2$ oxidation products have been characterized as *N*-(3,5-di-*O*-acetyl-2-deoxy-β-D-*erythro*pentofuranosyl)cyanuric acid and 9-(3,5-di-*O*-acetyl-2-deoxy-β-D-*erythro*pentofuranosyl)-4,8-dihydro-4-hydroxy-8-oxoguanine (Figure 5.3). The formation of the cyanuric acid derivative is due to the initial addition of $^1O_2$ across the 4,5 double bond of the purine ring, followed by a rearrangement and subsequent photooxidation of the enamine intermediate. The second oxidation product in which the purine ring remains intact and which exhibits significant UV absorption arises from an unstable transient 1,4-endoperoxide.

The main type I photodegradation products are the α and β anomers of 3,5-di-*O*-acetyl-2-deoxy-D-*erythro*pentofuranose (Figure 5.3). Their formation involves (see Chapter 4 for the mechanism of type I reactions) an initial electron transfer reaction from the purine ring to the photosensitizer in its triplet state and subsequent reaction of the resulting purinyl radical with oxygen. Alkaline lesions at guanine produced in DNA when exposed to 313-nm light in the presence of acetone are likely due to *apurinic sites* generated by type I photoprocesses. The relative importance of type I and type II photodynamic mechanisms depends on the photosensitizers. Thus photosensitizers with good electron acceptor properties (riboflavin, methylene blue) preferentially give type I prodtype I product   type II products

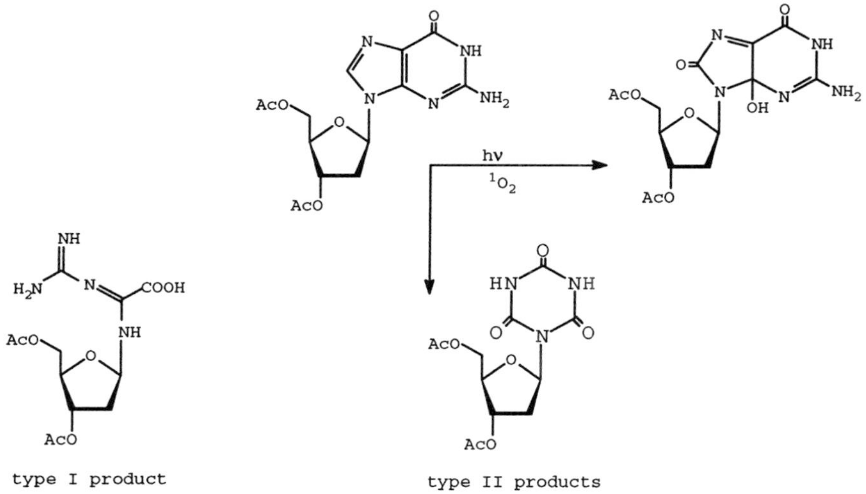

FIGURE 5.3 Specific products for type I and type II photodynamic reactions of purines.

ucts. On the other hand, hematoporphyrin and rose bengal produce type II products.

In addition to damage to guanine, DNA strand breakages are induced by various photosensitizers in the presence of oxygen. Illumination of an aerated aqueous solution of DNA produces six times more single-strand breaks (SSBs) when rose bengal is present. The involvement of singlet oxygen in the production of DNA strand breaks is ruled out on the basis of the lack of any detectable isotopic effect when D_2O is used as the solvent (see Chapter 4). Single- and double-strand scissions of supercoiled DNA produced by riboflavin photosensitization are not prevented by the addition of specific scavengers or quenchers of superoxide radical, hydrogen peroxide, hydroxyl radical, and singlet oxygen. The major conclusions drawn from these two studies is that the triplet states of photosensitizers can be involved in the photoinduced formation of DNA SSBs, without the intervention of activated oxygen species.

5.2 DIRECT AND SENSITIZED PHOTOCHEMISTRY OF AMINO ACIDS AND PROTEINS

5.2.1 Direct Photochemistry of Simple Amino Acids and Peptides at Room Temperature in Aerated Solutions at Neutral pH

Proteins are the most abundant macromolecules in cells. Functional proteins may be highly sensitive to light because their biological properties are controlled by relatively small regions of the entire macromolecule, e.g., the active center of an enzyme. The absorption of light by a protein is localized at chromophoric groups with significant extinction coefficients at the irradiation wavelengths. However, as shown in Chapter 4, the subsequent events may not be localized to the initial target because electrons can migrate within the macromolecule. Protein photochemistry has been investigated with many techniques, including assays of permanent residue destruction, laser flash photolysis, electron spin resonance detection of photochemical intermediates, and circular dichroism studies on conformation changes. As noted at the beginning of this chapter we will only consider the photochemistry of amino acids that are excitable by solar UV, e.g., the amino acids tyrosine (Tyr) and tryptophan (Trp), and also cystine, which has a small residual absorption at 300 nm. Tryptophan is the main chromophoric residue of proteins at the onset of the solar UV at 290 nm, where its molar extinction coefficient is still 3935 M^{-1} cm^{-1}. For comparison, the molar extinction coefficients of Tyr and Cys at this wavelength are 113 and 50 M^{-1} cm^{-1}, respectively. It follows that under physiological conditions, the Trp photolysis plays a key role in protein photoinactivation by solar UV light.

5.2.1.1 Tryptophan

In Chapter 4 it was shown that the excitation of Trp in its first absorption band (absorption maximum at 280 nm in neutral aqueous solution) leads to photoionization of the indole ring, with formation of the neutral indolyl radical and of the strongly reducing hydrated electron. In aerated solutions, in the absence of disulfide bridges, the latter is immediately scavenged by oxygen to yield the superoxide radical anion ($O_2^{\bar{\cdot}}$). The subsequent reactions of the indolyl radical yield many photoproducts, including hexahydropyrroloindole (HPI), N-formylkynurenine (NFK), ammonia, and melanine-like high-molecular-weight polymers. The formation of NFK is particularly important because NFK, which absorbs light at longer wavelengths (absorption maximum at 320 nm) than does Trp, is a photodynamic photosensitizer.

Figure 5.4 illustrates the degradation of Trp in an air-saturated neutral aqueous solution by solar UV light, in the middle of May in Paris, France. The increase in the rate of Trp photolysis (and thus in the quantum yield of destruction) during the course of the irradiation is due to the production of NFK. Once formed, NFK, whose molar extinction coefficient is 3180 M^{-1} cm^{-1} at 320 nm, acts as a photodynamic photosensitizer toward Trp, an excellent type I and type II photodynamic substrate (see Chapter 4). The NFK is also formed by direct Trp photolysis in peptides such as Gly-Trp, Trp-Gly, Trp-Lys, and Ala-Gly-Trp-Leu and by reaction of $^1O_2$ with Trp (see below).

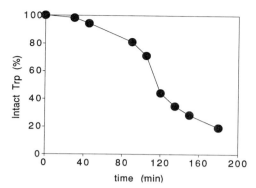

FIGURE 5.4 Actual data showing the degradation of 230 μM Trp in 3 ml of air-saturated, 50 mM, pH 7 phosphate buffer illuminated with solar light in Paris, France, on May 17, 1993, between 10 a.m. and 2 p.m. (solar time). The irradiation was carried out in a quartz cell (1-cm light path) and the weather was mostly sunny.

5.2.1.2 Tyrosine and Cystine

Photolysis of aqueous tyrosine gives hydroxylated aromatics [especially 3,4-dihydroxyphenylalanine (DOPA), the precursor of melanine], bityrosine, aliphatic amino acids, and ammonia. The reaction mechanism is not well established. The formation of the bityrosine photoproduct is interesting because it is strongly fluorescent at about 410 nm on excitation at 320 nm, and can be used to probe tyrosine oxidation. Furthermore, tyrosine can be oxidized by the neutral indolyl radical (Trp·) produced by Trp photolysis. This reaction can be written as follows:

$$Trp\cdot + TyrOH \longrightarrow Trp + TyrO\cdot \qquad (5.1)$$

In Eq. (5.1), the notation TyrOH is used for tyrosine instead of the conventional Tyr to emphasize the oxidation of the hydroxyl group of the phenolic ring.

Although cystine is a weakly absorbing amino acid, it is especially important in protein photochemistry because cystine photolysis has a high quantum yield ($\Phi \sim 0.1$) and may lead to the splitting of interchain disulfide bridges. The probable primary events are the splitting of S—S bonds, leading to RS·-type radicals, and the splitting of C—S bonds, leading to S—S·-type radicals. In air-saturated solutions, one cysteic acid molecule and one alaninesulfinic acid molecule are formed from each S—S bridge split.

5.2.2 Photosensitized Degradation of Amino Acids in Photodynamic Reactions

Only a few amino acids are susceptible to the photodynamic action via type II and type I reactions. These are, at neutral pH, His, Cys, Met, and Trp.

Histidine is one of the most important target amino acids for singlet oxygen. It is rather reactive and the first isolable products are usually products of extensive degradation of the heterocyclic ring. The primary product of histidine and various related imidazoles is the endoperoxide (Figure 5.5). The excellent specificity of the reaction of His with $^1O_2$ makes it possible to use this amino acid as a probe for singlet oxygen formation in aqueous media.

Cysteine (RSH) undergoes type I and type II photodynamic reactions. Under certain conditions, sulfinic (RSO_2H) and sulfonic (RSO_3H) acids have been isolated; in other cases cystine is the isolable product.

Methionine undergoes photooxygenation, both as the free amino acid and in peptides. The type II product (singlet oxygen) is the sulfoxide (Figure 5.5); in type I processes (a minor process), loss of the carboxyl group and formation of methional are observed.

Tryptophan reacts by type II and type I reactions; the products are extremely complex as was observed under direct UV photooxidation. A 1,2-dioxetane intermediate has been characterized. The primary isolable product is the same as that obtained with direct photolysis, e.g., *N*-formylkynurenine (Figure 5.5). However, the HPI derivative is also formed, as well as melanin-type polymers, because of type I reactions. The similarity of reaction products in direct and photosensitized photolysis is most probably due to the fact that NFK is a photosensitizer by itself.

5.2.3 Photochemistry of Proteins

Direct or sensitized photolysis of an aerated aqueous protein solution leads to extensive changes in almost all of the solution properties. These are changes in the absorption spectrum, molecular weight, electrophoretic pattern, solubility, and heat sensitivity, changes in acid–base titration curves, increased digestibility by proteolytic enzymes, modification or destruction of residues, and changes in the catalytic activity of enzymes. Attempts to attribute the overall damage to specific initial photochemical reactions are difficult because of the structural complexity in the amino acid composition of proteins, in addition to the occurrence of complex secondary, tertiary, and quaternary structures.

In Chapter 4, the inactivation of proteins by photodynamic reactions was exemplified with lysozyme. Proteins are the main targets of many crucial photodynamic reactions *in vivo*, but the methodology presented

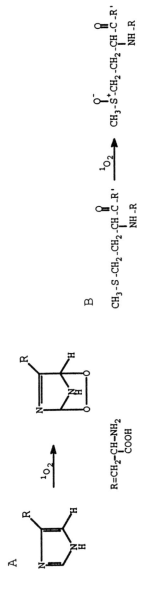

FIGURE 5.5 Products of reaction of $^1O_2$ with (A) His, (B) Met, and (C) Trp.

for lysozyme can be generalized to other proteins. Consequently, in this section we will concentrate our attention on the direct photoinactivation of proteins.

It must be stressed that direct excitation of Trp and Cys residues by wavelengths corresponding to the onset of solar UV radiations can, in principle, lead to protein photoinactivation. However, the resulting photochemistry of an amino acid residue in a protein may differ significantly from the amino acid in solution, for the following reasons:

1. Electrons generated by the photoionization of Trp residues may migrate and react at different sites. For instance, they may be trapped by disulfide bridges of cystine (see Chapter 4).
2. The splitting of disulfide bridges may lead to large changes in the conformation of the protein, favoring reaction of primary radical species formed by the irradiation with residues at remote places in the primary structure.
3. Tyrosinyl residues can be oxidized by intramolecular electron transfer from an intact tyrosine residue to an oxidized tryptophan.

The identification of pathways leading from the initial photochemical reactions in a protein to permanent damage is therefore difficult. Studies are generally carried out on enzymes, hormones, serum proteins, and lens proteins, especially when data about the primary and three-dimensional structures are available. Quantum yields for the UV photoinactivation of enzymes are usually independent of oxygen, which does not rule out the attack of singlet oxygen on a "nonessential" site. It should be noticed, however, that most of the earlier work has been carried out with the strongly energetic 254-nm radiation, which notably changes the primary photochemical events. Under irradiation with wavelengths greater than 280 nm, photoinactivation of carbonic anhydrase was much faster under oxygen (Figure 5.6). In this protein, direct photoinactivation by near-UV light in the presence of oxygen produces loss of the enzyme activity, the tryptophanyl, histidyl, and, to a lesser extent, tyrosyl residues being destroyed. In nitrogen-saturated solutions, a dramatic drop is observed in the photoinactivation yield, whereas histidyl residues remain intact. This is explained by *an internal photodynamic action of N-formylkynurenine in the protein core produced by the UV photooxidation of tryptophanyl residues.* Photoinactivation of oxygenated enzyme solutions by this internal photodynamic action correlates with histidyl residue destruction via singlet oxygen. The sequence of photochemical reactions characterizing the internal photodynamic effect in proteins can be summarized as follows:

$$^1\text{Trp}_0 \xrightarrow{h\nu > 290\ \text{nm}} {}^1\text{Trp}_1 \xrightarrow{{}^3\text{O}_2} {}^1\text{NFK}_0 + \text{other products}$$

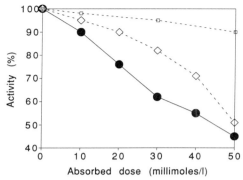

FIGURE 5.6   Photoinactivation of bovine carbonic anhydrase under irradiation with wavelengths >280 nm. ●, O_2-saturated pH 7.6 phosphate buffer; ◇, O_2-saturated pH 7.6 phosphate buffer + 10 mM NaN_3, a good quencher of $^1O_2$, showing that singlet oxygen, and thus type II photodynamic reactions, are involved in the photoinactivation, at least at the beginning of the irradiation; □, N_2-saturated pH 7.6 phosphate buffer.

$$^1NFK_0 \xrightarrow{h\nu > 290 \text{ nm}} {}^3NFK_1 \xrightarrow{{}^3O_2} {}^1NFK_0 + {}^1O_2$$

$$^1O_2 + Trp \longrightarrow \text{products}$$

Formation of N-formylkynurenine by photooxidation of Trp residues occurs in many proteins, including human lens proteins and bovine γ-cristallin. This photooxidation may be due not only to direct light absorption by Trp residues of proteins, but also, as shown above, to $^1O_2$-mediated reactions induced by photodynamic photosensitizers. As a result, photosensitized inactivation of proteins carried out with photosensitizers absorbing in the near-UV may be amplified by the transformation of Trp residues into NFK.

5.3 PHOTOCHEMISTRY OF STEROLS AND POLYUNSATURATED LIPIDS

5.3.1 Photorearrangement of 7-Dehydrocholesterol

With the exception of 7-dehydrocholesterol found in the malpighian cells of human skin, all the other biologically important sterols and lipids do not absorb UV radiation above 290 nm. The 7-dehydrocholesterol is the photochemically active precursor of vitamin D_3, but the key intermediate in its biosynthesis is the provitamin D, which is thermally trans-

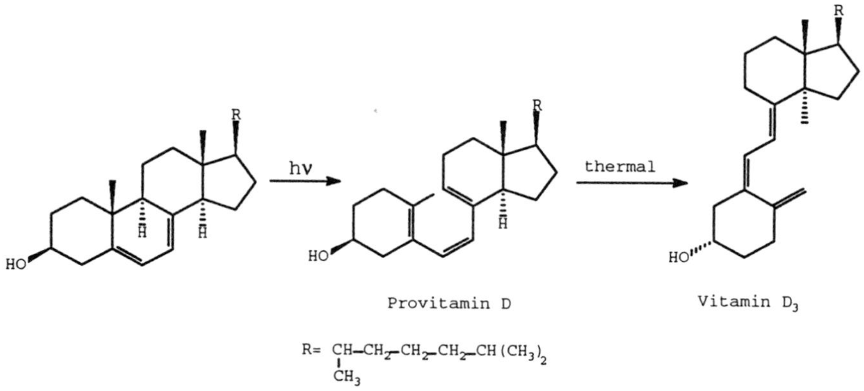

Provitamin D   Vitamin D$_3$

R= CH–CH$_2$–CH$_2$–CH$_2$–CH(CH$_3$)$_2$
    |
    CH$_3$

FIGURE 5.7   Photosynthesis of vitamin D$_3$.

formed into vitamin D$_3$. This photorearrangement is perhaps the only beneficial photochemical reaction in normal human skin (Figure 5.7).

5.3.2 Photosensitized Oxidation of Lipids

Biological membranes are important targets of the photodynamic action of many sensitizers. Two possible reaction centers are encountered. Essential amino acids of membrane proteins, such as Met, His, Trp, and Cys, can be readily photooxidized with ensuing protein denaturation, as shown in Section 5.2.2. Polyunsaturated fatty acids (PUFAs) and cholesterol can be also oxidized, not only by singlet oxygen ($^1O_2$) but also by other activated oxygen species produced secondarily during the course of photosensitization, especially in living systems. Photosensitization at the membrane level results in a decrease in enzyme activities, morphological changes, and transport activities. All the cellular membranes that are susceptible to photosensitized damage (plasma membrane, mitochondria, lysosomes, Golgi apparatus) contain unsaturated lipids ($\sim$50%) in relatively constant amounts (Table 5.1). In addition, cholesterol, found in all membranes, is at much greater concentration in the plasma membrane.

5.3.2.1 Reaction of $^1O_2$ with Cholesterol

This is the typical example of the so-called ene reaction of $^1O_2$ with double bonds. It has already been detailed in Chapter 4 (see Figure 4.9). The bond in cholesterol (or derivatives) is shifted to the allylic position while a hydroperoxide is formed: 3β-hydroxy-5α-hydroperoxy-Δ^6-cholestene. On the other hand, radical oxidation yields complex mixtures, including 7α- and β-hydroperoxides.

TABLE 5.1   Unsaturated Fatty Acid Content of Organelles

Fatty acids[a]	Mole(%)			aH[b]
	Mitochondria	Lysosomes	Microsomes	
Oleic acid (C18:1)	7	8	7	2:0
Linoleic acid (C18:2)	21	22	16	3:1
Arachidonic acid (C20:4)	24	23	25	5:3
Linolenic acid (C18:3)	—	—	—	4:2

[a] The number of carbon atoms and double bonds is given in parentheses after the name of each fatty acid.

[b] Allylic hydrogens (aH) are hydrogen atoms on a carbon adjacent to a double bond. The data give the respective number of mono and biallylic hydrogens in the fatty acid.

5.3.2.2 Oxidation of PUFAs

The poor reactivity of $^1O_2$ with unsaturated lipids was illustrated in Chapter 4 (Table 4.4). Reaction rate constants for the reaction of PUFAs with $^1O_2$ are of the order of $10^4-10^5\ M^{-1}\ sec^{-1}$, depending on the number of double bonds and allylic hydrogens. However, this apparently poor reactivity is counterbalanced by the *radical chain reaction of lipid peroxidation*. The propagation of the latter requires the *presence of allylic hydrogen*.

It is not possible to review in detail all the reactions encountered in the peroxidation of the lipids mentioned in Table 5.1. The reader is therefore invited to read the key reference given at the end of this chapter. However, there are a few clues that can be given to understand the body of reactions (Frankel, 1985). Thus, under controlled laboratory conditions, unsaturated fatty acids undergo the ene reaction with $^1O_2$. Another reaction path, as shown in Figure 5.8, consists of hydrogen abstraction by radicals, giving allylic free radicals that can react at either allylic position to give two different allylic hydroperoxides. Both pathways constitute the initiation step of the radical chain reaction.

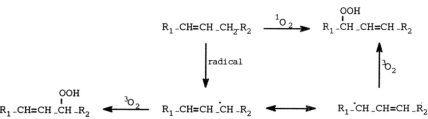

FIGURE 5.8   Hydroperoxide formation in unsaturated fatty acids under $^1O_2$ or radical reaction.

The chain reactions of the peroxidation of PUFAs (RH) are propagated by redox metal ions. Ferrous ions are excellent propagators by reducing lipid hydroperoxides (ROOH) via the following reaction:

$$ROOH + Fe^{2+} \longrightarrow Fe^{3+} + OH^- + RO\cdot \tag{5.2}$$

This reaction is called a Fenton-like reaction because it is formally equivalent to the decomposition of hydrogen peroxide by ferrous ions (Fenton reaction), which produces hydroxyl radicals ($OH\cdot$).

$$H_2O_2 + Fe^{2+} \longrightarrow Fe^{3+} + OH\cdot + OH^- \tag{5.3}$$

It should be noted that H_2O_2 can be produced during type I photodynamic reactions by the dismutation of O_2^- radical anions (see Chapter 4).

$$2H_3O^+ + O_2^- + O_2^- \longrightarrow H_2O_2 + O_2 + 2H_2O \tag{5.4}$$

This reaction is catalyzed by the superoxide dismutase (SOD). The $OH\cdot$ radical is the strongest oxidant on earth. *It can initiate lipid peroxidation by abstracting allylic hydrogens of PUFAs*, leading to hydroperoxide formation as shown in Figure 5.8.

The alkoxyl radical ($RO\cdot$) formed by the Fenton-like reaction can, in turn, initiate reactions such as

$$RO\cdot + ROOH \longrightarrow ROH + ROO\cdot \tag{5.5}$$

$$ROO\cdot + RH \quad (or\ R'H) \longrightarrow ROOH + R\cdot \quad (or\ R' \tag{5.6}$$

(where R'H is another PUFA).

$$R\cdot (or\ R') + {}^3O_2 \longrightarrow ROO\cdot (or\ R'OO\cdot) \tag{5.7}$$

The $ROO\cdot$ and $R\cdot$ species are called peroxyl and alkyl radicals, respectively. Reactions such as

$$R\cdot + R\cdot \longrightarrow R—R \tag{5.8}$$

are called termination reactions because they consume radicals that could otherwise initiate the chain reactions.

The complexity of secondary products can be illustrated with the photosensitized oxidation of linoleate. Singlet oxidation of methyl linoleate by the concerted "ene addition" mechanism produces a mixture of four isomeric hydroperoxides, two conjugated 9-hydroperoxy-*trans*-10-*cis*-12- and 13-hydroperoxy-*cis*-9-*trans*-11-octadecadienoates, and two unconjugated 10-hydroperoxy-*trans*-8-*cis*-12- and 12-hydroperoxy-*cis*-9-*trans*-13-octadecadienoates.

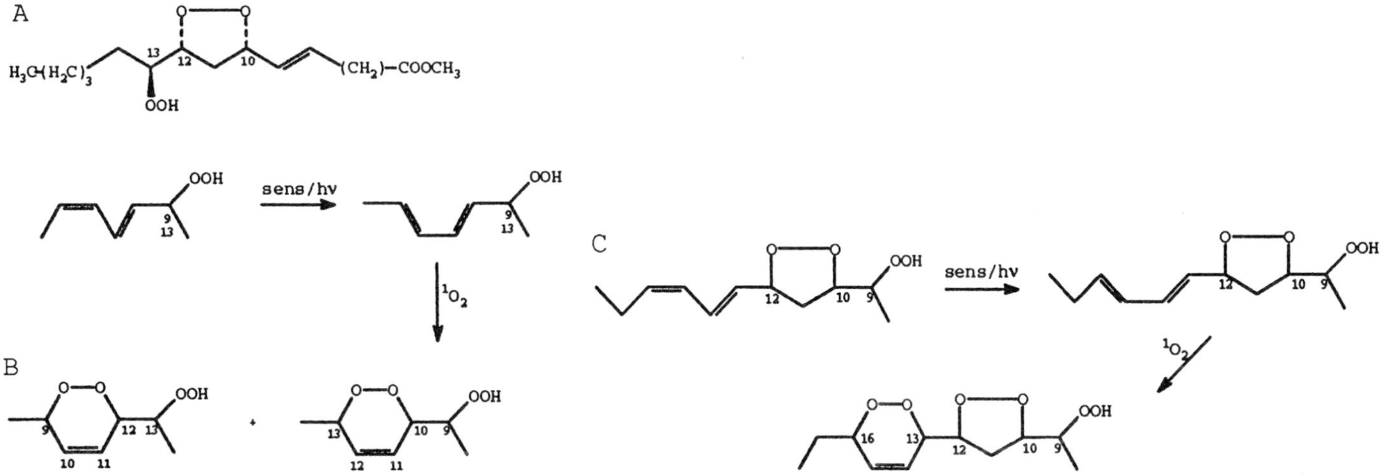

FIGURE 5.9 (A and B) Scheme for the photosensitized peroxidation of linoleate hydroperoxides by $^1O_2$. Note that the photosensitizer (sens) is responsible for the isomerization of the double bonds. This isomerization is required for addition of $^1O_2$. (C) Same as above but linolenic acid is the fatty acid.

Secondary oxidation products include epoxy esters, di- and trioxygenated compounds, dihydroperoxides, and hydroperoxy-cyclic peroxides. Five-membered cyclic peroxides (Figure 5.9A) are formed from the monohydroperoxides, such as those shown in Figure 5.8, by oxygen addition in the dark. The reaction of singlet oxygen with the hydroperoxide, the primary product of the reaction of linoleate with $^1O_2$, is possible because these hydroperoxides still have reactive double bonds. The photosensitized oxidation of a mixture of 9- and 13-hydroperoxides from methyl linoleate produces six-membered cyclic peroxides by 1,4 addition of singlet oxygen to the conjugated diene systems (Figure 5.9B). Therefore complex chemical structures such as those shown in Figure 5.9C can be found for *linolenate,* which contains three double bonds, allowing the simultaneous formation of five- and six-membered cyclic hydroperoxides. One leaves the reader to imagine structures that can be obtained with arachidonic acid!

These intermediates are responsible for the formation of decomposition products, which include pentane and many aldehydes, ketones, and carboxylic acids. The reaction of these decomposition products with reactive groups of proteins (free NH_2 groups, thiols) can lead to their inactivation and/or functional alterations, especially in membranes, where a close contact between oxidized lipids and proteins is possible. Thus *triggering of hydroperoxide formation is a quite harmful event* for the cell. However, the cell antioxidant defenses can, in principle, cope with the oxidative stress. This antioxidant system (enzymes such as peroxidases, catalase, superoxide dismutase, and the well-known low-molecular-weight antioxidants — vitamins E, C, and glutathione) will be examined in detail in Chapter 12.

5.4 THE MAIN CELL PHOTOSENSITIZERS

A major conclusion that can be drawn from the discussions in this chapter is that photosensitized reactions are particularly important in the photochemical alterations of biomolecules. One can therefore suggest the occurrence of photosensitized processes in living cells, especially in cutaneous cells or bacteria. In mammalian cells, tetrapyrrolic derivatives (Figure 5.10) are commonly involved in photosensitization reactions. Thus, as will be shown in Chapter 16, porphyrins that accumulate in the skin of porphyric patients suffering from impaired heme biosynthesis or metabolism are responsible for the phototoxic response of areas of skin exposed to visible or UV light.

Genetic or acquired impairment of Trp metabolism is also known to induce a strong phototoxic response. For instance, exaggerated produc-

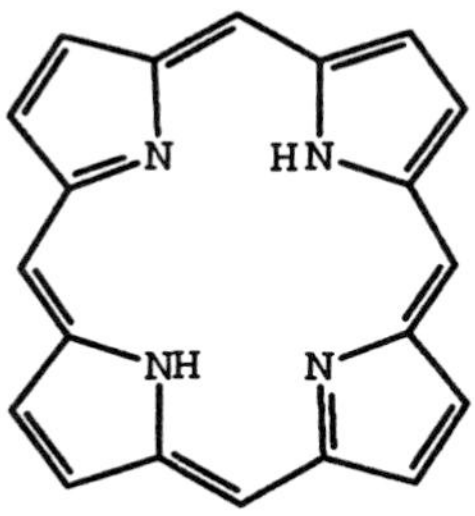

Porphyrins 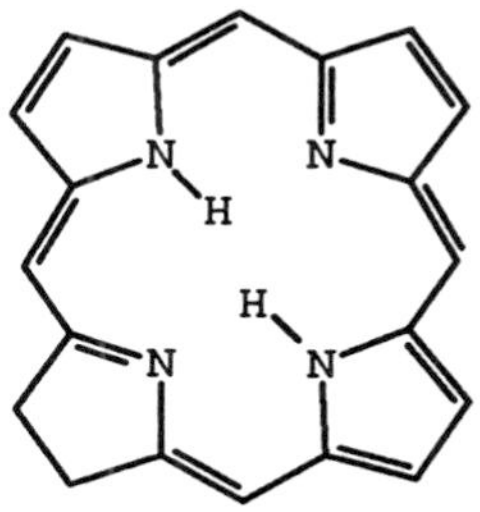   Chlorins

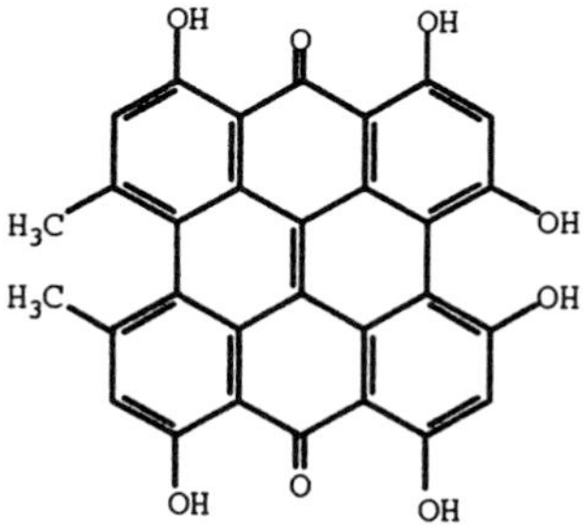

Hypericin 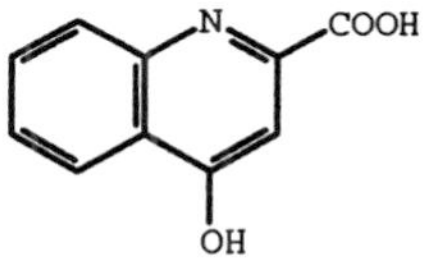   Kynurenic acid

FIGURE 5.10   Chemical structure of some natural photosensitizers.

tion of kynurenic acid is responsible for actino-reticulosis disease (Figure 5.10). Plants also contain photosensitizers, such as psoralens or hypericin, which induce abnormal skin responses to light when ingested by cattle or deposited on hands. Furthermore, mycotoxins can induce an accumulation of phylloerythrin, a product of chlorophyll (Figure 5.10) metabolism, in the blood of sheep. It is shown that many drugs have adverse or curative photosensitizing properties (see Chapters 16, 18, and 19).

Whereas the above example deals with abnormal or excess photosensitizers found in mammalian skin cells, all normal cells contain precise amounts of natural photosensitizers that are potentially able to trigger type I and type II photodynamic effects and/or lipid peroxidation. Thus oxidized flavins (Figure 5.11) are excellent photodynamic agents whereas NADH or NADPH are readily oxidized by UV light at

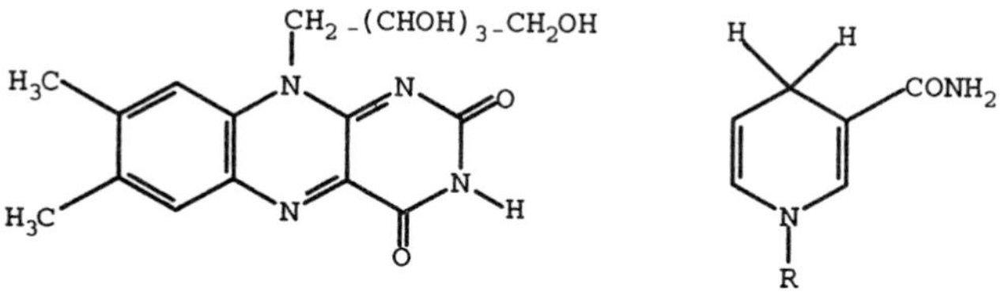

CH_2–$(CHOH)_3$–CH_2OH

H_3C

H_3C

Riboflavin

CONH$_2$

R=adenine dinucleotide (NADH)

R=adenine dinucleotide phosphate (NADPH)

FIGURE 5.11   Chemical structure of flavins and nicotinamide adenine dinucleotides.

neutral pH:

$$O_2 + NAD(P)H + UV\ light + H_2O \longrightarrow NAD(P)\cdot + O_2^- + H_3O^+$$

$$NAD(P)\cdot + O_2 \longrightarrow NAD(P)^+ + O_2^-$$

The O_2^- radical-anion thus produced can induce lipid peroxidation as described above. Also, ferritin, the cell store for ferric ions, can be activated by UV or visible light, thereby releasing the harmful Fe^{2+} ions that readily reduce molecular oxygen. These examples are not intended to be an extensive review of all natural photosensitizers. They are only a minute illustration of the many photosensitizers encountered in the animal and plant kingdoms. In conclusion, it can be stated that our skin cells are undoubtedly excellent photochemical reactors. It can also be anticipated that the products of harmful photochemical reactions are unavoidable contributors to the shortening of our cell life expectancy.

Bibliography

Foote, C. S. (1981). Photooxidation of biological model compounds. In "Oxygen and Oxy Radicals in Chemistry and Biology" (M. A. J. Rodgers and E. L. Powers, eds.), pp. 425–440. Academic Press, New York.

Frankel, E. N. (1985). Chemistry of free radical and singlet oxidation of lipids. *Prog. Lipid Res.* **23**, 197–221.

Smith, K. C. (1989). "The Science of Photobiology." Plenum, New York.

6

Bioluminescence

6.1 INTRODUCTION

Firefly bioluminescence is the archetype of the various enzymatic processes that produce light which occur in organisms ranging from marine bacteria and phytoplankton to the large luminous beetles from South America. Photosensitized reduction and phosphorylation must have been the primary energy sources necessary to synthesize the molecules metabolized by the first living organisms (presumably nitrogen-fixing blue–green algae in shallow pools). The advent of the manganese–enzyme complex for the photosensitized oxidation of water provided an inexhaustible supply of replenishment of electrons from water in order to balance the electrons lost by photoreduction of $NADP^+$.

There were two adventitious consequences of the photosensitized oxidation of water by green plants:

1. The diffusion of plant-released O_2 into the stratosphere, where a balance between the photochemical production of ozone, O_3, and chemi-

cally catalyzed production of O_2 resulted in a vital concentration of O_3, equivalent to only 3 mm at standard atmospheric pressure.

2. The evolution of significant concentrations of O_2 and the transformation of the surface of the earth from a reducing to an oxidizing environment.

The bioenergetically wasteful bioluminescence may represent a vestigial survival mechanism for primitive anaerobes, to eliminate O_2 evolved from photosynthesis, until selection of aerobic electron transport systems. However, there were very low levels of dissolved O_2 in primitive earth waters, and the plant-released O_2 would have created an impossible metabolic burden on the bacteria for oxidizable substrates.

An alternative hypothesis links the origin of bioluminescence with mixed-function oxygenases, light emission being as in present-day hydroxylase systems, an adventitious event. Much later, subsequent to the advent of vision, these might have been a selective advantage for bioluminescence in terms of biological efficiency.

6.2 LOW QUANTUM YIELD CHEMILUMINESCENCE

The mitogenetic radiation of Gurwitch and the microsomal chemiluminescence of metabolites of carcinogenic polycyclic aromatic hydrocarbons are nonfunctional very low quantum yield biological chemiluminescences (CLs).

Chemiluminescence in biological function is called bioluminescence (BL). The acronym ABC referes to "adventitious" biological chemiluminescence. CL ranges from bioluminescent reactions with quantum yields nearing 1 to those with yields of 10^{-15}. BL is customarily reserved for the function of light signaling by interacting organisms, and all other light emissions by cells, tissues, and organisms are biological CL.

6.3 CHEMICALLY INITIATED ELECTRON EXCHANGE LUMINESCENCE

Besides the absorption of photons, ground state molecules may be raised to excited states by high-temperature electrical discharges, electroluminescence, rupturing of crystalline bonds (crystallo- or triboluminescence), frictional turbulence (sonoluminescence), back reactions of previously separated charges (delayed luminescence of chloroplasts), and exergonic chemical reactions (ion or free radical recombinations involving oxidations by molecular oxygen or decomposition of peroxides). With rapid relaxation of molecules to thermal equilibrium (10^{-13}–10^{-12}

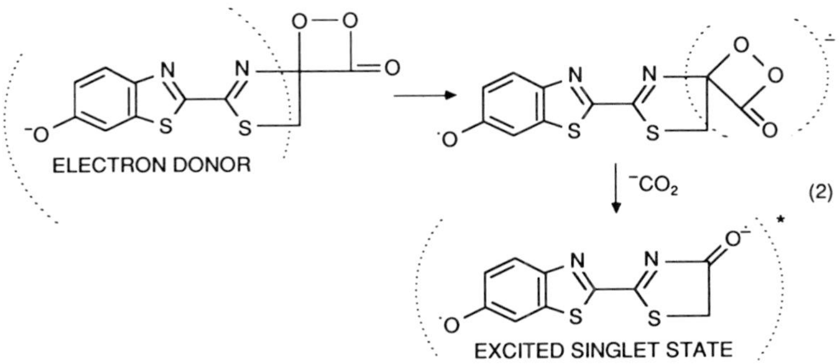

FIGURE 6.1 Scheme of chemically initiated electron exchange luminescence (CIEEL). An intramolecular CIEEL followed by rapid decarboxylation will generate a charge-transfer resonance structure of the product, which will be in a singlet excited state. Courtesy of H. H. Seliger (1989), Biology Department, The Johns Hopkins University, Baltimore, Maryland. Fig. 1 in Cell Structure and Function (Kohen, E. and Hirschberg, J. G. ed.) Chapter 23, p. 422; with permission of Academic Press, Orlando, Florida.

sec), *whatever the initial distribution of excited states formed by chemical reaction, the emission will always be from the first excited state.*

The high bioluminescence quantum yields of the firefly reaction are explained by the formation of a dioxetanone intermediate, via the addition of oxygen to luciferin. The dioxetanone intermediate most likely dissociates into CO_2 and dicarboxyketoluciferin *(excited state)*. An intramolecular electron exchange from the easily oxidized heterocycle portion of luciferin to the high-energy dioxethane moiety, followed by rapid decarboxylation, will generate a *charge-transfer structure* that will recombine to a high quantum yield singlet excited state (instead of a low quantum yield triplet). This process is called *chemically initiated electron exchange luminescence* (CIEEL) (Figure 6.1).

If we assume that the chemiluminescence reaction is

$$A + X \longrightarrow {}^1S_1 \longrightarrow {}^1S_0 + h\nu$$

with k_c the rate constant of S_1 formation, then the intensity of chemiluminescence is given by

$$I_{ch} = k_f[{}^1S_1] = [k_F/(k_F + k_{FQ} + k_{ISC})]k_c[A][X] \qquad (6.1)$$

where I_{ch} is the intensity of chemiluminescence, k_F is the rate constant of fluorescence, k_{fIC} is the rate constant of internal quenching, and k_{isc} is the rate constant of intersystem crossing. Using the Eq. (2.16) for ϕ_F, the quantum yield of fluorescence, the above equation can be simplified to

$$I_{ch} = \phi_F k_c[A][X] \qquad (6.2)$$

Thus, the quantum yield is defined in a manner analogous to that for fluorescence, except that it is related to the rate of substrate molecules reacting.

6.4 METHODS FOR DETECTION OF SINGLET OXYGEN IN BIOLOGICAL REACTIONS BY CHEMILUMINESCENT PROBES

Chemiluminescence attributed to singlet oxygen was first reported during studies of lipid peroxidation by rat microsomal extracts and phagocytic activity of polymorphonuclear leukocytes. Direct emission of singlet oxygen (1268 nm) has a low quantum efficiency, and the emission of dimol singlet oxygen has major bands at 634 and 703 nm.

Favorable quantum yields are obtained using chemiluminescent probes of $^1O_2$ that form dioxetane. The overall quantum yield of probe chemiluminescence Q_{CL} is

$$Q_{CL} = A Q_{exc} Q_f \tag{6.3}$$

where A is the chemical yield of dioxetane in the reaction with $^1O_2$, Q_{exc} is the relative yield of the excited state product of dioxetane, and Q_f is its chemiexcited fluorescence yield.

Following a brief exposure to $^1O_2$, *trans*-1(2-methoxyvinyl)pyrene (*trans*-MVP) is 180 times more chemiluminescent than 7,8-diol-7,8-dihydrobenzo[a]pyrene, a metabolite of benzo[a]pyrene (BaP) involved in microsomal luminescence on interaction with $^1O_2$.

With a 10% yield of dioxetane (having a CL quantum yield, 0.027) by reaction of $^1O_2$ with *trans*-MVP (Figure 6.2), the overall quantum yield (0.027×0.10) is about 0.003. Thus small amounts of cellular $^1O_2$ can be detected with trace amounts of *trans*-MVP, which may not interfere with cellular processes.

6.5 CHEMILUMINESCENCE OF A PROXIMATE CARCINOGEN 7,8-DIOL-BENZO[a]PYRENE

The specificity of the microsomal 7,8-diol-BaP CL mechanism can be substantiated by coincidence between the CL spectrum (quantum yield, 10^{-8}) and the fluorescence spectrum of the reaction product. The most probable chemiluminescent intermediate is a dioxetane and the excited-state product is expected to be 7,8-diol-9,10-dialdehyde-BaP (which has a spectrum similar to that of the monomer of 7,8-diol-BaP).

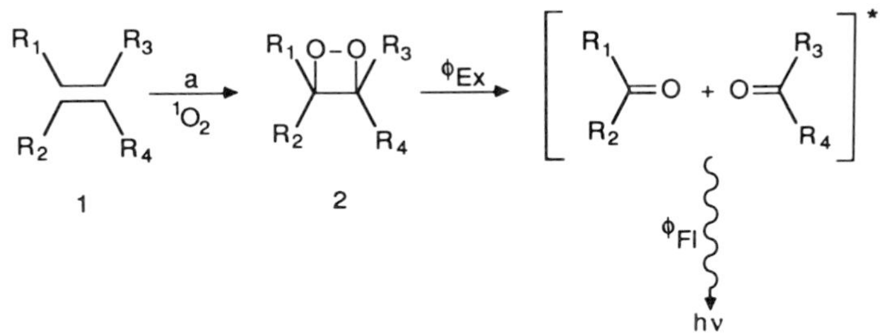

FIGURE 6.2 Reaction of $^1O_2$ with the vinyl probe's (1) double bond leads to production of dioxetane (2) (see Chapter 4); the "a" over the first arrow is the yield of dioxetane production in the reaction with the probe. This chemical yield of dioxetane is the product of the second-order rate constant of the probe with $^1O_2$ and the efficiency of the dioxetane formation relative to competing $^1O_2$ pathways, such as endoperoxide and hydroperoxide formation. Decomposition of dioxetane leads to formation of an excited state carbonyl product, which emits the chemiluminescence. Φ_{Ex} is the relative yield of the excited state carbonyl product molecule (denoted by an asterisk) produced by the decomposition of a dioxetane. Φ_{Fl} is the chemiexcited fluorescence yield of the resultant excited product. Courtesy of H. H. Seliger (1989), Biology Department, The Johns Hopkins University, Baltimore, Maryland. Fig. 3, with permission of Academic Press, Orlando, Florida.

6.6 THE ORIGIN OF BIOLUMINESCENCE

Bioluminescence is the result of metabolic oxidations of saturated phenanthrene ring systems and n-alkanes by mixed-function oxidases, yielding product molecules in excited electronic states. An increased metabolism of energy sources not easily accessible to the primitive oxidases could counterbalance the initial energetically wasteful oxygenase reactions.

Oxygen partial pressures of 0.0007 torr, at which suspensions of luminous bacteria produce just visible luminescence, are many orders of magnitude below the oxygen requirement for the transition from anaerobic fermentation to aerobic oxidation. Before the advent of photosynthetically produced oxygen and efficient stepwise electron transport, oxidations may have been carried out by mutant flavoproteins, with production of a superoxide radical, according to the following reactions:

$$FH_2 + O_2 \longrightarrow FH—OO^- + H^+$$

$$FH—OO^- \longrightarrow FH \cdot + O_2^-$$

$$2FH \cdot \longrightarrow F_{ox} + FH_2 \text{ (dismutation)}$$

where F_{ox} and FH_2 are the oxidized and reduced flavins, respectively.

The production of radicals O_2^- or $HO \cdot$ can induce chemiluminescent reactions in suitable substrates. O_2^- has a long lifetime in solution due to extremely low dismutation reactions at pH values >4.5, and in the absence of redox metal ions.

The substrate and enzyme molecules in BL are called by their respective generic names, luciferin and luciferase. The emission colors of BL range from the blue of the marine bacteria and deep sea fish to the red of the railroad worm *Phrixothrix*. There are three cases of BL emission— direct, enzyme-complex, and sensitized.

6.6.1 Direct Bioluminescence

The emission spectrum of direct BL is identical with the fluorescence spectrum of the excited product molecule free in solution, and with nonenzymatic CL of the substrate molecule.

6.6.2 Enzyme-Complex Bioluminescence

The energy of the tight enzyme–product complex is modified compared to the free excited state product, with emission shifted in peak, but essentially identical in spectral shape and half-width.

6.6.3 Sensitized Bioluminescence

A "fluorescent protein" that binds tightly to the luciferase–product complex serves as a "wavelength shifter" by absorbing the electronic excitation energy of the luciferase–product excited state and emitting its own characteristic fluorescence. *In sensitized BL the emitter molecule is not a participant in the chemical reaction leading to the product excited state.*

6.6.4 Bioluminescent Systems

The bioluminescent systems for which substrate and product molecules have been identified are depicted in Figure 6.3 for the firefly, a crustacean (*Cypridina* sp.), and the Seapen *(Renilla reniformis)*. The luciferin substrate is different in each case, but in all cases molecular oxygen is consumed stoichiometrically, resulting in a *decarboxyketoluciferin* product in an excited state, and free carbon dioxide. The decarboxy keto oxygen atom comes from molecular oxygen. In the *hydroxyl pathway* (Figure 6.4), intermediate VII requires that one of the oxygen atoms of CO_2 comes from water, whereas in the *dioxetane pathway,* intermediate VIII requires that it comes from molecular oxygen.

The luciferin–luciferase oxidation reaction constitutes a *model system for mixed-function oxygenase reactions,* the chemical reaction fundamental to all organisms.

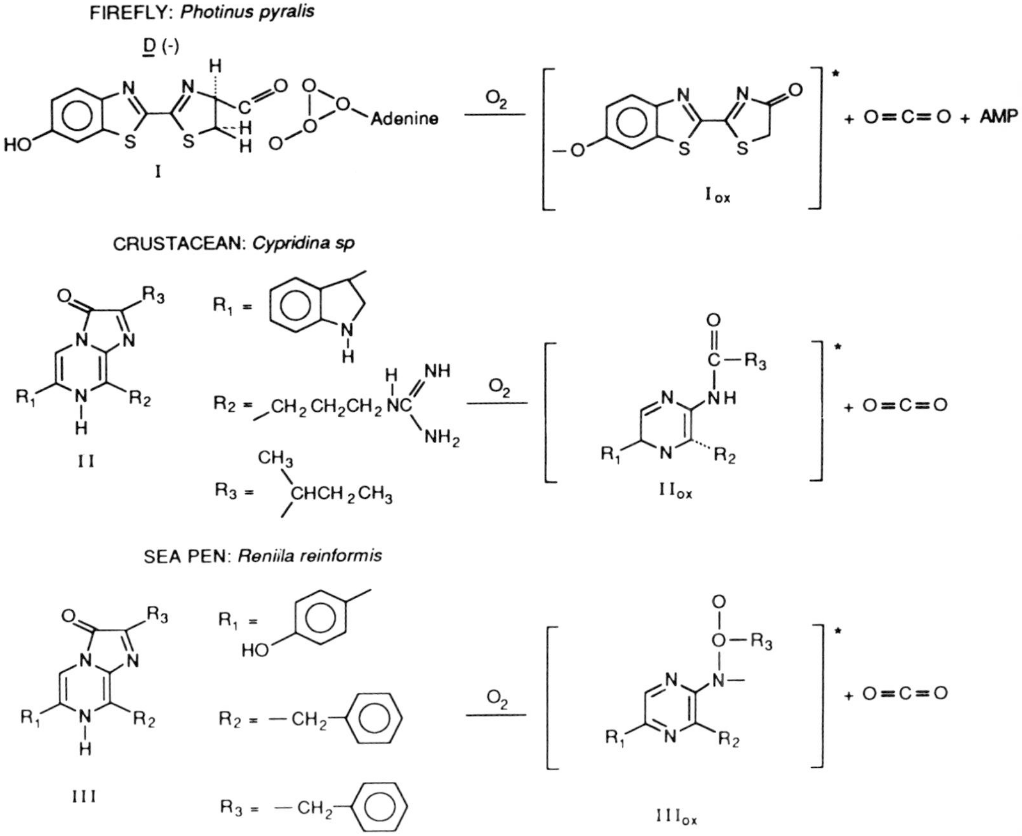

FIGURE 6.3 Bioluminescent systems in the firefly *(Photinus pyralis)*, a crustacean (Cypridina sp.), and the sea pen *(Renilla reniformis)* have a common mechanism, even though the luciferase enzymes and luciferin substrates are not identical. The product and substrate molecules have been identified in all three cases. In each case molecular oxygen is consumed stoichiometrically and two ketones are produced: a decarboxy keto luciferin product in an excited state (responsible for the bioluminescence) and a free carbon dioxide molecule. Courtesy of H. H. Seliger, Biology Department, The Johns Hopkins University, Baltimore, Maryland, with permission of Academic Press, Orlando, Florida.

6.7 BIOLUMINESCENCE OF EUKARYOTES

In *eukaryotes*, relatively large molecules serve as luciferins. *Luciferase catalyzes both an oxidation and a decarboxylation.* BL is manufactured within specialized cells called *photocytes*. Many higher vertebrates have a highly organized collection of these cells, making up a light organ, or *photophore*.

On land, in addition to the firefly, there exist luminous fungi (foxfire), glowworms, freshwater snails (*Latia* in New Zealand), earthworms, and

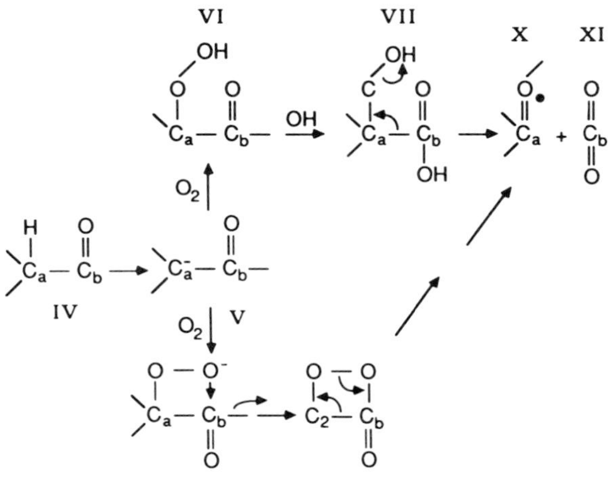

VI VII X XI

IV

VIII IX

FIGURE 6.4 The similar steps in the oxidation of the three luciferins are represented by the pathways shown here. In all cases the decarboxy keto oxygen atom comes from molecular oxygen. Intermediate VII, the hydroxyl ion pathway, requires that one of the oxygen atoms of CO_2 comes from water, whereas in the dioxetane pathway, intermediate VIII requires that an oxygen atom of CO_2 comes from molecular oxygen. Courtesy of H. H. Seliger (1989) Fig. 8, with permission of Academic Press, Orlando, Florida.

insects, including the railroad worm (South America). In the ocean, protozoa and the dinoflagellates are mainly responsible for BL, but the coelenterates comprise the greatest number of bioluminescent species, i.e., the soft corals (Anthozoa), jellyfish (Hydrozoa), and comb jellies (Ctenophora). Some of the luminous fish derive their BL from symbiotic bacteria. There are bioluminescent marine worms *(Chaetopterus, Balanoglossis, Odontosyllis)*, a clam *(Pholas)*, crustacea *(Cypridina* or *Vargula)*, squid *(Watasenia)*, shrimp *(Holophorus)*, and echinoderma (sea stars and sea urchins).

6.8 TYPES OF BIOLUMINESCENT REACTIONS

There are five different types of BL reactions:

1. *Substrate oxidation.* This BL reaction is accomplished by *Cypridina hilgendorfii,* an ostracod crustacean, 2–3 mm in size, living along the Japanese coast; it expels the enzyme and substrate in separate granules into the seawater, where the reaction occurs.

2. *Substrate activation* followed by oxidation. This BL reaction occurs in the firefly and the sea pansy, *Renilla.*

3. *Reduction followed by oxidation.* As examples, luminous marine bacteria and bioluminescent fungi are hosts for this type of BL reaction. In fungi, NADH (or NADPH) and a soluble dehydrogenase reduce an intermediate electron carrier X to XH_2; XH_2 reacts with luciferin (L), luciferase (E), and structural protein (B) (L-E-B) to yield LH_2-E-B, which in the presence of the structurally bound luciferase is oxidized by O_2 to L-(O)*-E-B, the excited state of L(O)-E-B.

4. *Peroxidation.* An example of this type of BL reaction is the H_2O_2 requirement for the blue luminescence in extracts of the acorn worm, *Balanoglossus,* a marine hemichordate. Luciferase functions as a peroxidase and molecular oxygen is not required.

5. *Precharged systems.* Prototypes of this BL reaction are the *scintillon luminescence* of marine dinoflagellates such as *Gonyauilax polyedra* and the luminescence of certain hydromedusae such as *Aequorea* and *Halistaura* and the marine annelid *Chaetopterus.* Precharged systems are biochemically poised, so that light emission can be readily triggered. The first of these systems involves a particulate structure, termed a *scintillon,* which may be a cell organelle; experimentally, simply lowering the pH from 8 to 5.7 will cause this structure to emit light. In other systems the emission observed is not that of the luciferin but that of the excited state of a protein (*photoprotein* associated with *sensitized bioluminescence*) following energy transfer from the luciferin. In *Aequora* and *Halistaura,* on addition of calcium, a luminescent reaction occurs. A purified protein isolated from the marine annelid *Chaetopterus* luminesces when mixed with H_2O_2 and ferrous iron in the presence of oxygen.

6.9 GENERALIZATIONS ABOUT BIOLUMINESCENCE REACTIONS

A general sequence of reactions leads to the emission of light. Usually some preliminary steps occur to "prepare" the luciferin for the BL reaction. The steps are activation, oxygenation, excitation, and turnover.

6.9.1 Activation

Activation is the first step after binding of the substrates to luciferase E:

$$E:LH_2B + A \longrightarrow E:LH_2B + C$$

where A is a substrate, B is binding protein, and C is some product, e.g., ATP and pyrophosphate, respectively (firefly BL).

6.9.2 Oxygenation

In the *Cypridina* BL reaction, with no activation step, oxygenation of luciferin is the first reaction:

$$E:LH_2B + O_2 \longrightarrow E:LBHOOH$$

6.9.3 Excitation

The peroxide formed in the oxygenation step breaks down through the decomposition of dioxetane and the CIEEL process to yield the excited state product:

$$E:LBHOOH + M \longrightarrow E:LBO^* + P + M$$

where M is another cofactor that may be needed to initiate the process (e.g., Ca^{2+} in the jellyfish aequorin reaction) and P is other products (e.g., H_2O or CO_2 for the firefly). Light emission results from deexcitation of $E:LBO^*$.

6.9.4 Turnover

Finally, the luciferase is released, ready to act again:

$$E:LBO \longrightarrow E + LO + B$$

where LO is oxidized luciferin. The firefly does not need the turnover step *in vivo*, because it has all the luciferase it needs for a lifetime of one-time reactions. *In vitro*, firefly luciferase can be made to turn over.

6.10 SOME EXAMPLES OF BIOLUMINESCENCE

6.10.1 *Cypridina* Luminescence

When stimulated, the ostracod secretes luciferin and luciferase *from separate glands* into the surrounding seawater. Luciferase catalyzes oxygenation, excitation, and release for turnover:

$$E + LH_2 \longrightarrow E:LO^* + CO_2$$

The BL emission maximum is about 465 nm. Although the mechanism of oxidation is the same, the substituent groups (R) differ for the *Cypridina*-type luciferin and the coelenterate-type luciferin, called coelenterazine.

Luciferin, chemically identical to that from *Cypridina,* has been found in luminous organs of teleost fish (*Apogon ellioti* and *Parapriacanthus beryciformes*). Although cypridinas are found in the fish gut, gel filtration and the immunological and kinetic behavior of fish luciferase suggest its complete *de novo* synthesis by the fish.

The fireworm *Odontosyllis* (Bermuda), a marine polychaete annelid, releases bioluminescent materials into the water during mating. This is a very spectacular and precisely timed event, beginning just 55 min after sunset on the second, third, and fourth days after a full moon. The purified luciferin (λ_{abs} 235, 285, and 330 nm) serves as the substrate in the enzymatic reaction; the product of this reaction (λ_{abs} 250 and 445 nm) evidently acts as the emitting molecule; the spectrum of its fluorescence (λ_{max} 507 nm) closely matches that of the BL. An unusual property of the *Odontosyllis* system is its stimulation *in vitro* (and *in vivo* as well) by small concentrations (between 10^{-3} and 10^{-6} M) of cyanide, a known product of millipedes.

Latia is a gastropod freshwater mollusc that exudes a brightly luminescent mucus (λ_{max} 520 nm). The BL reaction requires only oxygen, no cofactors, with a reaction mechanism generally similar to that of *Cypridina*.

6.10.2 Firefly Bioluminescence

Firefly luciferin (LH_2) is D($-$)-2-(6'-hydroxy-2'-benzothiazolyl)-Δ^2-thiazoline-4-carboxylic acid (Figure 6.3). The first step in the sequence (Section 6.9.1) is activation, by a luciferase-catalyzed reaction:

$$E:LH_2 + ATP \longrightarrow E:LH_2AMP + PP_i$$

Oxygenation proceeds in several steps, the first possibly being

$$E:LH_2AMP + O_2 \longrightarrow E:LHOO\ AMP$$

The excitation step is an intramolecular electron transfer, required for the CIEEL mechanism.

A single protein, firefly luciferase, possesses all of the required catalytic properties to generate light: activation, oxygenation, and excitation. The different colors of firefly BL, ranging from green to orange–red, are apparently produced by slightly different species-specific luciferases that form different complexes with the same excited state decarboxyluciferin product.

As examples of biological adaptation, species active during twilight exhibit yellow BL against the green background of foliage; dark-active species exhibit green BL; in two yellow bioluminescent species studied, the electroretinogram (ERG) action spectra were also shifted to yellow— the widths of these yellow-peaked spectra were sharply narrowed relative to the green-peaked ERG action spectral of the green bioluminescent species.

Optimization of the signal-to-noise (S/N) ratio for the detection of BL in the presence of "noise" due to ambient light can be expressed in terms of a dimensionless ratio, the biologically effective adaptation (BEA). The optimization model postulates that (1) visual sensitivity in nonbioluminescent ancestors of fireflies was conditioned by the predominance of green light reflected from foliage and (2) in bioluminescent dark-active species the color green was selected to match the green visual spectral sensitivity.

Optimization of BEA during twilight proceeded in steps: (1) selection for a screening pigment that preferentially absorbed green light, (2) extension of behavioral patterns to include a different photic environment, (3) selection for modification of visual sensitivity in the new photic environment, and (4) selection for modification of the color of BL to maximize signal detection by the already modified visual sensitivity.

Optimization can occur in terms of screening pigment properties (screening pigment pathway) or visual pigment (i.e., opsin) sensitivity (opsin pathway) (see Figure 6.5).

The predictions of the optimization model were compared with experimental measurements of action spectra of visual sensitivities (Figure 6.6), BL emission spectra, spectral intensity distributions of ambient light, microspectrophotometric absorption curves of rhabdomeres, and observations of diurnal timing of flashing activities. The model is based on three assumptions:

1. The visual spectral sensitivity curves of all North American fireflies, with varying and selective attenuation of the short-wavelength regions by different concentrations of the screening pigment, while long-wavelength regions of all species are superposable (Figure 6.6).
2. The similar shapes of the BL emission spectra independent of the shifts in the color of BL.
3. A highest mean efficiency for the detection of BL when the λ_{max} of BL is equal to the λ_{max} of visual sensitivity.

The model shows that a narrowing of visual spectral sensitivity via screening pigment in order to discriminate against green ambient light is more efficient than a shift in visual spectral sensitivity via a change in the opsin photoprotein.

6.10.3 Bioluminescence of Coelenterates

Within the Cnidaria (Coelenterata) are the classes Anthozoa, which include the soft coral, *Renilla reniformis* (sea pansy), and Hydrozoa, which includes the jellyfish, *Aequorea aequorea*.

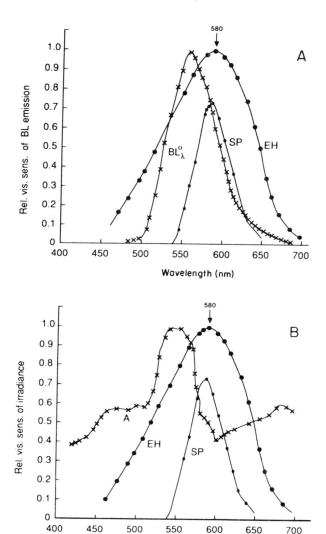

FIGURE 6.5 (A) Relationship between the spectral emission of green bioluminescence, $\lambda_{max} = 550$ nm ($\times$), versus the yellow visual spectral sensitivities of a firefly produced via the opsin pathway (EH), with $\lambda_{max} = 580$ nm, and the screening pigment (SP) pathway, implying a greater detection efficiency for BL for the opsin pathway. (B) Relationship between the spectral radiance of foliage-reflected sunlight (curve A) versus the yellow visual spectral sensitivities of a firefly produced via the opsin pathway (EH) and the screening pigment (SP) pathway, implying a significantly lower detection efficiency for ambient light for the screening pathway than for the opsin pathway. Courtesy of H. H. Seliger (1989) Figs. 9A, 9B) Academic Press, Orlando, Florida.

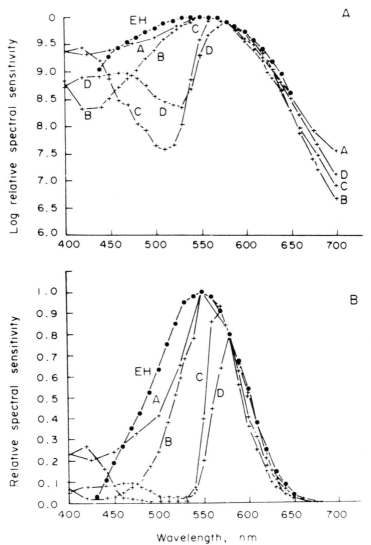

FIGURE 6.6 (A) Best fit of the alignment along the vertical axis of the log of the visual spectral sensitivities of four representative firefly species, A, B, C, and D, and the model pigment absorption curve [Ebrey–Honig (EH) nomogram]* for λ_{max} = 550 nm, illustrating the decreases in visual spectral sensitivities in the short-wavelength region of the spectrum (which corresponds to the role of the screening pigment) relative to the EH nomogram. The curves are for firefly A, *P. versicolor;* firefly B, *P. lucicrescens;* firefly C, *P. pyralis;* and firefly D, *P. scintillans.* (B) Visual spectral sensitivities of *P. versicolor* (A), *P. lucicrescens* (B), *P. pyralis* (C), and *P. scintillans* (D) relative to the Ebrey–Honig nomogram for λ_{max} = 550 nm, drawn from the curves in Figure 6.5A. Courtesy of H. H. Seliger (1989) Figs. 12A, 12B, copyright Academic Press.

*The Ebrey Honig nomogram is the arbitrarily reconstructed absorption spectrum of firefly rhodopsin for λ_{max} set at 550 nm. For calculation of the visual pigment absorption

6.10.3.1 Bioluminescence of *Renilla*

The first step of *Renilla* BL is activation:

$$E_s:LH_2S + PAP \longrightarrow LH_2 + PAPS + E_s \qquad (6.5)$$

Luciferyl sulfate LH_2S (LH_2 is luciferin) is complexed with a sulfokinase, E_s. Activation requires 3',5'-diphosphoadenosine (PAP), and liberates PAPS.

Free luciferin is released from binding protein B by Ca^{2+}:

$$B:LH_2 + Ca^{2+} \longrightarrow B:Ca^{2+} + LH_2 \qquad (6.6)$$

If Ca^{2+} is added to *Renilla* luciferin bound to B and luciferase, the BL intensity rises in a few tenths of a second and then falls very slowly. The light-giving organ, or photocyte, of *Renilla* is made of lumisomes, membrane-enclosed vesicles (0.2 μm), containing B, LH_2, luciferase E, and a "green fluorescent protein." On adding Ca^{2+} to lumisomes, a flash very much like that of the *in vivo* BL is observed.

After luciferin complexation to Ca^{2+}, oxidation and excitation take place on the anthozoan luciferase, E:

$$E:LH_2 + O_2 \longrightarrow E:LHOOH$$

$$E:LHOOH \longrightarrow E:LO^* + CO_2$$

to yield the excited final product, $E:LO^*$.

The *Renilla* BL observed *in vivo* is green. The product of the luciferin–luciferase reaction *in vitro* (i.e., $E:LO^*$) has a blue emission. The green *in vivo* BL of *Renilla* is sensitized BL produced by a "green fluorescent protein" (GFP). The purified luciferin–luciferase reaction *in vitro* is blue emitting, as can be determined by carrying out the purified luciferin–luciferase reaction. Addition of extracted and purified GFP (with green fluorescence properties) to the luciferin–luciferase *in vitro* demonstrates the transition from blue BL to green BL. For natural selection, the blue to green change is significant: in the shallow coastal water habitats of *Renilla*, blue light is strongly absorbed and green light is transmitted maximally.

The mechanism of the sensitization process is explained by (1) dipole–dipole energy transfer and (2) GFP functions as a fluorescent activator in

---

spectrum (nomogram) three columns are set from left to right: (1) absorption coefficient (percent of maximum); (2) (λ[nm]) scale; (3) (λ_{max}[nm]) scale. A straight edge set at the λ_{max} in the right hand column will intersect the central column (λ[nm]) at a chosen wavelength for which the percent absorption (absorption coefficient) can be read from the column on the left side.

a CIEEL mechanism, i.e.,

$$E:LO^* + GFP \longrightarrow E:LO + GFP^*$$

6.10.3.2 Bioluminescence of Hydrozoa

In the Hydrozoa Ca^{2+} interacts with the product of oxygenation (LHOOH) complexed to a "photoprotein" E_A, i.e., $E_A:LHOOH$. This photoprotein, called either aequorin or mnemiopsin, replaces luciferase in the excitation step, with formation of the excited state product. What is extracted from the jellyfish is the oxygenated species, $E_A:LHOOH$. Adding Ca^{2+} to photoprotein triggers the BL:

$$EA:LHOOH + Ca^{2+} \longrightarrow E_A:LO^* + CO_2$$

The stable oxygenated complex may be regenerated by adding LH_2 to apoaequorin.

The system isolated from hydromedusae (*Aequorea* and *Halistaura*) specifically involves only calcium ions and a single proteinlike molecule. On addition of Ca^{2+} to the highly purified photoprotein, a BL emission (peak around 460 nm) occurs with a half-life of about 1 sec at 20°C.

The molecular basis for cellular flashing seems to involve either a transient change in the permeability to calcium or its active transport. The *in vivo* photogenic system is localized in granules 0.5 μm or smaller, which would allow calcium compartmentalization and movement.

6.10.4 Bioluminescence of Ctenophores

There is a remarkable difference between the *in vivo* BL spectra of the Ctenophora and some of the Cnidaria (i.e., Anthozoa, Hydrozoa). The Cnidaria have a narrow, structured emission, identical to the fluorescence of GFP, with an emission maximum around 509 nm. The GFP of Cnidaria is very fluorescent (quantum yield around 0.8, no luciferase activity) and it evidently sensitizes BL.

The Ctenophora, on the other hand, have a broad BL (maximum around 490 nm), with no evidence of sensitized BL in the ctenophores or the scyphozoans.

6.10.5 Bioluminescence of Bacteria

BL enzymes (luciferases) are mixed-function oxygenases. Protobioluminescence (proto-BL) would be a consequence either of fortuitous chemiluminescent substrates in food or of an oxidase mutant that also provided a fortuitous CL of an enzyme–product complex.

The following sequence of events has been proposed:

1. The α-oxidation cycle of exogenous fatty acids could provide chain lengths appropriate for incorporation into the bacteria's own acetyl carrier protein β-synthetic pathway.

2. Transient severe hypoxia in poorly mixed bottom waters could have selected for a flavoprotein able to react with toxic aldehydes formed by oxidation of fatty acids (retained by contemporary luminous bacteria).
3. With sufficient brightness of proto-BL, a new function of BL signaling might have been a factor in further selection for optimization of interspecies interactions in the echo system of the bacteria.

The above hypothesis can be tested by searching for luminous colonies of luciferase mutants in large numbers of colony plates of α-oxidizing nonluminescent bacteria subjected to a variety of mutagenic factors.

In contemporary bioluminescent bacteria, an $FMNH_2$-oxygenase (bacterial luciferase) can catalyze the oxidation of saturated aldehydes of chain length C_8 to C_{18} to their corresponding acids. There are two steps in the cyclic flavoprotein degradation pathway in bacterial BL: (1) α-oxidation of C_n fatty acid to C_{n-1} aldehyde, by operation of cytochrome oxidase-mediated electron transport in the presence of oxygen and (2) oxidation of C_{n-1} aldehyde to C_{n-1} fatty acid by luciferase-$FMNH_2$. The enzyme-bound 4α-hydroxy-FMNH excited state product [luciferase-4α-OH-FMNH]* is responsible for the BL emission, with the resulting free luciferase, water, and oxidized flavin.

Mutants with $FMNH_2$ aldehyde oxidases would gain two selective advantages over other aerobes: (1) accessibility to food sources, i.e., fatty acids, at oxygen concentrations where cytochrome oxidase-mediated electron transport is inhibited, and (2) the ability to degrade long-chain fatty acids to chain lengths appropriate for incorporation into their own acetyl-acyl carrier protein β-synthesis pathway.

Under sunlight-limited conditions the "appearance" of a biological light source such as proto-BL, with flavoproteins ($FMNH_2$) acting as "protoluciferases," could, if it particularly mimicked a sunlight stimulus, elicit photic responses from nearby organisms. If such responses resulted in increased food availability, survival, or reproductive potential for the organisms which are emitters of proto-BL, such fortuitous BL could serve as a new bioluminescent signaling. There would be three requirements for the adaptation of the new function in protobioluminescent bacteria:

1. A proto-BL sufficiently bright to be detected, with an emission spectrum overlapping the absorption of the visual photopigments of phototactic organisms.
2. Presence of such organisms within the ecosystem of the bacteria and within the effective illumination volume of the protobioluminescent colony.
3. As a result of phototaxis, increase in local concentrations of detrital food sources for the protobioluminescent bacteria or ingestion

of these saprophytic bacteria into the nutrient-rich digestive system of the phototactic omnivorous predators.

Selection for phototaxis in marine waters is optimized by blue emission. Efficient chemiluminescent reactions involving O_2 limit the maximum energy of excited state levels to blue or longer wavelength emissions. Bacterial colonies of 10^8 cells would emit 10^{11}–10^{12} photons sec^{-1} of blue light. At a distance of 10 cm, this corresponds to 10^{11}–10^{12} photons emitted through a surface of $4\pi \times 10^2$ cm^2 corresponding to about 0.1 nW/cm^2 for blue emission. In contemporary luminous bacteria the biosynthesis of luciferase is induced only at the time they achieve high concentrations, i.e., substrate oxidation occurs only when the *combined* BL of the individuals will be the brightest. The ubiquity of luminous bacteria in marine waters, on the surface of marine animals, and as symbionts in the light organs of fish species has resulted from environmental selection and optimization of original proto-BL. In any bacterial–fish association, precise overlapping of bacterial BL and host visual spectral sensitivity can occur in two ways: (1) mutation of luciferases or flavin cofactors and (2) a fluorescent lumazine–protein complex, coupled to the luciferase (alteration of the emission color or increased photon yield by sensitized fluorescence).

The Eberhard–Hastings mechanism of bacterial BL with FMN 4α-hydroxide in the excited state as emitter is now challenged by the finding of a blue fluorescent protein isolated from luminous bacteria [lumazine protein (LP)]. Its addition, in the cell-free luciferase reaction solution results in an increase in the quantum yield of BL, and a blue shift of emission spectrum from 490 to 476 nm.

The blue shift cannot be attributed to Forster energy transfer from the excited FMN 4α-hydroxide emitter to the lumazine acceptor. Thus the primary excited species generated in the reaction may not be the 4α-hydroxyflavin.

A chemically initiated electron exchange luminescence mechanism for chemiluminescent reactions of some organic peroxides in the presence of a catalytic "activator" has been proposed. The key steps of this mechanism, are as follows:

1. A one-electron transfer from the activator (A) to the peroxide (P) to form a radical pair $P^{\mp}$ and A^+.
2. A chemical transformation of $P^{\mp}$ to a new anionic radical $D^{\mp}$, which is a powerful electron donor.
3. A second step of one-electron transfer from $D^{\mp}$ to A^+ leading to the formation of an excited state A*
4. Radiative relaxation of A*.

The initial proposal identifies A* as the excited state but in principle either A* or D* could be formed at step 3.

According to the CIEEL mechanism, bacterial luciferase energy transfer may take place from an excited acylium cation $(R\text{-}C{\equiv}O^+)^*$ to flavin 4α-hydroxide or to a second chromophore X (such as the lumazine protein) to produce HFl-4α-OH* or X*. The latter can account for the blue-shifted emission. A 4α-hydroxyflavin radical cation is a key intermediate in the proposed CIEEL scheme.

In summary, in all phylla studied, the luciferins fall into three classes: (1) aldehydes (bacteria, gastropods, earthworms), (2) fluorescent bile pigments (dinoflagellates), and (3) fluorescent pigments related to melanin metabolites—pyrazines (jellyfish, ostracods, some fish species) and thiazole (firefly).

The common aspect of all bioluminescent reactions is that they involve metabolic oxidations by molecular oxygen, resulting in excited state products that are highly fluorescent. With the exception of the aldehyde oxidases, the monooxygenase reactions appear to be oxidative decarboxylations via dioxetane intermediates. In the dioxetane reactions the oxidized substrate is fluorescent, whereas in bacterial BL, fluorescence is emitted by the enzyme-bound flavin cofactor or the lumazine protein acting as an energy trap for the free energy released.

Bibliography

Ebrey, T. G., and Honig, B. (1977). New wavelength dependent visual pigment monograms. *Vision Res.* **17,** 47–151.

Gast, R., and Lee, J. (1978). Isolation of the *in vivo* emitter in bacterial bioluminescence. *Proc. Natl. Acad. Sci. U.S.A.* **75,** 833–837 (in Tu and Mager, 1991).

Hader, D.-P., and Tevini, M. (1987). "General Photobiology," pp. 90–111. Pergamon, Oxford.

Hastings, J. W. (1968). *Bioluminescence Ann. Rev. Biochem.* **37** pp. 597–630, from library of Professor Marcel Bessis, Laboratoire de Pathologie Cellulaire, Kremlin Bicetre, France.

Hastings, J. W. (1991) Bioluminescence. In: *"Neural and Integrative Animal Physiology"* (Prosser, C. L.), pp. 131–170. Wiley, Interscience, New York.

Koka, P., and Lee, J. (1979). Separation and structure of the prosthetic group of the blue fluorescence protein from the bioluminescent bacterium *Photobacterium phosphoreum*. *Proc. Natl. Acad. Sci. U.S.A.* **76,** 3068–3072. (in Tu and Mager, 1991).

Lee, J. (1989). Bioluminescence. *In* "The Science of Photobiology" (K. C. Smith, ed.) 2nd ed., pp. 391–417. Plenum, New York and London.

Mager, H. I. X., and Adink, R. (1984). On the role of some flavin adducts as one electron donors. *In* "Flavins and Flavoproteins" (R. C. Bray, P. C. Engel, and S. G. Mayhew, eds.), pp. 37–40. de Gruyter, Berlin (in Tu and Mager, 1991, pp. 319–328).

O'Kane, D. J., Karle, V. A., and Lee, J. (1985). Purification of lumazine proteins from *Photobacterium leiognathi* and *Photobacterium phosphoreum*: Bioluminescence properties. *Biochemistry* **24,** 1461–1467 (in Tu and Mager, 1991, pp. 319–328).

Schuster, G. B. (1979). Chemiluminescence of organic peroxides. Conversion of ground-state reactions to excited-state products by the chemically initiated electron-exchange luminescence mechanisms. *Acc. Chem. Res.* **12,** 366–373 (in Tu and Mager, 1991, pp. 319–328).

Seliger, H. H. (1989). The measurement of light emitted by living cells. *In* "Cell Structure and Function by Microspectrofluorometry" (E. Kohen and J. G. Hirschberg, eds.), pp. 417–449. Academic Press, San Diego.

Shimomura, O., Johnson, F., and Saiga, H. (1963). Extraction and properties of Halistaurin, A bioluminescent protein from the Hydromedusan Halistaura *J. Cell. Comp. Physiol.* **62,** 915 (in Hastings, 1968, pp. 621–623).

Tu, S.-C., and Mager, H. I. X. (1991). Recent advances in chemical modeling of bacterial bioluminescence mechanism. *In* "Photobiology: The Science and Its Applications" (E. Riklis, ed.), pp. 319–328. Plenum, New York and London.

CHAPTER 7

Environmental Photobiology

7.1 THE PROBLEM OF OZONE DEPLETION

Occurrence of a "hole" in the ozone layer has been observed and attributed to the presence of man-made chlorofluorocarbons (CFCs). The chlorofluorocarbons contained in many coolants, electronic solvents, foam, etc. are slowly released into the atmosphere, where they can exist for up to 100 years. In the stratosphere, via short-wavelength UV radiation, they produce free chlorine, which destroys the ozone. Each CFC molecule acts in a catalytic fashion to destroy about 100,000 ozone molecules. Even if harmless chemicals are now substituted for CFCs, the long lifetimes of CFCs already present in the environment mean that the ozone layer will continue to be adversely affected until the middle of the twenty-first century. The resulting increase in the UV spectrum reaching the earth's surface will be largely confined to the wavelength region 295–315 nm. Ozone consumption is believed to result from photochemical reactions such as photodissociation of chlorofluoromethane (CCl_2F_2)

or CCl_3F, resulting in the release of chlorine. Subsequently,

$$O_2 + Cl\cdot \longrightarrow O\cdot + ClO\cdot \qquad (7.1)$$

There is also an O_3-consuming reaction of ClO with atomic oxygen produced by the O_3 photodissociation equilibrium:

$$ClO\cdot + O\cdot \longrightarrow Cl\cdot + O_2 \qquad (7.2)$$

Changes in total column atmospheric ozone are measured in Dobson units by satellite sensing using backscattered UV radiation [with the total ozone mapping spectrometer (TOMS)]. A Dobson unit is 0.01 mm of the thickness of the ozone layer, if condensed to standard pressure and temperature.

7.2 OZONE REDUCTION AND INCREASED SOLAR ULTRAVIOLET RADIATION

Concern about ozone reduction revolves around ozone in the stratosphere, the location of most of the atmospheric ozone. In addition to ozone, effective UVB light is affected by solar elevation and cloud cover.

Apart from the Antarctic ozone hole, there have already been general reductions in stratospheric ozone. Computations from satellite observations show that effective UVB light on clear days may have increased by as much as 10% at temperate latitudes during cooler months of the year.

UVA light (315–400 nm) is not affected by ozone radiation, whereas UVB (280–315 nm) is. UVC light (<280 nm) is completely absorbed by ozone and oxygen in the atmosphere.

7.2.1 Factors Affecting the Present Solar UV Climate

Apart from local cloud cover there are two primary factors that determine the solar UV radiation:

1. *The total ozone column:* more than 90% of the ozone is in the upper atmosphere (stratosphere) of the earth, although considerable ozone exists in the lower atmosphere (troposphere) in areas where there is air pollution. The stratospheric ozone layer is thinnest in the tropics and thickest toward the poles.

2. *The sun angle* determines the length of the path that the sun rays take as they penetrate the atmosphere (the shortest path occurs when the sun is directly overhead; longer paths and increased filtering action of the air occur when the sun is lower in the sky).

The change in ozone thickness with latitude and the differences in prevailing sun angles combine to produce a very large gradient in solar UV radiation, from the intense UV-irradiated climate of the tropics to the very benign UV radiation received at high latitudes. Cloud cover, haze, dust, and air pollution (e.g., sulfur dioxide and tropospheric ozone) are particularly effective UV attenuators.

Increases of CO_2, methane, and chlorofluorocarbons are generally causing a warming of the planet (greenhouse effect). Because the planet will warm unevenly, differences in atmospheric pressure patterns will change the way storms track around the earth, with an obvious bearing on sunlight climate.

7.2.2 Nature of Biological Photoreactions: The Radiation Amplification Factor

With reduction of ozone or its redistribution toward lower latitudes, shorter wavelengths with greater destructive influences are intensified in regions beyond lower latitudes. This amplification of the radiation effect is expressed as a *radiation amplification factor (RAF), which is the percentage increase of biologically effective radiation caused by a given percentage change of total ozone column.*

Action spectra (Figure 7.1) expressing the relative effect of radiation of the same photon flux density at different wavelengths (Section 7.3.1) evaluate (1) the response of higher plants to UV radiation, (2) the damaging effects on microorganisms mediated by DNA, (3) the erythemal (sunburning) response of Caucasian skin, (4) UV photocarcinogenesis, and (5) the photolysis of ozone in the troposphere. In most of these

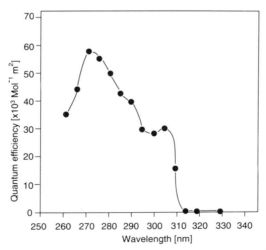

FIGURE 7.1 An example of an action spectrum. Quantum efficiency of the inhibition of motility at different wavelengths in the flagellate *Euglena gracilis* by ultraviolet radiation. From Hader *et al.* (1991), p. 36.

examples, the increasing effectiveness of the radiation with decreasing wavelength is quite apparent.

Using action spectra as *weighting functions*, the relative effectiveness is multiplied by the spectral irradiance at each wavelength. The weighted radiation values summed over the appropriate wavelength range yield a biologically effective dose. Many of the RAF values range between 1.0 and 2.5.

A 1% increase of effective irradiance leads to a greater than 1% increase of biological effect *(biological amplification factor)*.

7.2.3 Future Solar UV Climate

Future levels of atmospheric ozone (Figure 7.2) will be influenced by quantities of ozone-depleting chemicals: CFCs, CO_2, methane, and nitrogen oxides (NO_x). When stratospheric ozone is reduced, UVB radiation penetrates more effectively into the troposphere and affects the hydrocarbon / NO_x photochemistry, and therefore the formation of tropospheric ozone. While tropospheric ozone increases in industrial regions as a result of pollutants, generally it cannot compensate for stratospheric ozone depletion.

In the tropics, solar UVB, already intense now, would surpass the intensities that have been experienced on the Earth's surface in recent geological history.

7.3 THE ROLE OF BIOLOGICAL ACTION SPECTRA

Biologically effective radiation is calculated by the three weighting functions—plant damage, DNA damage, and the Robertson–Berger meter. All of these functions decrease with increasing wavelength in about the same region. With ozone reduction RAFs are much greater for effective radiation calculated with steeply dropping action spectra (i.e., DNA damage) than with less steep spectra (the Robertson–Berger meter erythema-like spectrum). It is noteworthy that a single action spectrum used as a weighting function for biological organisms still does not exist.

7.3.1 Biological Action Spectra

Action spectroscopy (Figure 7.1) is most simply defined as the measurement of a biological effect as a function of wavelength (λ). A carefully constructed action spectroscopy (AS) can identify the chromophore if the AS corresponds closely to the absorption spectrum of a molecule that can be shown to be affected by exposure to radiation in the wavelength

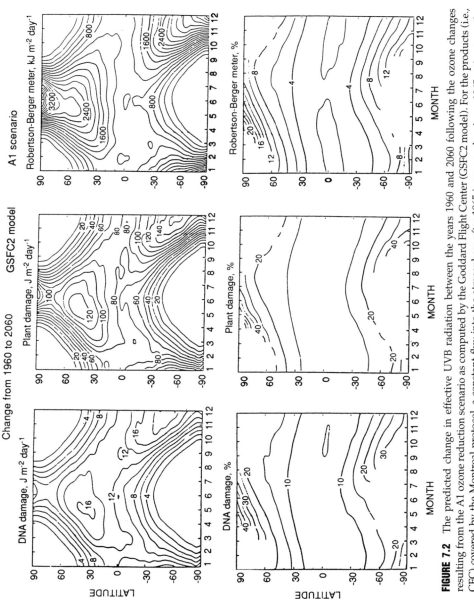

FIGURE 7.2 The predicted change in effective UVB radiation between the years 1960 and 2060 following the ozone changes resulting from the A1 ozone reduction scenario as computed by the Goddard Flight Center (GSFC2 model). For the products (i.e., CFC) covered by the Montreal protocol, a constant flux into the atmosphere after 1965 is assumed, using a 1965 estimate as the annual release into the atmosphere. From Madronich *et al.*, in Draft UNEP Atmospheric Science Panel Report 1989, reproduced by Caldwell *et al.* (1989), p. 8.

region tested. *Six fundamental conditions are necessary for any analytical action spectrum (AAS):*

1. At no wavelength should more than half of the incident radiation be absorbed by the sample before every participating chromophore is exposed (e.g., less than a third of the radiation at 260 nm will be absorbed before it reaches the nucleus of a flattened cell).

2. Scattering and absorption of radiation in front of the target chromophore should either be negligible, or amenable to a "correction factor."

3. *In vitro/in vivo* absorption spectra of chromophore should be identical.

4. The quantum yield (i.e., *the probability of a photochemical change in a chromophore that has absorbed a photon*) should be the same for all the wavelengths tested. Otherwise the effect can vary with the wavelength even without a change in the absorption of radiation by the chromophore.

5. It should be possible to multiply each separable wavelength survival curve by a fluence modification factor such that the curves are reasonably superimposable on one another (a crucial factor in determining whether one is looking at the same mechanism of action throughout the spectral region being employed).

6. The effect for a given amount of total radiation dose should be independent regardless of the rate at which the exposure is given (reciprocity law to be tested over a fluence range of at least a factor of three).

There are experimental variables that can limit the reliability of any AS: (1) spectral purity of the radiation source, (2) accuracy of the dosimetry measurements, (3) placement of the dosimeter (ideally at the sample), (4) presence of exogenous and/or endogenous "nonparticipating" chromophores, (5) ambient (even microenvironmental) conditions, (6) time in the life cycle (cellular) or growth cycle (developmental) of the exposed organism, (7) physical state of the target molecule (e.g., DNA extended or coiled in chromatin), (8) extent of repair of photobiological damage during irradiation and before assay.

Using the spores of *Sphaerocarpus donnelli* and the fungus *Trichophyton metagrophytes*, UV radiation in the region of about 265 nm (coincident with absorption of nucleic acids) is the most effective wavelength for cell mutation. The fungal action spectrum for mutation shows another maximum at 218 nm, due to substantial UV absorption (75%) by fungi at 218 nm, before the radiation reaches the center of the cell.

With the unique flattened geometry that mammalian cells assume in monolayer cultures, the determination of the AAS in such preparations has assumed more significance. The first AAS for mammalian cell killing showed a peak effect (270 nm) between the nucleic acid (260 nm) and protein (280 nm) peaks, giving rise to the belief that both compounds might contribute to reproductive death. With the AS for pyrimidine dimer production it can be shown that DNA alone is responsible for mammalian cell lethality by UVC radiation.

7.3.2 Polychromatic Action Spectra

The system using a polychromatic source is closest to natural conditions. A major advantage of polychromatic action spectra (PAS) is that biological interactions at and responses to different wavelengths (usually unknown) can be empirically defined, e.g., photorepair systems at longer wavelength radiation that might mitigate the damaging effect of shorter wavelength radiation.

7.3.3 Ultraviolet B Photobiology

The percentage of UVB radiation in the solar output reaching the earth's surface is less than 0.3% (2 W m^{-2}) due to the filtering effect of atmospheric chemicals. Even so, there is enough ambient UVB especially due to ozone depletion reaching earth to produce some damage to cellular DNA (absorption peaks at 260 nm and drops by three orders of magnitude at 320 nm). Plastoquinone and plastoquinol, both important in photosynthesis, also absorb strongly at a wavelength below 310 nm. Plant cells are often highly pigmented and especially able to shield target molecules such as DNA.

7.3.4 Effectiveness Spectra

Action spectra are obtained for a given effect, and these may be combined with the known (or estimated) ambient solar radiation expected due to various ozone depletion scenarios (e.g., solar erythemal effectiveness spectrum *vs.* O$_3$-depleted solar erythemal effectiveness spectrum).

Attempts to construct an effectiveness spectrum (ES) from a generalized plant damage AS, or considerations weighted for photosynthesis, allow a preliminary estimate of a 1% decrease for each 1% decrease in ozone layer.

7.4 HUMAN HEALTH AND UVB RADIATION IN THE ENVIRONMENT

Most of the responses to UVB radiation show a relationship between dose and effect that is nonlinear in certain areas of the dose–response curve (e.g., the formation of vitamin D$_3$ in the skin by UVB radiation). In other cases, adaptation of the skin to UV radiation will modify the effects, e.g., thickening of the most superficial layers of the skin and pigmentation; both responses provide the skin with an efficient protection against sunburn. However, hyperproliferation of skin cells may render the skin more susceptible to cancer.

7.4.1 Ocular Damage: Effects on the Cornea, Lens, and Retina, Regardless of Ozone Exposure

7.4.1.1 Damage to the Cornea

The initial response is photokeratitis, commonly seen in skiers (snowblindness). *The eyes do not develop a tolerance to UV light but become more sensitive with repeated exposures.* Below 290 nm, damage is principally to the corneal epithelium. Between 290 and 315 nm, the corneal stroma and endothelium begin to show damage as well. UV radiation damage is cumulative and additive for all wavelengths.

7.4.1.2 Damage to the Lens

Cataracts are prevalent, and may be nuclear, posterior subcapsular, or cortical in nature, the last two being etiologically more related to UVB radiation. In nuclear cataracts, one sees a yellowing of nuclear proteins. In posterior subcapsular cataracts, abnormal and degenerate cells migrate and accumulate at the posterior surface of the lens. In cortical cataracts, gaps form in the cortex and fill with water and debris. UVA radiation also may contribute to cataract formation.

In Nepal, cataract prevalence increases from 1.26 to 4.63% when the average daily sunlight hours increase from 7 to 12. A doubling of the cumulative exposure increases the risk of cortical cataract by a factor of 1.6. Individuals ranking in the upper quarter for average annual UVB exposure had a three-fold increased risk of cortical cataract compared to those in the lowest quarter.

7.4.1.3 Damage to the Retina

In the aphakic (lens removed), the UVA (325 nm) threshold for retinal damage is 10 times lower, compared to nonaphakic individuals.

7.4.1.4 Intraocular Melanoma

Individuals born in the southern United States show nearly a threefold increased risk for introcular melanoma. Blue-eyed individuals constitute the phenotype with the highest risk.

7.4.2 Immunologic Effects/Infectious Diseases

In UV-exposed skin the loss of the antigen-presenting Langerhans cell (Chapter 15) is associated with the subsequent appearance of suppressor T cells (T_s cells). These cells normally serve a regulatory function, preventing an inappropriate autoimmune response.

In UV-treated skin, T_s cells prevent a response to antigens immediately or shortly after the cells appear in the skin. Thus, in animal models, UV radiation-induced tumors are not recognized as foreign and thus are not controlled by the skin's immune system. The skin cancer effects of UV radiation occur mainly in white-skinned races. However, the effect on antigen-presenting cells at very low UV dose is seen in all, i.e., Caucasians, Africans, Asians, and Australian Aborigines.

Impaired local immunity at the site of irradiation occurs following doses of UV radiation received in a normal sun-exposure scenario (a week-long sunny vacation). Several months of daily sun exposure could induce impaired systemic immunity with a diminished host response to foreign organisms.

7.4.2.1 Herpes Viruses

A variety of stimuli can cause reactivation of viruses that exist in a latent state. Herpes simplex reoccurrences are noted after exposure to UV light (multiple suberythemal doses or single doses equivalent to a mild sunburn). UV irradiation of the herpes-infected site leads to the development of T_s cells that decrease the activity of those T cells that would normally respond to herpes infection.

7.4.2.2 Protozoa Infections

In both leishmaniasis (parasitic protozoan infection transmitted by the sandfly) and malaria, exposure to sunlight enriched with the shorter wavelength UVB is suspected to result in a suppressed immune response to the parasites. Immune responses to the four strain-specific malarial infections differ. T_s cells are probably induced during malarial infection, but the evidence is inconclusive.

7.4.2.3 Bacterial and Fungal Infections

In immunosuppression, *Candida albicans*, *Staphylococcus aureus*, and *Escherichia coli*, which are normal skin flora, can become a major cause of morbidity.

7.4.2.4 Tuberculosis and Leprosy

Immunity to the mycobacteria that cause tuberculosis is T cell dependent, has a cutaneous phase, and is best elicited via immunization of the skin. Studies in animal models of immunization with the bacillus of (BCG) suggest that moderate doses of UV light may impair development of an immune response and retard recovery. Leprosy also may be affected by UVB increases.

7.4.2.5 Vaccination Programs

Immunizations administered through UV-treated skin have shown that the potential exists for the treatment to render the individual more susceptible, rather than less susceptible, to the administered antigen.

7.4.3 Skin Cancer

There are two types of keratinocyte-affecting nonmelanoma skin cancer (NMSC): basal cell carcinoma (BCC) and squamous cell carcinoma (SCC). The causative role of sunlight in SCC is supported by the following data (see also Chapter 14): (1) the predominant occurrence is on the most sun-exposed parts of the skin, face, neck, and hands, (2) the highest incidence is in locations with the most sunlight, (3) it is predominant in fair-skinned persons, presumably due to a lack of protective pigment, and (4) the risk of developing SCC is strongly related to the cumulative dose of sunlight received throughout life.

In mice the carcinogenic effect of UVB radiation can be described as a power function of the doses of UVB regularly received.

The *"amplification factor"* gives the percentage increase of the SCC incidence caused by a 1% decrease of ozone. This factor is >1% as a result of the combined action of two amplifications: radiation amplification and biological amplification. *Radiation amplification is greatest for the shortest wavelength in the solar spectrum,* because the increased penetration in the case of ozone depletion is wavelength dependent. Especially an increase in irradiance in the wavelengths just above 300 nm will have a comparatively large effect. *The radiation amplification factor (RAF) has a* value of 1.6 for a 1% decrease of ozone.

Biological amplification reflects the fact that a 1% increase of the effective irradiance leads to a greater increase in the incidence of SCC. This is due entirely to the steepness of the power relationship between the doses of UV radiation and the incidence of skin cancer. The *biological amplification factor (BAF) for SCC* has a value of 2.9. Thus, with the two amplifications taken together, a 1% decrease of ozone will lead to an increase of the incidence of SCC by $1.6 \times 2.9 = 4.6$. It will take several decades for this increase to be fully realized.

An action spectrum specific to BCC has not been determined, because mice under UV lamps rarely develop BCC. Assuming that the induction of BCC has the same action spectrum as that of SCC, a biological amplification factor of 1.7 is obtained for BCC, leading to an overall amplification of $1.6 \times 1.7 = 2.7$.

Should total column ozone be reduced by 5%, the carcinogenically effective UVB irradiance would increase by $5 \times 1.6 = 8$. In the long run, this would lead to an increase of BCC by 14% and of SCC by 25% (RAF multiplied by BAF).

7.4.3.1 Cutaneous Malignant Melanoma

The two most common forms of cutaneous malignant melanoma are superficial spreading melanoma (SSM) and lentigo malignant melanoma (LMM) (see Chapter 14). LMM shows a relationship similar to the cumulative dose relationship of BCC and SCC. The relationship between solar exposure and SSM [and nodular melanoma (NM) and unclassified melanoma (UCM) as well] may be related to peak exposures or possibly exposures early in life. Evidence supporting a relationship between CMM and UVB radiation includes the following points: (1) there are higher CMM incidence rates in people lacking protective pigmentation, (2) there is a correlation of higher CMM incidence rates with decreasing latitude and increasing UVB, (3) there is a positive correlation between freckling and nevus formation (risk factors for CMM) and solar exposure, (4) the CMM rates for natives and immigrants to sunny climates are not the same, (5) there are high rates of CMM in xeroderma pigmentosum (XP) patients with genetically deficient DNA repair, and (6) intermittent and severe sun exposure at early ages results in higher CMM risks.

Recently it has been found that two animal models for malignant melanoma can be used: (1) post-UVB melanoma of a small hybrid fish and (2) post-UVB melanoma of the marsupial *Monodelphis domestica*.

7.5 TERRESTRIAL PLANTS

Reductions of leaf area, fresh and dry weight, lipid content, and photosynthetic activity were found in UVB-sensitive plant species, resulting from destruction of the stratospheric ozone layer by chlorofluorocarbons. Additional alterations of leaf surface, epicuticular waxes, UV-absorbing pigments, and water vapor diffusion through the stomata have been reported.

7.5.1 Artificial and Solar Radiation

UVB damage is accentuated by low levels of white light (less than a microeinstein $m^{-2} sec^{-1}$ between 400 and 700 nm). Most plant researchers use *the generalized plant action spectrum* and/or *the DNA action spectrum* as weighting functions.

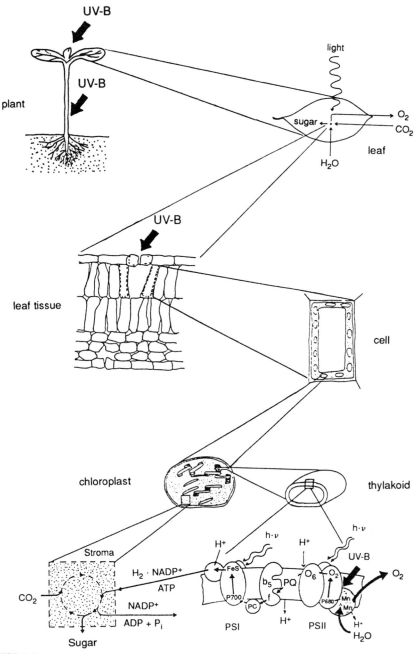

FIGURE 7.3 A generalized model for the effects of UVB radiation on plants. Presumed targets of UVB on growth, stomatal function, and thylakoid structures are indicated by arrows; PSI and PSII, photosystems I and II. From Tevini *et al.* (1989), p. 31.

7.5.2 Effects on Plant Growth, Competition, and Flowering

Plant height and leaf area are reduced by UVB radiation (Figure 7.3), with a clear dose–response relationship demonstrated for cucumber seedlings in growth chambers. For studies on plants, the ozone filter technique utilizes two identical growth chambers covered with a UV-transmitting filter. Ambient UVB radiation is attenuated in one growth chamber by passing ozone through the cuvette on top. The second growth chamber utilizes ambient air in the cuvette on top. This technique is adaptable for simulating natural solar UVB radiation at many latitudes and altitudes. The ozone filter technique is used to simulate a 25% increase in solar UVB irradiance (approximately equivalent to a 12% ozone depletion): in soybean (*Glycine max* L. cv. Essex), UVB radiation is most effective during the transition time between the reproductive and vegetative stages. Reductions in plant height and leaf are not associated with reduced photosynthesis and they are not always correlated with total biomass reduction. The reasons for large response differences among cultivars of a single species might be differential accumulation of UV-screening pigments or morphological alterations.

One of the post-UVB metabolic products of the growth regulator indole-3-acetic acid (IAA) reduces hypocotyl growth when applied exogenously. Action spectra indicate that growth inhibition is not associated with DNA damage.

7.5.2.1 Pollination and Flowering

The anther walls filter out over 98% of incident UVB radiation, and thus pollen is well protected against UVB. The pollen wall, which contains UV-absorbing compounds, is also well protected during pollination. *However, after transfer to the stigma, the pollen tube is susceptible to UVB radiation,* which has been shown to impact pollen germination. Ovules are well hidden in the ovaries and therefore may be sufficiently protected.

UVB-altered timing of flower induction may affect natural ecosystems when plants flower earlier or later than the arrival of their natural insect pollinators.

7.5.3 Effects on Plant Function: Photosynthesis and Transpiration

The action spectra for stomatal closure lead to a reduction of photosynthesis below 290 nm, whereas radiation longer than 313 nm is nearly ineffective. Transpiration is reduced in UV-sensitive cucumber and sunflower seedlings. Photosystem II activity in chloroplasts is also reduced.

7.5.3.1 Effects on Plant Composition

Generally stable chlorophylls and carotenoids are affected in UV-sensitive plants such as beans and cucumbers. The distribution of the main surface waxy compounds is shifted toward shorter chain lengths.

7.5.3.2 UV-Protective Pigments

Most higher plants have accumulations of UV-absorbing pigments (i.e., phenylpropanoids, such as flavonols in the epidermal portion of their leaves) in concentrations that are linearly dependent on UVB dose. Flavonol accumulation is due to an increased activity and biosynthesis of phenylalanine ammonia lyase (PAL), which in rye is activated within minutes (by a decrease in the inhibitor, *trans*-cinnamic acid), through its wavelength-dependent isomerization to the cis form. The trans–cis shift may be responsible for regulation of genetic transcription and may protect seedlings within minutes of their breaking through the soil surface; the UV absorption of isolated epidermal layers is shifted to shorter wavelengths by cis isomerization. In parsley leaves, each individual cell synthesizes mRNA that encodes the key enzyme of flavonoid synthesis, the enzymes themselves, and the end product. Carotenoids protect the photosynthetic apparatus by quenching singlet oxygen or free radicals. UVA/blue photoreactivation of DNA photolyase is responsible for the repair of UV-induced pyrimidine dimers.

7.5.4 Effects of Combinations with Other Stresses

Water stress in combination with enhanced UVB radiation adversely affects water loss in cucumber seedlings unless UVB-absorbing leaf flavonoids are induced. Enhanced UVB radiation reduces net photosynthesis by 20% in plants well-supplied with phosphorus.

7.6 AQUATIC ECOSYSTEMS

Photosynthetic organisms on our planet are estimated to fix about 10^{11} tons of carbon annually (Figure 7.4), more than 50% of it by planktonic organisms in the oceans.

In marine ecosystems (Figures 7.5–7.6) the different sensitivities of the

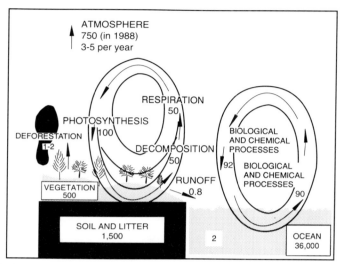

FIGURE 7.4 Major biosphere processes, pools, and fluxes in the carbon cycle (pools and fluxes are in the gigatons of carbon). From Hader *et al.* (1991), p. 35.

primary producers (e.g., phytoplankton) and of the primary consumers (e.g., zooplankton) to UVB radiation result in altered species instabilities in ecosystems, thus affecting the biological food chain.

Phytoplankton are restricted to the upper layers of the oceanic environment. The penetration of light into a body of water strongly depends on marine characteristics, ranging from turbid coastal to clear oceanic water. Shortwave violet and longwave red light are more strongly absorbed than is blue–green light. The penetration of even partially attenuated solar UVB into the photic zone creates a potential hazard for planktonic organisms, which do not possess the protective epidermal UV-absorbing layers of higher plants and animals. The annual biomass production by phytoplankton is about 6×10^{14} kg. A loss of 10% would far exceed the gross national product of all countries in the world, assuming any reasonable price for biomass on the market.

7.6.1 Primary Producers: Phytoplankton

In the water column, planktonic organisms are oriented with respect to light, gravity, chemical gradients, temperature, and the earth's magnetic field. Solar UVB affects the general metabolism, photosynthetic energy production, and nitrogen fixation and assimilation in many species.

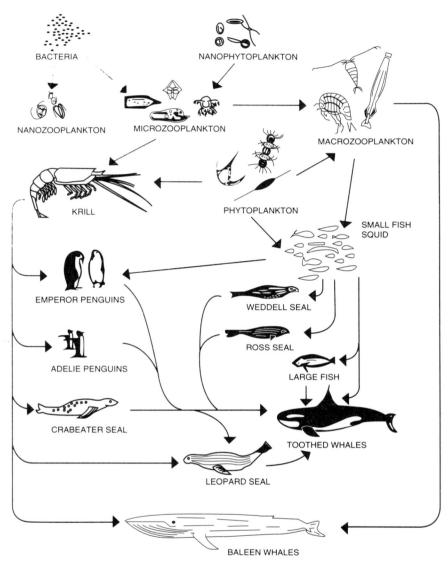

FIGURE 7.5 Example of the biological food chain in a marine ecosystem, which starts with the primary producers; the biomass they produce is utilized by the primary consumers, which in turn serve as food input for the next level in the food chain. From Hader *et al.* (1989), p. 40.

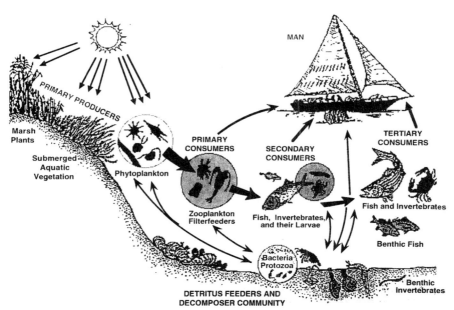

FIGURE 7.6 A representation of the Antarctic marine food chain. Adapted from Voytek (1990), by Hader *et al.* (1991), p. 37.

7.6.1.1 Orientation

Most motile microorganisms use various bands in the visible and long-UV range for photoorientation. Any UVB-induced decrease in orientation of motile phytoplankton prevents the necessary constant adaptation to changing environmental conditions and possibly hazardous situations. The organisms have a repertoire of several responses to light. Flagellates move toward the surface at low light intensities, but at high intensities they escape from the light source, which may bleach their photosynthetic pigments. Planktonic organisms are known to undergo daily vertical movements of up to 12 m.

7.6.1.2 Development and Physiology

UVB inhibition of development may drastically alter the species composition of phytoplankton, and in turn may "fuel" transfer through the food chain. Predominance of toxic or nonpalatable organisms may interfere with growth and multiplication of higher organisms in the chain.

High UVB doses affect general metabolic processes and cellular membrane permeability, and induce irreversible protein damage, eventually causing death.

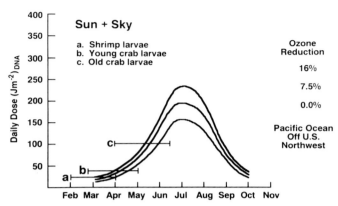

FIGURE 7.7   Estimated effective UVB solar daily dose (active on DNA) at various atmospheric ozone concentrations. *Upper curve:* ozone reduction 16%; *middle curve:* ozone reduction 7.5%; *lower curve:* ozone reduction 0.0%. Also shown are approximate thresholds of the UVB daily dose for shrimp and crab larvae, in their natural seasonal position. Adapted from Damkaer (1981), by Hader *et al.* (1989), p. 44.

7.6.1.2.1 Photosynthesis   Enhanced UVB reduces the concentration of photosynthetic pigments (the photosystem II reaction center is damaged) and decreases ATP and NADPH production and CO_2 fixation. There are also structural changes in the membranes, a decrease in the lipid content, and changes in nucleic acid and protein composition, affecting enzyme activity and production.

7.6.1.2.2 Nitrogen Assimilation   Through the action of the enzyme nitrogenase, cyanobacteria, which are capable of utilizing atmospheric nitrogen dissolved in water, provide nitrogen for higher plants in rice paddies. The annual worldwide nitrogen assimilation (35 million tons) by cyanobacteria exceeds the amount of artificial fertilizer (30 million tons) produced annually. Visible light activates nitrogenase, whereas UVB radiation inactivates nitrogenase.

7.6.1.3 Pigmentation and Fluorescence

Following a short exposure to UVB radiation, a strong increase in red fluorescence by the phytoplankton demonstrates that the photosynthetic apparatus does not effectively use the photosynthetic light, but rather wastes it in fluorescence. After longer exposures, the fluorescence decreases, indicating that radiation increasingly destroys the absorbing pigments.

7.6.1.4 UVB Targets

In green flagellates, after inducing UVB damage, no recovery of motility could be found in white, dim light, which argues against an involvement of DNA governed by photolyase. The primary targets of UVB could be receptor molecules involved in photoorientation or specific components of the motor apparatus.

7.6.2 Consumers: Marine Animals

7.6.2.1 Effects on Zooplankton

One small crustacean suffers 50% mortality from UVB dose rates that are smaller than those currently present at the sea surface. By contrast, some shrimp larvae tolerate sea surface dose rates greater than those forecast for a 16% ozone depletion.

With a 16% ozone depletion over temperate pelagic waters (Figure 7.7), a lethal (59% mortality) cumulative radiation dose for about half the zooplankton examined would be reached at the depth of 1 m in less than 5 days in the summer.

7.6.2.2 Effects on Fisheries

A 16% ozone reduction would result in increases in larval fish mortality of 50, 82, and 100% for anchovy larvae at a 5-m depth at ages 2, 4, and 12 days, respectively.

7.6.2.3 Changes in Species Composition

The generation time of marine phytoplankton ranges from hours to days, against the potential increase of solar UVB spread over decades. The question is whether the gene pool within species is variable enough to adapt to this relatively gradual change in exposure to UVB radiation. Indirect effects also may occur in the form of altered patterns of predation, competition, diversity, and trophic dynamics if UVB-resistant species were to overtake UVB-sensitive species.

7.6.2.4 Atmospheric Carbon Dioxide Concentration and Global Climate Change

Fluxes of CO_2 show that the oceans take up a major proportion of the CO_2 released annually. A 10% decrease in CO_2 uptake by the oceans due to UV-induced depletion of phytoplankton would leave about the same amount of CO_2 in the atmosphere as is produced by fossil fuel burning. This would have long-term consequences for the global climate.

UVB inhibition of dimethyl sulfide production by phytoplankton could result in a reduction of atmospheric cloud condensation nuclei, decreased cloudiness, and increased solar UVB—a positive destructive feedback loop. An alternative prediction argues that the increased evaporation due to a substantial warming of the atmosphere would cause more massive cloud formation.

7.7 UPDATE ON ULTRAVIOLET RADIATION CHANGES AND BIOLOGICAL CONSEQUENCES

Despite increases in tropospheric ozone that more than compensate for stratospheric ozone losses (unlike in the United States), an increase in UVB (larger than expected, i.e., $10 \pm 5\%$ per decade) is seen in data from the high-altitude European observatory at Jungfraujoch. Although the enhanced irradiances exist at wavelengths where the absolute energy flux is small, living cells are quite sensitive to damage in the spectral region 300–315 nm.

ENVIRONMENTAL EFFECTS OF OZONE DEPLETION AND ULTRAVIOLET RADIATIONS

Examples of Radiation Amplification Factors at 30°N

Example	RAF
DNA related	
Mutagenicity and fibroblast killing	2.0–2.2
Fibroblast killing	0.3–0.6
Generalized DNA damage	1.9
HIV-1 activation	3.3–4.4
Plant effects	
Inhibition of growth of cress seedlings	3.0–3.8
Isoflavanoid formation in beans	2.3–2.7
Anthocyanin formation in sorghum	0.9–1.0
Photosynthetic electron transport	0.1–0.2
Membrane damage	
Membrane-bound K^+-stimulated ATPase inactivation	1.6–2.1
Skin	
Elastosis	1.1–1.2
Photocarcinogenesis	1.4–1.6
Melanogenesis	1.6–1.7
Eyes	
Damage to cornea	1.1–1.2
Cataract	0.8–0.7
Movement	
Inhibition of motility in *Euglena gracilis*	1.5–1.9
Immunological	
Immune suppression	0.8–1.0

Two additional cases where the ocular system is presumably affected by UVB are (1) presbyopia and (2) deformations of the anterior lens capsule. An association is found between the early onset of presbyopia and living in areas of high levels of sunlight (and high temperature). In the population of Somalia, it was found that deformation of the anterior lens capsule in the central pupillary area shows a strong association with climatic keratopathy and, by inference, reflected UVB radiation. The deformations found interfere severely with vision.

The viruses activated by UVB irradiation *in vitro* include HIV-1 and a variety of papilloma viruses, associated with a variety of hyperplastic, dysplastic, and malignant lesions of the squamous epithelium.

Only 1 out of 12 skin cancer patients reacted by contact hypersensitivity to challenge by dinitrochlorobenzene whereas there was a reaction from 22 of the 34 healthy volunteers. Thus there is an indication of UV-induced immunosuppression in cancer patients(?) Salivary gland cancers may be related to UVB radiation exposure. With updated RAF and BAF values, a 1% depletion in ozone will eventually increase the BCC incidence by 2.0% and the SCC incidence by 3.5%.

When UV irradiation is added to chemical exposure, melanomas occur sooner. The acceleration of the appearance of melanomas by UV irradiation is due to a local effect at the site where the melanoma develops; possibly that is an immunologically mediated effect. Human melanomas occurring in sun-exposed skin areas exhibit a high frequency of point mutations in the N-*ras* gene.

Bibliography

Caldwell, M. M., Madronich, S., Bjorn, L. O., and Ilyas, M. (1989). Ozone reduction and increased solar ultraviolet radiation. *In* "Environmental Effects Panel Report," U.N. Environ Programme (UNEP), pp. 1–10. U.S. Environ. Prot. Agency, Washington, DC.

Coohill, T. P. (1992). Action spectroscopy and stratospheric ozone depletion. *In* "UV-B Monitoring Workshop: A Review of the Science and Status of Measuring and Monitoring Programs, pp. C89–C112. Alternative Fluorocarbons Environmental Acceptability Study (AFEAS), Washington, DC.

Cooper, K. D., Oberhelman, L., LeVee, G., Baadsgard, O., Anderson, T., and Koren, H. (1991). UV exposure impairs contact hypersensitivity in humans; correlation with antigen presenting cells. *1991 Annu. Meet. Am. Soc. Photobiol.,* San Antonio, TX (in UNEP, 1991, p. 17).

Damkaer, D. M., Dey, D. B. and Heron, G. A. (1981). *Oecologia,* Dose/dose rate responses of shrimp larvae to UVB radiation. 48, 178–182.

Donawho, C. K., and Kripke, M. L. (1991). Photoimmunology of experimental melanoma. *Cancer Metastasis Rev.* 10, 177–188 (in UNEP, 1991, p. 19).

Gates, F. L. (1930). A study of the bactericidal action of the ultraviolet light. III. The absorption of ultraviolet light by bacteria. *J. Gen. Physiol.* 14, 31–42 (in Coohill, 1992, p. C90).

Gery, M. M. (1989). Trophospheric air quality. *In* "Environmental Effects Panel Report," U.N. Environ. Programme (UNEP), pp. 41–43, 49–54. U.S. Environ. Prot. Agency, Washington, DC.

Hader, D.-P., Worrest, R. C., and Kumar, H. D. (1989). Aquatic ecosystems. *In* "Environmental Effects Panel Report," U.N. Environ. Programme (UNEP), pp. 39–48. U.S. Environ. Prot. Agency, Washington, DC.

Hader, D.-P., Worrest, R. C., and Kumar, H. D. (1991). Aquatic ecosystems. *In* "Environmental Effects of Ozone Depletion," U.N. Environ. Programme (UNEP), pp. 33–43. U.S. Environ. Prot. Agency, Washington, DC.

Hollaender, A., and Emmons, C. W. (1941). Wavelength dependence of mutation production in the ultraviolet, with special emphasis on fungi. *Cold Spring Harbor Symp. Quant. Biol.* **9**, 179–186 (in Coohill, 1992, p. C90).

Knapp, E., Reuss, A., Rise, O., and Schreiber, H. (1939). Quantitative analyses der mutation-sauslosenden wirkung monochromtischen UV-lichtes. *Naturwissenschaften* **27**, 304 (in Coohill, 1992, p. C92).

Longstreth, J. D., de Gruijl, F. R., Takizawa, Y., and van der Leun, J. C. (1991). Human health. *In* "Environmental Effects of Ozone Depletion," U.N. Environ. Programme (UNEP), pp. 15–24. U.S. Environ. Prot. Agency, Washington, DC.

Madronich, S., Bjorn, L. O., Ilyas, M., and Caldwell, M. M. (1991). Changes in the biologically active ultraviolet radiation reaching the earth's surface. *In* "Environmental Effects of Ozone Depletion: 1991 Update" (J. C. van der Leun and M. Tevini, eds.), pp. 1–13. United Nations, Rome.

McKenzie, R. L., Ilyas, M., Frederick, J. E., Filyushkin, V., Wahner, A., Mathusubramanian, P., Roy, C. E., Stammes, K., Blumthaler, M., and Madronich, S. (1992). UV radiation changes. *In* "UV-B Monitoring Workshop: A Review of the Science and Status of Measuring and Monitoring Programs," pp. C1–C22. Alternative Fluorocarbons Environmental Acceptability Study (AFEAS), Washington, DC.

Scotto, J., Cotton, G., Urbach, F., Berger, D., and Fears, T. (1988). Biologically effective ultraviolet radiation: Surface measurements in the United States, 1974 to 1985. *Science* **239**, 762–764 (in Caldwell *et al.*, 1989).

Setlow, R. B. (1974). The wavelengths in sunlight effective in producing skin cancer: A theoretical analysis. *Proc. Natl Acad. Sci. U.S.A.* **71**, 3363–3366 (in UNEP, 1991, p. 2).

Stadler, L. J., and Uber, F. M. (1941). Genetic effects of ultraviolet radiation in maize. IV. Comparison of monochromatic radiation. *Genetics* **27**, 84–118 (in Coohill, 1992, p. C92).

Teramura, A. H., Tevini, M., Bornman, J. F., Caldwell, M. M., Kulandaivelu, G., and Bjorn, L. O. (1991). Terrestrial plants. *In* "Environmental Effects of Ozone Depletion," U.N. Environ. Programme (UNEP), pp. 25–32. U.S. Environ. Prot. Agency, Washington, DC.

Tevini, M., Teramura, A. H., Kulandaivelu, G., Caldwell, M. M., and Bjorn, L. O. (1989). Terrestrial plants. *In* "Environmental Effects Panel Report," U.N. Environ. Programme (UNEP), pp. 25–37. U.S. Environ. Prot. Agency, Washington, DC (UNEP, 1989, pp. 61–64).

United Nations Environment Programme (UNEP) (1989). "Environmental Effects Panel Report." U.S. Environ. Prot. Agency, Washington, DC.

United Nations Environment Programme (UNEP) (1991). "Environmental Effects of Ozone Depletion." U.S. Environ. Prot. Agency, Washington, DC.

van der Leun, J. C., Tevini, M., and Worrest, R. C. (1989). Executive Summary United Nations Environment Programme (UNEP), "Environmental Effects Panel Report," pp. i–iii. U.S. Prot. Agency, Washington, DC.

van der Leun, J. C., Talizawa, Y., and Longstreth, J. D. (1989). Human health. *In* "Environmental Effects Panel Report," U.N. Environ. Programme (UNEP), pp. 11–24. U.S. Prot. Agency, Washington, DC.

Voytek, M. A. (1990). Addressing the biological effects of decreasing ozone in the Antarctic environment. *Ambio* **19**, 52–61.

Yoshikawa, T., Rae, V., Bruin-Slot, W., van den Berg, J. W., Taylor, J. R., and Streilein, J. W. (1990). Susceptibility to effects of UV-B radiation of contact hypersensitivity as a risk factor for skin cancers in humans. *J. Invest. Dermatol.* **95**, 530–536.

CHAPTER 8

Marine Photobiology

8.1 PHYTOPLANKTON BIOLUMINESCENCE

8.1.1 Circadian Rhythms

John Murray, one of the naturalists on the Challenger expedition about 120 years ago, wrote "*Pyrocystis* . . . is the chief source of diffuse phosphorescence of the sea in equatorial regions . . . the most brilliant display of phosphorescence observed during the cruise."

The dinoflagellate marine algae *Pyrocystis noctiluca* and *Pyrocystis fusiformis* emit a thousand times more light than is emitted by another dinoflagellate, *Gonyaulax*. Photoinhibition and photoenhancement accurately predict the vertical distribution of bioluminescence during daylight and darkness. *Large variations occur with depth within the euphotic zone.* With vertical excursions of phototactic organisms, the intense bioluminescence observed at the sea surface at night is observed at greater depths during the day.

The *in situ* diurnal rhythm of blue-green light flashes due to the dinoflagellate luciferin–luciferase reaction may be endogenous or exoge-

nous, controlled by diurnal migration. During a solar eclipse, bioluminescent organisms respond to the eclipsing sun much as they normally
respond to the setting sun. Thus, the exogenous factor overrides such
endogenous rhythms as may exist.

The only microplankton constituents capable of luminescence are dinoflagellates. *The rate of luminescent flashing following stimulation (e.g., a
single bright light flash) is greatest at night and is inhibited by light.*

The emission spectrum of the large armored dinoflagellate, *Pyrodinium
bahamense,* peaks at 476 nm. This dinoflagellate exhibits both a diurnal
physiological rhythm in bioluminescence capacity and a diurnal vertical
migration.

In *Gonyaulax* no synchronization by means of intercellular communication is detectable between cells that are even slightly out of phase. In
Pyrocystis the persistance of the circadian rhythm is observed during
24 hr of darkness, but the rhythm quickly damps out in 24 hr of light. In
different *Pyrocystis* species, either stimulated luminescence only or both
spontaneous and stimulated luminescence are linked to circadian cycles.

In *P. noctiluca,* a rhythm of chloroplast expansion and contraction is
visualized by autofluorescence of chlorophyll together with a concomitant rhythm in the intracellular distribution of bioluminescent microsources on stimulation of cells by acid. The intracellular changes of
bioluminescence apparently involve (1) the distribution of microsources
and (2) the amount of light emitted. Periodicity of stimulated bioluminescence persists in constant darkness; thus the rhythm cannot only be a
consequence of exogenously varied photoinhibition.

8.1.2 Scintillons

The regulation of luminescence is linked to interactions between luciferin, luciferase, and a luciferin-binding protein (LBP). There is a 10-fold
increase in the concentration of LBP, luciferin, and luciferase from the
day cycle to the night cycle. A higher concentration of LBP mRNA at
night indicates that the circadian clock is acting at the DNA transcription
level.

The basic reaction is the LBP binding to luciferin (LH_2) at pH greater
than 7.

$$LBP + LH_2 \longrightarrow LBP \cdot LH_2$$

$$LBP \cdot LH_2 + 4H^+ \longrightarrow (H^+)_4 LBP + LH_2$$

$$LH_2 + O_2 \xrightarrow[\text{luciferase}]{} L{=}O + H_2O$$

LBP acts to prevent autooxidation of LH$_2$. Production of LBP is highest during the onset of night. The concentration of LBP remains constant for approximately 6 hr, and then steadily decreases to 10% of its nighttime concentration at daylight.

During night phase, light is emitted as 100 msec flashes from tiny luciferase-containing organelles, the *scintillons*. These are extremely fragile and may be isolated as intact vesicles that emit flashes when the pH is rapidly lowered to 6.

8.2 FLOW CYTOMETRY OF PHYTOPLANKTON

Fluorescence intensity measurements from at least 1000 *Gonyaulax* cells yield a histogram (i.e., log of fluorescence intensity *vs.* number of cells). Fluorescence emissions in several wavelength regions (the "signatures" of plankton cells) enable discrimination between groups of phytoplankton and pigment fluorescence resulting from changes in the population structure.

Distinct subpopulations of the cyanobacterium *Synechococcus* (less than 1 μm in diameter) with fourfold differences in phycoerythrin (PE) fluorescence (580 nm) may represent different strains, or possibly different stages of photoadaptation among cells of a single strain.

8.2.1 Study of Picoplankton

A laser-based flow cytometer provides sufficient sensitivity and resolution to detect autofluorescence, forward light scatter, and mythramycin-stained DNA fluorescence signatures in marine phytoplankton (e.g., starved marine bacteria, freshwater *Synechococcus*, and *Escherichia coli*).

Flow cytometry has been used to detect and quantify sexual differentiation in the centric diatom *Thalassiosira weissfloggi* (Grun.) by measuring size (light scattering) and chlorophyll, protein, and DNA content for each cell throughout the process of differentiation. *Sexuality can be induced experimentally by shifts in light or temperature under nutrient-replete conditions or by nutrient depletion.*

8.3 RELATIONSHIP BETWEEN NATURAL FLUORESCENCE, PHOTOSYNTHESIS, AND CHLOROPHYLL CONCENTRATION IN THE SEA

An optical instrument designed by Kiefer *et al.* (1989) was used to measure the natural or solar-induced fluorescence of chlorophyll *a*, within the phytoplankton.

8.3.1 Relationship between Chlorophyll Concentration and Natural Fluorescence

The relationship between chlorophyll concentration $[Chl(t, z)]$ (mg m^{-3}) and natural fluorescence (at a given depth z in a parallelipedic water column of 1 m $\times$ 1 m base area at a time t) is determined by the natural fluorescence coefficient F (m^{-1}), the specific light absorption coefficient of phytoplankton A (m^2 mg^{-1}) and the quantum yield of fluorescence Q (einsteins fluoresced/einsteins absorbed):

$$[Chl(t, z)] = F/AQ \qquad (8.1)$$

Even in the extremely oligotrophic waters of the South Pacific gyre, *natural fluorescence is easily measurable* through the euphotic zone at depths greater than 6 m. As found in previous studies, the value of the fluorescence varies spatially and temporally with ambient scalar irradiance of photosynthetically available radiation (PAR) and the concentration of chlorophyll *a*.

The results of measurements between depths of 2 and 150 m and over a 1500-fold range in production indicate that photosynthesis is highly correlated with natural fluorescence. The quantum yield of photosynthesis decreases more rapidly than the quantum yield of natural fluorescence with increasing irradiance.

8.3.2 The Two Methods of Fluorescence Measurements

Two methods to measure fluorescence are available: *natural* and *flashed*.

Natural fluorescence provides a measure of the rate of light absorption by photosystem II with the assumption that *the ratio of photosynthesis to fluorescence* yields a predictable entity.

In the *flashed fluorescence method* the probe, or the measuring, flashes are sufficiently short and infrequent so that they do not perturb the steady state condition of the cells in the sample. When the fluorescence induced by a *probe flash* is measured shortly after a high-intensity *exciting flash*, the signal depends both on quenching and on the photosynthetic reaction centers that are closed at that instant (see Chapter 9, Section 9.9). If the exciting flashes are large enough to close reaction centers, variations in fluorescence induced by the probe flashes can be interpreted in terms of photochemical quenching, which provides detailed information related to the quantum yield of photosystem II. The estimated yield is then multiplied by an estimate of the light absorption by the cells to obtain a prediction of the photosynthetic rate.

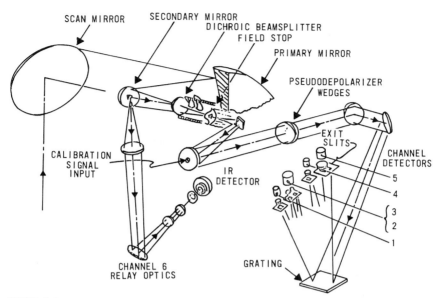

FIGURE 8.1 The CZCS optical system. Courtesy of H. R. Gordon, from "Advances in Geophysics," Vol. 27, p. 304. Academic Press, New York, 1985.

8.4 SATELLITE MEASUREMENTS OF PIGMENT CONCENTRATIONS

8.4.1 The Coastal Zone Color Scanner

The Coastal Zone Color Scanner (CZCS) on the Nimbus-7 satellite, launched in October, 1978, is a sensor in orbit that is specifically designed to study living resources (Figure 8.1). A scanning radiometer views the ocean in six coregistered spectral bands, five in the visible and near-infrared (443, 520, 550, 670, and 750 nm) and the sixth a thermal IR band (10.5–12.5 μm). From a nominal height of 955 km the sensor produces a ground resolution of 825 m at nadir.

The purpose of the CZCS experiment is to provide estimates of the near-surface concentration of phytoplankton pigments by measuring the spectral radiance backscattered out of the ocean. *The radiance backscattered from the atmosphere and/or sea surface (having the quality of a mirror) is typically at least an order of magnitude larger than the desired (i.e., from pigment) radiation scattered out of the water (i.e., L_w).* The process of retrieving L_w from the total radiance measured at the sensor (i.e., L_t) is usually referred to as *atmospheric correction*. The phytoplankton pigment concentration is estimated from the retrieved spectral radiances by the application of an in-water and biooptical algorithm.

CHLOROPHYLL <u>a</u> + PHAEOPIGMENTS <u>a</u> (MG/M^3)

FIGURE 8.2   A map of phytoplankton pigments in the Gulf of Mexico on November 2, 1978, derived from the Nimbus-7 CZCS data of orbit 130. Open-water picture elements (pixels) are coded according to the scale to show estimated chlorophyll *a* + phaeopigments *a*. Courtesy of H. R. Gordon, from *Science* **210,** 62 (1980).

CZCS data can be processed to a level that reveals subtle variations in water color, which show direct relationships to changes in the concentration of phytoplankton pigments. For spectra from 400 to 700 nm, up-welled radiance (expressed in microwatts per square centimeter per steradian per nanometer) was measured just beneath the sea surface at different pigment concentrations. From a map of phytoplankton pigments in the Gulf of Mexico (Figure 8.2), information was obtained pixel by pixel on the distribution of chlorophyll *a* and phaeopigments *a* (mg m^{-3}). The CZCS spectral bands 1 through 4 used in spectral scans were 433–453 (chlorophyll *a* absorption), 510–530 (scattered light), 540–560 (fucoxanthin), and 660–680 nm (chlorophyll absorption). The signal/noise ratio was in the range of 118–200/1.

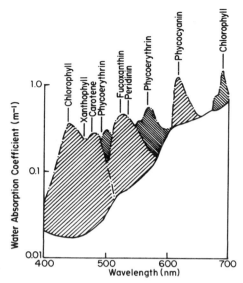

FIGURE 8.3 The absorption spectra of different algal pigments in the "window of clarity" in water absorption. The spectra for the pigments approximate those measured *in vivo*. For the purpose of this illustration, fucoxanthin and pteridinin absorptions are considered identical. Courtesy of H. R. Gordon, "Advances in Geophysics," Vol. 27, p. 300. With permission of Academic Press, Orlando, Florida.

8.4.2 The Signal Measured by the CZCS: The Phytoplankton Pigments

The signal measured by the CZCS is related to the pigment concentration in the water through the scattering and absorption properties (Figure 8.3) of the phytoplankton, which contains the photosynthetically active chlorophyll *a* (Chl *a*), strongly absorbing near 443 nm. Thus the solar radiation backscattered out of the ocean at 443 nm decreases with increasing Chl *a* concentration. The Chl *a* absorption is much weaker at 520 and 550 nm. *Therefore, an increase in Chl* a *causes the backscattered radiance to increase relatively at 520 and 550 nm as a result of the scattering associated with the phytoplankton.* Water that is poor in Chl *a* will appear a deep blue in sunlight, whereas water rich in Chl *a* will appear green.

Phytoplankton supports all higher forms in the sea. Further observations of water color over large geographical areas can lead to improved understanding of the state of the standing crop of phytoplankton. This can, in turn, lead to improved methods for managing and exploiting fisheries. Another potential application of water-color data is the early

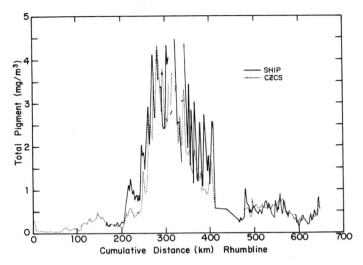

FIGURE 8.4 Comparison between the ship-measured surface pigments and the CZCS-derived pigments from orbit over the Gulf of Mexico. Courtesy of H. R. Gordon, from "Advances in Geophysics," Vol. 27, p. 322. With permission of Academic Press (1985), Orlando, Florida.

detection of massive phytoplankton bloom due to excessive nutrient influxes: if unchecked, such conditions can deplete the oxygen content of the water and result in massive fish kills.

Comparisons of the phytoplankton pigment concentrations derived from the uncorrected CZCS radiances with surface measurements agree to within less than 0.5 log C, where C is the sum of concentrations of chlorophyll a plus phaeopigments a (in milligrams per cubic meter) (Figure 8.4).

8.4.3 Relationship between Radiance Measurements and Pigment Concentration

The pigment concentrations are related to ratios of radiances at various wavelengths rather than to absolute radiances. This procedure compensates for the influence of nonorganic suspensoids in water, as well as the masking effects of organic disturbances.

From a comparison of Nimbus satellite measurements of subsurface spectral radiance and shipboard measurements (aboard the R.V. *Athena*

II) of the actual pigment concentration (Figure 8.4), this technique shows considerable promise for estimating pigment concentrations in regions where surface measurements are not available.

8.4.4 An Indicator of Phytoplankton Biomass

Typically, the concentration of chlorophyll *a* is used as an indicator of phytoplankton biomass because of the ease with which this quantity can be measured fluorometrically. In work with color imagery, however, the sum of the near-surface concentrations of Chl *a* and phaeopigments (i.e., pigment concentration) is used as an index of phytoplankton biomass.

The new satellite technology can provide a comprehensive overview of the surface temperature and pigments in several regions of an open ocean simultaneously.

8.5 PHYTOCHROMES IN MARINE ALGAE

Far-red (FR, λ greater than 700 nm) irradiation promotes and near-red irradiation inhibits rapid stipe (short plant stalk) elongation in the alga *Neurocystis luetkana* (Mertens). A reversible red/far-red (R/FR) reaction in cell division has been demonstrated in four marine phytoplanktons. Sporangial branch formation is inhibited by red light and promoted by FR light in the red alga *Porphyria tenera* Kjellman. In the same species, monospore formation was stimulated by R and inhibited by FR irradiation. Partially purified phytochrome was extracted recently from the red alga *Acrochaetium daviesii* (Dillwyn) Nageli. A number of marine algal photoperiod studies also suggest possible phytochrome activity.

Measurements with the Scripps Institute underwater spectroradiometer in clear oceanic waters show measurable irradiation in the R region (640–660 nm) as deep as 20 m and in the FR region (above 700 nm) as deep as 9 m.

Phytochrome may function in nature to detect changes in the ratio of quantum flux density QFD at 660 and 730 nm ($QFD_{660}/QFD_{730} = \sigma$). It has been shown that plants grown under equal photosynthetically available radiation (PAR), but with different σ, showed a number of different developmental responses, including stem elongation. In an aquatic environment the water affects σ at any given depth. FR (far infrared) is attenuated more rapidly, resulting in a rapid increase of σ with depth. These factors, together with phytoplankton absorption in the NR, determine changes in σ with depth.

There are some striking dissimilarities in spectrophotometric properties of plants compared to marine algae:

1. The absorption maxima for algal P_R and P_{FR} forms of phytochrome (see Chapter 10) are blue shifted by 16 and 10 nm, respectively, compared with those of the etiolated higher plant phytochromes.
2. Algal phytochrome exhibits significantly larger values for the ratios $A(P_{FR,\,\lambda_{max}})/A(P_{FR,\,\lambda_{shoulder}})$ and $A(P_{FR,\,\lambda_{max}})/A(P_{R,\,\lambda_{max}})$.
3. Algal phytochrome exhibits a significantly lower molar absorption coefficient compared with those of higher plant phytochromes.

With regard to the spectral differences compared to higher plants, these may reflect a higher percentage of P_{FR} at photoequilibrium under red light for the algal phytochrome, differences in the spectral overlap between P_R and P_{FR} forms, or the occurrence of apophytochrome–holophytochrome heterodimers.

8.6 FLUORESCENT PTERINS AND FLAVINS IN MARINE PHYTOPLANKTONS

Kalle (1937) described the "yellow substance—*substance jaune—gelbstoff*" dissolved in seawater. Among the compounds discovered to be dissolved in seawater are vitamin E, thiamin and biotin, auxins, carbohydrates, proteins, peptides, amino acids, sterols, hydrocarbons, fatty acids, and nucleic acids. In addition, humic type substances have been identified. These compounds are possibly derived from excretion products of living organisms or the post-mortem decomposition of organisms, or possibly from materials originating on land.

Phytoplankton and zooplankton are interlinked by *trophodynamic* relationships (as primary or secondary producers). In terms of carbon content, the dissolved organic matter (DOM) is 100-fold more abundant than the living organic matter. The greatest producers of DOM are the algae, which, in addition to pigment substances, yield glycolic acid and proteins.

According to the hypothesis of Lucas (1947), the species within an ecosystem find themselves in a network of complex relationships where the described molecules (DOM) can have different modalities of action, such as trophic, sensory, and/or hormonal (which can modify species development, growth, and behavior). It has been proposed that these external metabolites be named *ectocrine* substances.

Momzikoff studied the repartition of pterins and flavins in the upper layers, where the essential biomass of the phyto- and zooplankton is to be found.

8.6.1 Flavins

The flavins have different roles. First, there is their role as *vitamins;* for example, riboflavin is vitamin B_2. It is constantly present in phytoplankton, fish, and crustaceans. In salmonides B_2 avitaminosis is characterized by weaker growth, anemia, vascularization of the cornea, intraocular hemorrhages, photophobia, reduced vision, and/or abnormal coloration of the iris.

The second role of flavins is as *coenzyme;* riboflavin functions as a prosthetic group in FMN and FAD, which both participate in oxidation and reduction reactions catalyzed by enzymes involved in respiratory sugar metabolism. Flavins take part in bacterial bioluminescence.

Third, flavins play a role as *type I and type II photosensitizers* (see Chapters 4 and 5). In particular, photoreduction of riboflavin leads to formation of a superoxide ion intervening in oxidation and reduction reactions.

Flavins have been identified in the phytochrome effect, as blue-absorbing phytochromes.

8.6.2 Pterins

The pterins and the lumazines form a group called pteridins. There are about 40 natural derivatives. In living organisms the pterins are present in the free form. Pterin dimers are known as well as an example of an unstable complex (riboflavin–formicapterin). The only known conjugated form is folic acid (pteroylglutamic acid).

Pterins share with riboflavins the property of reversible oxido-reduction and can participate in electron transport mechanisms. They are rapidly degraded in the presence of light. In living organisms they are found only in trace amounts, which, however, should not prevent their participation in catalytic reactions. Insects, crustaceans, fishes, and amphibians do accumulate significant amounts of pteridin in their integuments, skin, and eyes. Folic acid, which contains one molecule of pterin, intervenes as a coenzyme in the synthesis of purine bases, thymidylic acid, methionine, and interconversion of serine into glycine. Pterin derivatives with a side chain on C-7 (erythropterin) account for red coloration of certain planktonic marine copepods. These pigments serve for protection, as inter- and intraspecies recognition signals, or as screening pigment and luminous screens in the eyes. Biopterin is a growth factor for *Crithidia fasciculata,* which cannot synthesize this compound.

There is some evidence that pterins are associated with the fundamental processes of respiration. At the level of mitochondria they could participate in electron transfer in the chain of cytochromes. The stoichiometry of pteridins in mitochondria fits that of cytochromes.

The pteridin content of *Euglena* sp., *Rhodospirillum rubrum,* and photosynthetic bacteria is augmented when these cells are exposed to light. Pteridins stimulate photosynthetic phosphorylations. The pteridin antagonist 4-phenoxy-2,6-diaminopyridine (PAD) inhibits both the biosynthesis of pteridins and that of the normal photosynthetic system; the inhibition is released by addition of biopterin.

It is hypothesized that a hydrogenated pteridin in the semiquinone form is the primary acceptor of the electrons produced by the quanta of light energy absorbed by chlorophyll (its redox potential of -0.7 V is at the appropriate level for such a role).

Sepiapterin, which (together with riboflavin) colors the yellow spots of *Salamandra salamandra,* is more abundant in individuals placed in an aquarium with a yellow bottom compared to those placed in an aquarium with black bottom. Pterins in the eye have a screen effect, optimizing perception and filtering out the near-blue. The pterin liberated by *Phoxinus phoxinus* acts as a "fright substance" for individuals of the same species.

Both pterins and flavins strongly absorb UV light in the region of 340–370 nm. They are both fluorescent in their oxidized or semioxidized forms. Pterins fluoresce in the blue range and flavins in the yellow range, as indicated in Table 8.1

8.6.2.1 Pterins in Phytoplankton

Isosepiapterin and a biopterin glucoside have been isolated in the blue alga *Anacystis nidulans* and analogous compounds in nine other species of blue algae. In a phytoplankton sample largely composed of diatoms and obtained offshore from Boulogne in the British Channel, six fluorescent compounds were identified: (1) two pterins, one fluorescing in the blue region (biopterin glucoside), the other in the violet region (isoxanthopterin), (2) two flavins, riboflavin and lumiflavin, and (3) fluorochromes that could not be matched with known derivatives.

8.6.3 Pterins and Flavins in Zooplankton

Of the material collected from zooplankton fishing off Monaco, 90% was made of copepods. The *ecological efficiency* was determined; this is the ratio of the animal biomass to the vegetal biomass. For this calculation the respective carbon contents in the animal and vegetal biomasses were estimated by multiplying the dry weight of the animal biomass by 0.4 and the chlorophyll content of the seawater by 60. In the Monaco samples, the biological efficiency was 0.045, which is within the range of typical values (from 0.03 to 0.11) obtained in this region.

TABLE 8.1 Fluorescence of Pterins in Water

Compound	Wavelength of maximum (nm)	
	Excitation	Emission
Pterin	370	450
Biopterin	370	468
D-Neopterin	371	448
L-Monapterin	336	454
6-Hydroxymethylpterin	368	460
6-Carboxypterin	370	440
7-Hydroxybiopterin	360	422
Isoxanthopterin	358	425
Xanthopterin	408	525
Lumazine	318	420
6-Carboxylumazine	370	440
Xantholumazine	398	468
Riboflavin	465	525
Lumiflavin	470	530
Lumichrome	318	440
FMN	466	532
FAD	468	542

The zooplankton riboflavin content is statistically connected linearly to the abundance of vegetal biomass; it is independent of the biological efficiency value and it is very weakly connected to the animal biomass value. In contrast, the content of isoxanthopterin is directly connected to the value of the biological efficiency and to the animal biomass. On this basis, the riboflavin must be of vegetal origin; its variations and the amounts found within the organisms would reflect the abundance of the phytoplankton.

Though the riboflavin of the copepods proves to have an algal origin, isoxanthopterin is either biosynthesized or transformed from a precursor found in the diet. The content of riboflavin in copepods goes on increasing with the augmentation of the available phytoplanton biomass, but it decreases with the rise of the "consumer population" within the phytoplankton.

The high values (e.g., 2.3–16.0) of the ecological efficiencies that are sometimes found may be the result of vertical migrations, with populations from deeper layers migrating to concentrate near the surface. For species able to effect vertical migration, it is ecologically more beneficial to remain in deeper and colder layers, where metabolism is slowed down, enabling a more economical handling of available resources. At the very high values of the ecological efficiency that result from upward vertical migrations of the zooplankton, the phytoplankton biomass at -5 m is insufficient to meet the energetic requirements of the copepods.

As the vegetal biomass decreases, the metabolism is turned toward "economy." It may be that decreased excretion can explain the rise of isoxanthopterin (a product of pterin catabolism) levels in zooplankton. There could be also an adaptation to nutrient conditions, because isoxanthopterin can be synthesized from folic acid. The increase in isoxanthopterin may be an element of protection against UV light, a possible survival advantage. UV radiation does penetrate the upper layers of seawater, in which the copepods are concentrated. Pterins have a second UV absorption maximum at 340 nm for isoxanthopterin. This UV filtering property gains more significance in phytoplankton-impoverished waters, which are more transparent to the radiation.

8.6.4 Coral Samples from French Polynesia

The biomass essentially consists of a symbiosis between corals of the constructor type (scleratinar coelenterates) and a unicellular alga, the peridinian *Gymnodinium microadriaticum* (Freudenthal). The productivity of this alga is among the highest known. The corals proper are considered carnivorous; the study of the metabolism of the coral–alga symbionts reveals an exchange of materials, from which the coral seems to be the beneficiary, because it is capable of incorporating the metabolites produced by the alga.

The animal biomass here is one of the highest registered and simultaneously the vegetal biomass is one of the smallest. This apparent rise of the biological efficiency coefficient is associated with exceptionally high contents of isoxanthopterin (0.125 μg g^{-1} animal biomass) within the copepods as well as in the fishing waters.

Fluorescent substances were identified in three lots of corals, comprising three different species of *Aeropora*: *A. corymbosa, A. variabilis,* and *A. pulchra.* Analysis showed that riboflavin is the essential fluorochrome. The riboflavin/Chl *a* ratio varied from 0.053 to 0.068. In conclusion, the riboflavin of corals can be related to the symbiotic algae, as the riboflavin/Chl *a* ratio is close to that observed for algae in culture.

A progressive succession of populations in a phytoplanktonic ecosystem could manifest itself by a progressive succession of biochemical characters: e.g., first a high intraorganismal riboflavin/Chl *a* ratio and low concentration of dissolved riboflavin (young populations), then a low flavin/Chl *a* ratio and high dissolved flavin concentration (old populations). One consequence is that riboflavin can be reutilized by algae succeeding a preexisting population.

8.7 CORAL TISSUE FLUORESCENCE

The relationship between scleractinians (corals) and their endolithic algae was studied at depths ranging from 10 to 35 m in Curacao. *Endolithic algal concentrations are found in the coral skeleton under the living tissue of stony coral, and never in dead parts.* Endolithic algal concentrations were studied in fluorescent and nonfluorescent forms of two kinds of corals: *Agaricia agaricans* and *Meandrina meandrites*.

In *Agaricia* sp. and *Meandrina meandrites*, the green algal band is found deeper in the skeleton than in any other species living in about the same depth range. Depth-dependent light conditions may have an effect on the behavior of endolithic algal concentrations in living corals.

Living coral tissue fluorescence is excited by UV and blue light, both of which penetrate deeply into oceanic waters. Peak wavelengths of five bands present in fluorescence spectrograms were as follows:

Coral Species	Peak Wavelength (nm)				
A. agaricites (nonfluorescent)	—	—	—	605	—
A. agaricites	479	533	567	605	—
A. fragilis	479	533	567	605	—
M. cavernosa	450	540	—	601	—
M. meandrites	489	533	567	605	663
Algal tufts	—	—	—	605	663

The strongest fluorescence was found in the agaricids and in *M. meandrites*. Green endolithic algae displayed an orange–red fluorescence. In ground sections of the corals, endolithic algae can be seen as thalli having diameters of 6 μm and nodi at distances of 50 μm. They make contact with the soft coral tissue. The skeleton of *M. meandrites* is much more transparent than that of the agaricids.

Agaricia agarites and *M. meandrites* were selected to investigate the relationship between the distance and the width of the green zone (i.e., algae) in the skeleton versus the depth in the reef. For nonfluorescent *M. meandrites* a weak association was found between the width of the algal concentrations and depth. In fluorescent *M. meandrites* there is a strong negative association between the distance of the algal concentrations and depth.

There are three hypothetical advantages for algal concentrations to occur only under living coral tissue: (1) Corals resist sedimentation on their living surface, thus preventing blocking of light; (2) corals resist settling of sessile organisms on their living surface, preventing blocking of light and predation; and (3) in the case of fluorescence of the living coral tissue, the fluorescence itself can contribute as a light source to photosynthesis of endolithic algae.

An exchange of metabolites (transfer of nitrogen and phosphorus-containing compounds) is suggested by contacts of endolithic algae with the living coral tissue. From their position within coral skeletons, the possibility that algae take up nutrition from surrounding seawater in an environment poor in free nitrogen and phosphorus seems limited.

The brightest band of coral tissue fluorescence, around 479 nm, matches the broad absorption maximum around 450 nm in the photosynthetic action spectum of zooxanthellae (algae) from *Favia* sp. The emission peak around 479 nm coincides with the absorption maximum of the photosynthetic pigments β-carotene, diadinoxanthin, dinoxanthin, peridinin, and an unknown pigment of the zooxanthellae in *Pocillopora* sp. The bands around 533 and 577 nm in coral emission match the 540-nm maximum in the algal action spectrum. The bands around 603 and 663 nm are weak, and the latter band is probably caused by Chl *a* excitation.

For the coral *L. fragilis*, with maximal occurrence between depths of 100 and 145 m in the Red Sea, it is plausible that fluorescence from the tissue amplifies the photosynthesis in the zooxanthellae. *This supports the hypothesis that short wavelengths of the solar spectrum are converted to longer wavelengths by corals to enhance photosynthesis in the zooxanthellae.*

In the action spectrum of algae in *Favia* sp., the broad absorption features (340–480 and 645–720 nm) are due to the photosynthetic pigments Chl *a*, β-carotene, and siphonein. Thus, the coral emission band at 479 nm has a photosynthetic potential for endolithic algae, but more needs to be known as to the *in situ* intensities of this fluorescence.

Bibliography

Chamberlin, W. S., Booth, C. R., Kiefer, D. A., Morrow, J. H., and Murphy, R. C. (1990). Evidence for a simple relationship between natural fluorescence, photosynthesis and chlorophyll in the sea. *Deep-Sea Res.* **37,** 951–973.

Coble, P. G., Green, S. A., Blough, N. V., and Gagosian, R. B. (1990). Characterization of dissolved organic matter in the Black Sea by fluorescence spectroscopy *Nature (London)* **348,** 432–435.

Delvoye, L. (1992). Endolithic algae in living stony corals: Algal concentrations under influence of depth-dependent light conditions and coral tissue fluorescence in *Agaricia agaricites* (L.) and *Meandrina meandrites* (L.) (Scelaractinia, Anthozoa). *Stud. Nat. Hist. Caribb. Reg.* **71,** 24–41.

Duncan, M. J., and Foreman, R. E. (1980). Phytochrome-mediated stipe elongation in the kelp *Nereocystis* (Phaeophyceae). *J. Phycol.* **16,** 138–142.

Frankel, S. L., Binder, B. J., Chisholm, S. W., and Shapiro, H. M. (1990). A high-sensitivity flow cytometer for studying phytoplankton. *Limnol. Oceanogr.* **35,** 11674–11679.

Fritz, L., Milos, P., Morse, D., and Hastings, W. (1991). *In situ* hybridization of luciferin-binding protein antisense RNA to thin sections of the bioluminescent dinoflagellate *Gonyaulax polyedra. J. Phycol.* **27,** 436–441.

Gordon, H. R., Clark, D. K., Mueller, J. L., and Hovis, W. A. (1980). Phytoplankton pigments from the Nimbus-7 coastal zone color scanner: Comparisons with surface measurements. *Science* **210,** 63–66.

Gordon, H. R., Clark, D. K., Brown, J. W., Brown, O. B., Evans, R. H., and Broenkow, W. W. (1983). Phytoplankton pigment concentrations in the Middle Atlantic Bight: Comparison of ship determinations and CZCS estimates. *Appl. Opt.* **22**, 20–36.

Gordon, H. R., Austin, R. W., Clark, D. K., Hovis, W. A., and Yentsch, C. S. (1985). Ocean color measurements. *Adv. Geophys.* **27**, 297–333.

Hardeland, R., and Nord, P. (1984). Visualization of the free-running circadian rythms in the dinoflagellate *Pyrocystis noctiluca. Mar. Behav. Physiol.* **11**, 199–207.

Holdsworth, E. S. (1985). Effect of growth factors and light quality on the growth, pigmentation and photosynthesis of two diatoms, *Thalassiosira gravida* and *Phaeodactylum tricornutum. Mar. Biol.* **86**, 253–262.

Hovis, V. A., Clark, D. K., Anderson, F., Austin, R. W., Wilson, W. H., Baker, E. T., Ball, D., Gordon, H. R., Mueller, J. L., El-Sayed, S. Z., Sturm, B., Wugley, R. C., and Yentsch, C. S. (1980). Nimbus-7 coastal zone color scanner: System description and initial imagery. *Science* **210**, 60–63.

Johnson, C. H., Inoue, S., Flint, A., and Hastings, J. W. (1985). Compartmentalization of algal bioluminescence: Autofluorescence of bioluminescent particles in the dinoflagellate *Gonyaulax* as studied with image-intensified videomicroscopy and flow cytometry. *J. Cell Biol.* **100**, 1435–1446.

Kelly, M. G., and Katona, S. (1966). "An Endogenous Diurnal Rhythm of Bioluminescence in a Natural Population of Dinoflagellates," Contrib. No. 1745. Woods Hole Oceanographic Institution, Woods Hole, MA.

Kidd, D. G., and Legarias, J. C. (1990). Phytochrome from the green alga *Mesotaenium caldariorum. J. Biol. Chem.* **265**, 7029–7036.

Kiefer, D. A., Chamberlin, W. S., and Booth, C. R. (1989). Natural fluorescence of chlorophyll *a*: Relationship to photosynthesis and chlorophyll concentration in the western South Pacific gyre. *Limnol. Oceanogr.* **34**, 868–881.

Lipps, M. J. (1973). The determination of far-red effect on marine phytoplankton. *J. Phycol.* **9**, 237–242 (cited in Duncan and Foreman, 1980).

Momzikoff, A. (1977). Substances fluorescentes (Pterines et Flavines) dans les eaux de mer et les planktons marins. Essai d'interprétation écologique. Thèse de Doctorat d'Etat, 1077, Université Pierre et Marie Curie, Paris 6.

Moran, M. A., Wicks, R. J., and Hodson, R. E. (1991). Export of dissolved organic matter from a mangrove swamp ecosystem: Evidence from natural fluorescence, dissolved lignin phenols, and bacterial secondary production. *Mar. Ecol.: Prog. Ser.* **76**, 175–184.

Morse, D., Pappenheimer, A. M., and Hastings, J. W. (1989). Role of a luciferin-binding protein in the circadian bioluminescent reaction of *Gonyaulax polyedra. J. Biol. Chem.* **264**, 11822–11826.

Olson, R. J., Frankel, S. L., Chisholm, S. W., and Shapiro, H. M. (1983). An inexpensive flow cytometer for the analysis of fluorescence signals in phytoplankton: Chlorophyll and DNA distributions. *J. Exp. Mar. Biol. Ecol.* **68**, 129–144.

Olson, R. J., Vaulot, D., and Chisholm, S. W. (1985). Marine phytoplankton distributions measured using shipboard flow cytometry. *Deep-Sea Res.* **32**, 1273–1280.

Olson, R. J., Chisholm, S. W., Zettler, E. R., and Armbrust, E. V. (1990). Pigment size, and distribution of *Synecocchus* in the North Atlantic and Pacific Oceans. *Limnol. Oceanogr.* **35**, 45–58.

Richardson, N. (1970). Studies on the photobiology of *Bangia fuscopurpurea. J. Phycol.* **6**, 215–219 (cited in Duncan and Foreman, 1980).

Sweeney, B. M., and Folli, S. I. (1984). Nitrate deficiency shortens the circadian period in *Gonyaulax. Plant Physiol.* **75**, 242–245.

Swift, E., Biggley, H. W., and Seliger, H. H. (1973). Species of oceanic dinoflagellates in the genera *Dissodinium* and *Pyrocystis:* Interclonal and interspecific comparisons of the color and photon yield of bioluminescence. *J. Phycology*, 420–426.

Taylor, W. R., Seliger, H. H., Fastie, W. G., and McElroy, W. D. (1966). "Biological and Physical Observations on a Phosphorescence Bay in Falmouth Harbor, Jamaica, W. I.," *J. Marine Research* **24**, 28–43, also Contrib. No. 83. Chesapeake Bay Institute, Department of Oceanography, Johns Hopkins University, Baltimore, MD.

van der Vede, H. H., and Henrika Wagner, A. M. (1978). The detection of phytochrome in the red alga *Acrochaetium daviesii. Plant Sci. Lett.* **11**, 145–149 (cited in Duncan and Foreman, 1980).

Yentsch, C. S., Backus, R. H., and Wing, A. (1964). "Factors Affecting the Vertical Distribution of Bioluminescence in the Euphotic Zone," Contrib. No. 1471. Woods Hole Oceanographic Institution, Woods Hole, MA.

CHAPTER 9

Photosynthesis

SECTION I

9.1 INTRODUCTION

The photosynthetic process consists of three main stages: (1) the removal of hydrogen (electron and proton) from water with the release of oxygen; (2) the transfer of the hydrogen (basically electron) by energy from light trapped by the reaction center chlorophyll (Chl); and (3) the use of hydrogen (electron and H^+) to reduce carbon dioxide to carbohydrate.

9.2 THE HILL REACTION

Isolated chloroplasts with broken membranes provoke a release of oxygen in the presence of an oxidizing agent, called a *Hill reagent* (e.g., ferric ions reduced to ferrous salts). This reaction is inhibited by dichlo-

177

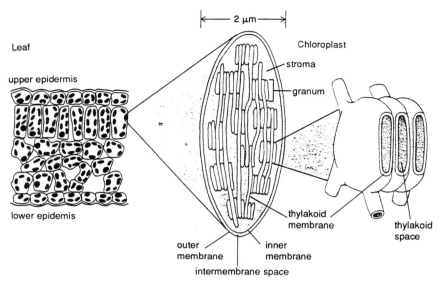

FIGURE 9.1 *Left*: the organization of the plant epidermis, showing cells filled with chloroplasts. *Middle*: enlarged structure of a single chloroplast. The chloroplast contains three distinct membranes (the outer, inner, and thylakoid membranes) that delimit three separate internal compartments—the intermembrane space, the stroma, and the thylakoid space. *Right*: enlarged structure of the thylakoid. The thylakoid membrane contains all of the energy-generating systems of the chloroplast. As indicated, the individual thylakoids are interconnected, and they tend to stack to form aggregates called grana. From Alberts *et al.* (1983), p. 511. Copyright, Garland Publishing, Inc., New York and London.

romethylurea (DCMU), a blocker of electron transport between the two photosystems. The overall reaction using the Hill reagent A can be written as

$$2H_2O + 2A \longrightarrow O_2 + 2AH_2$$

Thus, the oxidation of water is a light-dependent phenomenon that can occur without any reduction of CO_2. The photosynthetic process (Figures 9.1–9.3) is separated into a luminous phase and a dark phase. It has been confirmed by the use of labeled water ($H_2^{18}O$) that photosynthesis releases $^{18}O_2$.

9.3 THE PHOTOSYNTHETIC CHAIN: PHOTOSYSTEMS II AND I

9.3.1 Noncyclic Electron Transport

The release of one molecule of oxygen puts into action four photochemical stages, i.e., the primary donor Z of photosystem II (PSII) moves

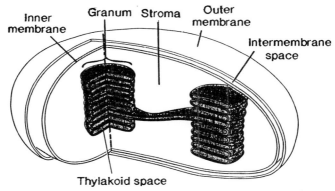

FIGURE 9.2   Diagram of a choroplast. After Stryer (1975), p. 434. This figure was adapted by Stryer with permission of Wadsworth Publishing Company Inc., Belmont, California, from Wolfe (1972).

through levels of oxidation indicated by the Zs in Fig. 9.4: Z^{+1} to Z^{+4}. The major structural components of the reaction center are the D_1 and D_2 polypetides bound to the inner surface of thylakoids, which serve as a stabilizing matrix for pigments and other molecules in PSII.

The *electron-transporting mechanism* in the reaction center of PSII acting as a light trap has a "special pair" of chlorophyll molecules, i.e., the original electron donor; the primary donor Z that replaces the electron donated by chlorophyll; pheophytin, a chlorophyll pigment with no Mg^{2+} that accepts the electron from chlorophyll; and a primary electron acceptor quinone (Q_A).

The reaction sequence is as follows: one electron of the excited chlorophyll trap (Chl a_{II}^*) is transferred to pheophytin and from there to the primary quinone (Q_A); Chl a_{II}^* receives an electron from Z. Q_A, located on the D_2 polypeptide and tightly bound to PSII (680-nm cut-off point for the activity of PSII, which drops sharply at longer wavelengths), passes its extra electron to Q_B (on D_1), which can diffuse freely when it has accepted two electrons. The "water-oxidizing clock" hypothesis holds that a cyclic mechanism supplies electrons to the P680 chlorophylls in PSII. *As each photon is absorbed by P680, the clock advances by one transient S state of oxidation, and releases one electron (e^-).* When the clock reaches S_4, it spontaneously releases an oxygen (O_2) molecule and reverts to S_0.

The more stable S_1 has one fewer electron than S_0. A clock that starts at S_1 goes to S_2, and one that starts at S_0 goes to S_1. The transition occurs because one electron is released from the clock to convert P680$^+$ back to P680. The second flash creates another P680$^+$ and boosts the S_2 to S_3, and so on with a third flash. When the clock reaches S_4, it has released four electrons and is ready to complete the water-splitting reaction. The clock then releases O_2 and drops back from S_4 to S_0. If, improbably, a photo-

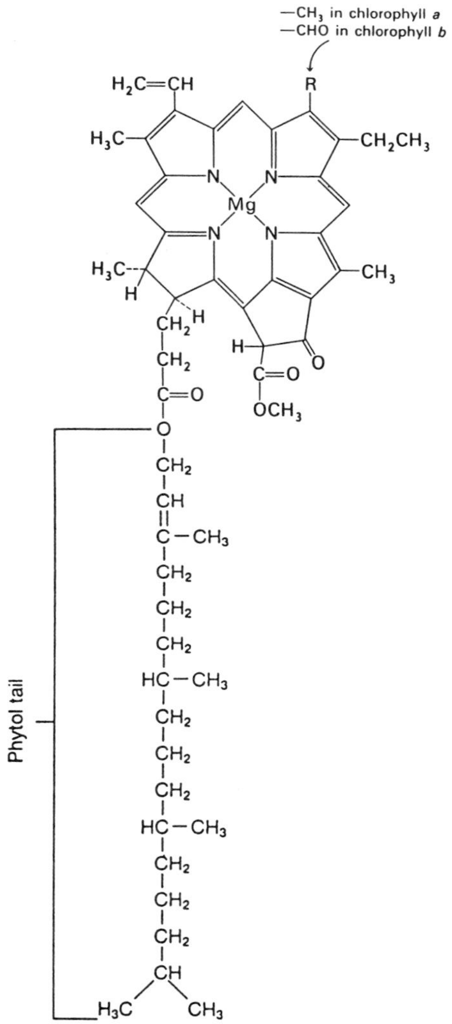

FIGURE 9.3   Structure of chlorophyll.

system absorbs two photons during a flash, the water-oxidizing clock advances in one step from S_1 to S_3.

Most of the clocks in a dark-adapted state are in S_1; thus, the maximum release of oxygen is after the third flash. The random "errors" that occur (clocks failing to advance or advancing by two S states) can account for the gradual damping-out in the oscillations of O_2 release and the slow desynchronization of the clocks.

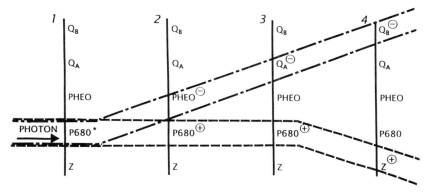

FIGURE 9.4 Stepwise electron transfer in the photosystem II reaction center, which stores some light energy in the form of separated positive and negative charges. Adapted from Govindjee and Coleman (1991), Scientific American **362**, 50–58. Copyright George K. Kelvin, Science Graphics, Great Neck, New York. With the permission of Govindjee.

9.3.2 The Role of Inorganic Ions in PSII

O_2 production does not take place unless there are four Mn ions in PSII for every P680. Mn ions can share two to seven electrons with other atoms. A four-point periodicity has been observed in Mn oxidation, in agreement with the above-described S_0–S_4 model (Kok model). Tentatively, S_0 has been identified with the presence of Mn(II), S_1 with Mn(III), and S_2 with Mn(IV). Both Mn(II) and Mn(III) appear to be stable and long-lived in PSII, which corroborates the prediction of stable S_0 and S_1 states. Mn(IV) associated with S_2 is a relatively transient intermediate. There is no discernible change in the Mn oxidation states between S_2 and S_3. The large polypeptides D_1 and D_2 are the most likely sites for Mn binding.

According to X-ray spectroscopy, in the S_1 state two Mn ions appear to be part of a binuclear complex and are separated by only 2.7 Å. The other pair of Mn ions is separated by a larger distance. Thus the four Mn ions can be visualized as being at the four corners of a trapezoid.

The water-splitting reaction also produces four protons released sequentially: one during the S_0-to-S_1 transition, one during the S_2-to-S_3 transition, and two during the S_3-to-S_4-to-S_0 transition. Protons may come directly from water, or from polypeptides: (1) water oxidation occurs prior to S_4, or (2) no water oxidation occurs until the final S_4-to-S_0 transition.

The removal of Ca^{2+} interrupts the S_3-to-S_4-to-S_0 transition and the fast reduction of P680$^+$ to P680. Ca^{2+} may act as a "gate" to bring H_2O molecules to the Mn complex. *The calcium ions in photosystem II may put the*

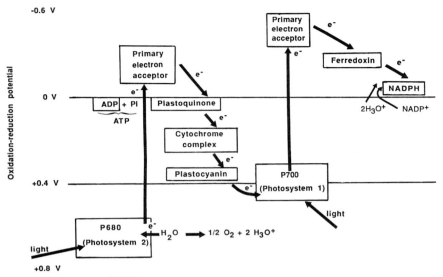

FIGURE 9.5   Scheme of the photosynthetic chain.

polypeptides of the water-oxidizing clock into the correct functional conformation.

9.3.3 The Electron Transport Chain from PSII to PSI

Electrons are transported from water to $NADP^+$ via two photochemical reactions that partake of the electron transport chain in three steps (Figures 9.5 and 9.6):

1. The primary donor Z moves the electrons from water to photosystem II, with release of oxygen.

2. The electrons of Chl a_{II}^* move by Q_A, the plastoquinone PQ, cytochrome f, and the plastocyanine. Electron movement to PSII is specifically inhibited by dichloromethylurea or cyanide. CO_2 (or HCO_3^-) has a unique stimulatory role in the conversion of plastoquinone to plastoquinol at the reaction center of PSII. CO_2 / HCO_3^- binding on Fe (iron–bicarbonate–protein complex) and the positively charged arginines in D_1 and D_2 are suggested to provide stability to the reaction center and stimulate electron flow and protonation.

3. The electrons of Chl a_I^*) move to the primary acceptor X of PSI, then

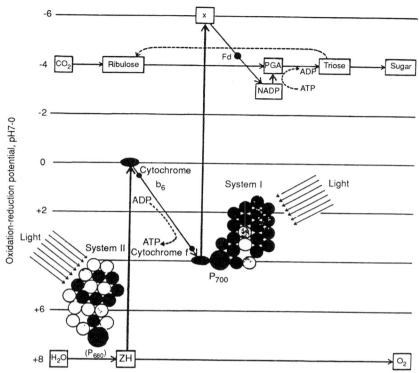

FIGURE 9.6 Scheme showing photosystem II, photosystem I "dark biochemistry," and their interconnection. ZH (which is currently referred to as Z) is suggested to be a tyrosine residue within a protein that donates electrons to the oxidized chlorophyll a. PGA is phosphoglyceric acid. X may represent an electron carrier (e.g., an iron–sulfur center F_X or FeS_X) that precedes $NADP^+$. Fd is ferredoxin. The numbers on the ordinate corresponding to the oxidation–reduction potential should be read as decimals, e.g., -0.6 for -6, -0.4 for -4, and so on. From E. I. Rabinowitch and Govindjee, Department of Plant Biology, UIUC, Urbana, IL, with kind permission of Govindjee.

to ferredoxin and to the flavin adenine dinucleotide (FAD) of the ferredoxin–$NADP^+$ reductase; $NADP^+$ is reduced to NADPH.

9.3.4 The Energy Balance of Photosynthesis

The electrons move from the constituent with the highest oxido-reduction potential E_0' to that with the lowest, i.e., from water to the chlorophyll trap a_{II} (PSII), from the primary acceptor Q_A of PSII to the chlorophyll trap a_I (PSI), and from the primary acceptor X to $NADP^+$.

The full oxidation of water is accomplished by eight photons, four 680-nm photons for movement of four electrons from Z to PSII, and four 700-nm photons for moving electrons from Chl a_I to the primary acceptor X of PSI. The energy furnished by a photon of wavelength 680 nm or shorter moves an electron from Chl a_{II} to acceptor Q_A of PSII. When graphed, the potential changes of electrons assume a *zig-zag shape*, which is so called "the scheme in Z" (Figure 9.7).

The phosphorylation of ADP to ATP is coupled to this noncyclic transport of electrons at sites localized in the first and second segments of the photosynthetic chain.

The electrons that move from water to $NADP^+$ cross the thylakoid membrane several times (Figure 9.8), successively from the luminal face to the stromal, then back toward the luminal before returning to the stromal.

With a pair of electrons transported from water to $NADP^+$ there is a translocation of *two protons from the stroma toward the intrathylakoid space via reduced PQ*. When $NADP^+$ is reduced, two protons are taken up from the stroma and transferred to the ferredoxin–$NADP^+$ reductase; one of these two protons is restored to the stroma when two electrons are transferred from $FADH_2$ to $NADP^+$. The balance is therefore three protons taken up from the stroma; two cross the membrane and one is consumed for the reduction of $NADP^+$.

Within the intrathylakoid space four protons are liberated — two when water is oxidized and two by translocation via fully reduced plastoquinone (PQH_2) formation:

$$PQ \cdot + H_3O^+ \longrightarrow PQH \cdot \quad \text{(first electron)}$$

$$PQH \cdot + (H_3O^+ + e^-) \longrightarrow PQH_2 \quad \text{(second electron)}$$

The primary donors of the two photosystems (Z for PSII, plastocyanin for PSI) are situated toward the luminal face of the thylakoid membrane whereas ferredoxin and ferredoxin–$NADP^+$ reductase are situated toward the stromal face. This vectorial arrangement of carriers has the consequence that the transport of electrons from water to $NADP^+$ is accompanied by a liberation of protons in the intrathylakoid space.

The return of three protons to the stroma by the hydrophobic base of an ATPase makes possible the phosphorylation of a molecule of ADP at the level of the spherical subunit CF_1 of the ATPase.

9.3.5 Energy and Reducing Power Supplied by the Dark Phase of Photosynthesis

ATP and NADP manufactured in PSI are fed into the final stage, involving carbon fixation. A molecule of ATP supplies about 2.3 kJ/mol

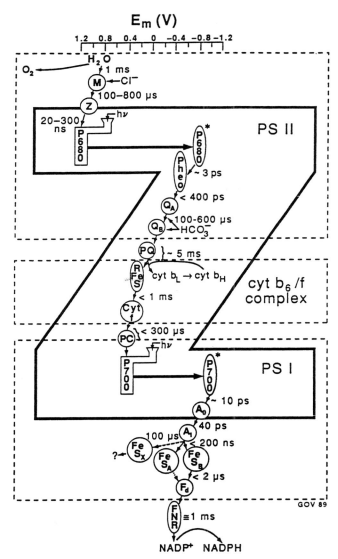

FIGURE 9.7   The Z scheme of photosynthesis. The time notations in ms, μs, and ps indicate the lifetimes of the respective steps. The complex FeS_X, FeS_A, FeS_B, F_d corresponds to X in Fig. 9.6; FNR, ferredoxin NADPH reductase. From Govindjee and W. C. Coleman (1993), *in* "Photosynthesis: Photoreactions to Productivity" (Y. Abrol, P. Mohanty, and Govindjee, Eds.) pp. 83–108. Kluwer Academic Publishers, The Netherlands. Courtesy of Govindjee.

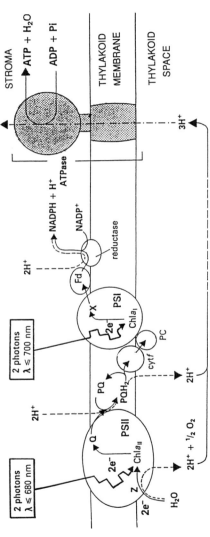

FIGURE 9.8 Proton movements across the thylakoid membrane. Taking into account the vectorial orientation of the constituents of the photosynthetic electron transport chain, the electrons that go from water to NADP$^+$ cross the membrane of the thylakoid several times: in effect, they move successively from the luminal face to the stromal face and then return toward the luminal face before coming back toward the stromal face. In addition, because the electron transport chain includes electron carriers alternating with hydrogen carriers (plastoquinone PQ and ferredoxin–NADP$^+$ reductase), some of the stages of electron transport are associated with a transport of protons. In the course of the translocation of a pair of electrons from water to NADP$^+$, the translocation of two protons takes place from the stroma toward the intrathylakoid space, because the reduced plastoquinone PQH$_2$ causes these to cross the thylakoid membrane at the time NADP$^+$ reduction is effected; two protons are also taken from the stroma and transferred to the ferredoxin–NADP$^+$ reductase, which has FAD for coenzyme. One of the two protons utilized is restored to the stroma when the two electrons are transferred from FADH$_2$ to NADP$^+$. The balance is therefore from three protons obtained from the stroma: two cross the thylakoid membrane and one is used for the reduction of NADP$^+$. Within the intrathylakoid space four protons are liberated: two by translocation and two by oxidation of water. The energy required for the noncyclic transport of a pair of electrons is provided by four photons: two photons (wavelength equal to or no longer than 680 nm) that are absorbed by chlorophyll a_{II} of the acceptor Q$_{II}$, and two photons (wavelength equal to or no longer than 700 nm) that are absorbed by photosystem I (PSI); these latter photons allow the transfer of two electrons from chlorophyll a_I to the acceptor X. This energy is converted in an electrochemical gradient that tends to make the protons return to the stroma. The return of three protons to the stroma through the hydrophobic base of an ATPase allows, at the level of the sphere CF$_1$, the phosphorylation of a molecule of ATP. Adapted from A. Berkaloff, J. Bourguet, P. Favard, N. Favard, and J.-C. Lacroix (1981). Biologie et Physiologie Cellulaires III C, Peroxysomes, Division Cellulaire, p. 43, Fig. 11.24. Copyright Hermann, 293 rue Lecourbe, 75015, Paris.

when it is hydrolyzed, and this is enough to provide the needed boost to the reducing power of the $NADP^+$/$NADPH$ system for CO_2 reduction to carbohydrate. ATP is also needed for the production of the "CO_2" acceptor ribulose phosphate.

9.3.6 The Emerson Effect: The Origin of the Concept of PSI and PSII

The fixation of one molecule CO_2 and *the liberation of one molecule of oxygen require a minimum of eight quanta of light energy.* The maximum quantum yield of photosynthesis, in number of O_2 molecules released per quantum of light absorbed, is 1/8, or 12.5%. It takes two light quanta to move one electron.

Constant at about 12.5% in most of the spectrum, the quantum yield of photosynthesis drops sharply near 680 nm (called the *red drop*). The quantum yield of photosynthesis can be brought to the full efficiency of 12.5% by simultaneously exposing the plant to a second beam of 650 nm (chl *b* absorption peaks at 650 nm; only Chl *a* absorbs at wavelengths longer than 680 nm). This relative excess in photosynthesis when a plant is exposed to two beams of light simultaneously, rather than separately, is known as the *Emerson effect,* or *enhancement.* The Emerson effect is explained by the coexistence of PSII and PSI with the two quanta required to move one electron (higher energy quantum acts on PSII).

Photosynthesis involves two photochemical processes: one supplied by Chl *a*, the other by Chl *b*, or some "accessory" pigment (i.e., Chl *b* in green cells, phycoerythrin in red algae, phycocyanin in blue–green algae, or fucoxanthol in brown algae).

In plants, Chl *a* is the main fluorescing pigment even when light is absorbed by other pigments. The initial absorber (i.e., chlorophyll *b*, phycoerythrin, or fucoxanthol) transfers its energy of excitation to Chl *a* by a resonance process (i.e., Förster energy transfer). This is known as *sensitized fluorescence.*

There apparently exist two forms of Chl *a in vivo;* one form absorbs mainly above 680 nm (mostly in PSI), and another at 670 nm (mostly in PSII and strongly assisted by accessory pigments). In Chl *a* extracts there is only one product. *In vivo* the two forms may differ in the way molecules of Chl *a* are clumped or associated with proteins.

9.4 CYCLIC TRANSPORT OF ELECTRONS IN PSI

In PSI (Figures 9.6 and 9.9), the electron, instead of moving from acceptor X to ferredoxin and $NADP^+$, is transferred to cytochrome b_6 (oxido-reduction potential near 0) and returns to Chl a_1 *by borrowing a*

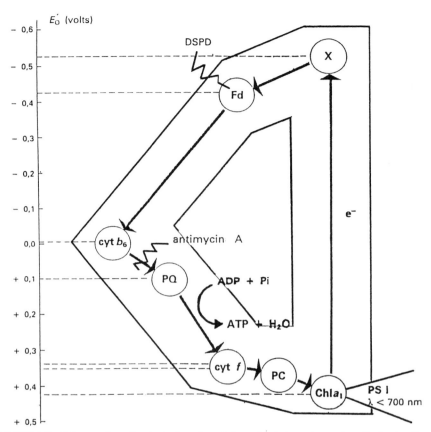

FIGURE 9.9   Scheme of cyclic photophosphorylation. A photon of wavelength shorter than or equal to 700 nm moves an electron from the chlorophyll trap a_1 to acceptor X of photosystem I. The photochemical reaction, having elevated the potential of this electron, returns it to chlorophyl a_1 through ferreroxin (Fd), cytochrome b_6, plastoquinone (PQ), cytochrome f, and plastocyanine (PC), with, respectively, decreasing oxido-reduction potentials E'_0. The phosphorylation of ADP to ATP is coupled to this cyclic electron transport, the coupling site being located between plastoquinone and cytochrome f, like one of the coupling sites associated with noncyclic electron transport. Cyclic electron transport is inhibited by antimycin A or disalicylidenepropane diamine; it is not inhibited by dichloromethylurea, which blocks the movement of electrons between the two photosystems in the case of noncyclic electron transport. From A. Berkaloff, J. Bourguet, P. Favard, N. Favard, and J.-C. Lacroix (1981). Biologie et Physiologie Cellulaires III Chloroplastes, Peroxysomes, Division Cellulaire, p. 42, Fig. 11.23. Copyright Hermann, 293 rue Lecourbe, 75015, Paris.

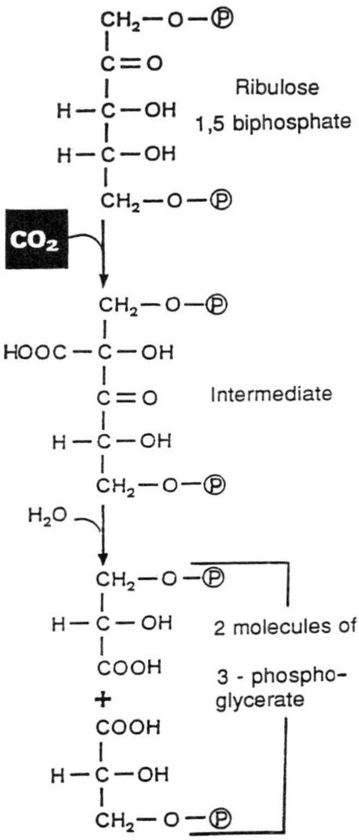

$$
\begin{array}{l}
CH_2-O-\text{\textcircled{P}} \\
| \\
C=O \\
| \qquad\qquad \text{Ribulose} \\
H-C-OH \quad \text{1,5 biphosphate} \\
| \\
H-C-OH \\
| \\
CH_2-O-\text{\textcircled{P}}
\end{array}
$$

CO_2

$$
\begin{array}{l}
CH_2-O-\text{\textcircled{P}} \\
| \\
HOOC-C-OH \\
| \\
C=O \qquad \text{Intermediate} \\
| \\
H-C-OH \\
| \\
CH_2-O-\text{\textcircled{P}}
\end{array}
$$

H_2O

$$
\begin{array}{l}
CH_2-O-\text{\textcircled{P}} \\
| \\
H-C-OH \qquad \text{2 molecules of} \\
| \\
COOH \qquad\quad \text{3 - phospho-} \\
+ \qquad\qquad\quad \text{glycerate} \\
COOH \\
| \\
H-C-OH \\
| \\
CH_2-O-\text{\textcircled{P}}
\end{array}
$$

FIGURE 9.10 CO_2 fixation and formation of two trioses from pentose by the activity of ribulose 1,6-phosphate decarboxylase.

segment of the photosynthetic chain between the two photosystems, i.e., PQ → cytochrome f → plastocyanine → Chl a_1. The coupling site for ADP phosphorylation to ATP, between plastocyanine and cytochrome f, is the same as in the case of noncyclic electron transport. Cyclic transport is inhibited by antimycin between cytochrome b_6 and PQ, and by disalicyli-denepropane diamine (DSPD) at the level of ferredoxin, but is not inhibited by DCMU. Only ATP (no NADPH) is produced in cyclic electron transport.

9.5 LIGHT-INDEPENDENT CARBON DIOXIDE FIXATION: THE CALVIN CYCLE

Both NADPH and ATP are needed for conversion of CO_2 to sugar (Figures 9.10 and 9.11). CO_2 is incorporated by fixation on *ribulose 1,5-diphosphate* (Figure 9.10), yielding a C_6 intermediate, which is cleaved into

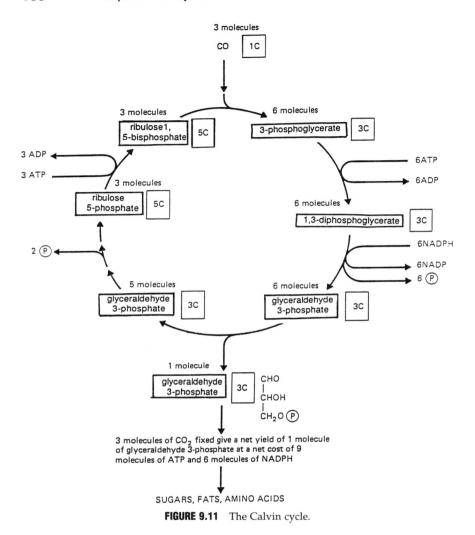

FIGURE 9.11   The Calvin cycle.

two molecules of *3-phosphoglyceric acid* (PGA), subsequently reduced to "oses" (Figure 9.12). The trioses contribute to the synthesis of phosphorylated hexoses and starch, and are also utilized to regenerate a C_5 acceptor.

The Calvin cycle effectively takes place within the stroma. The formation of two PGA molecules is catalyzed by *ribulose 1,5-diphosphate carboxylase*. Due to the mass of photosynthetic plants, *this enzyme is the most abundant protein on earth*. As a carboxylase it results in CO_2 fixation, forming two molecules of PGA (*photosynthesis*). As an oxygenase it cata-

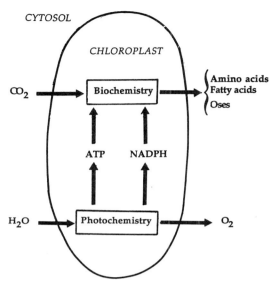

FIGURE 9.12   The photochemistry (water splitting) and biochemistry (carbon fixation, biosynthetic activity) of the chloroplast, with resulting synthesis of amino acids, fatty acids, and "oses."

lyzes the oxygenation of ribulose-1,5-diphosphate, which is then cleaved into one molecule of 3-phosphoglyceric acid and one molecule of phosphoglycolic acid (*photorespiration*).

From the carboxylation of three molecules of ribulose 1,5-diphosphate, six molecules of glyceraldehyde 3-phosphate result; five serve for the regeneration of three molecules of ribulose 1,6-diphosphate, and the last remains available for synthesis of organic molecules. The accumulation of trioses or hexoses in the stroma of the chloroplast, with a dangerous rise of the osmotic pressure, is prevented by the synthesis of starch macromolecules.

9.6 AN INSIGHT INTO THE PRIMARY PHOTOBIOCHEMISTRY OF PHOTOSYNTHESIS

9.6.1 Charge Separation through the Membrane of the Thylakoids

During photochemical reactions at the levels of PSI and PSII, an electron moves from Chl a_I^* and Chl a_{II}^* located near the intrathylakoid space to the primary acceptors Q and X located on the stromal side (Figures 9.5 and 9.6). As a consequence, through the thylakoid membrane, there is a separation of electrical charges lasting about 2.10^{-8} sec. The face of the thylakoid membrane facing the stroma is charged negatively (Q_A^-, X^-), and that facing the intrathylakoid space is charged positively (Chl a_I^+ and Chl a_{II}^+) (Figure 9.8). The membrane of the thylakoid then behaves as a capacitor, the dielectric of which is the lipidic phase of the membrane. There is then formed a potential difference with a value of about 100 mV between the two faces.

The gradient of protons establishes more slowly, i.e., within 10^{-2} to 10^{-3} sec; the influx of Cl^- and the efflux of Mg^{2+} that follow the movement of protons diminish the potential difference of photochemical origin.

9.6.2 Photophosphorylation

In chloroplasts (Figure 9.13), electron transport leads to a translocation of protons from the stroma toward the intrathylakoid space. The spherical subunit of the ATPase protrudes toward the stroma. Due to the gradient established, the protons tend to return toward the stroma through the hydrophobic base of the ATPase (Figures 9.8 and 9.13), in which case the enzyme is configured as an ATP synthase and ATP is formed.

In the mitochondria, electron transport is accompanied by proton translocation from the matrix toward the intramembranous space. Following the formation of an electrochemical gradient (Mitchell hypothesis), proton movement through the ATPase is so directed that the enzyme functions as an ATP synthase, and ATP is formed.

9.6.3 Photorespiration

In a high-O_2, low-CO_2 environment, glycolic acid (obtained by oxygenase activity of ribulose 1,5-dehydrogenase) is oxidized in the peroxisome to glyoxylic acid, which is converted into glycine that is oxidized in the mitochondria.

Photorespiration requires light energy that is necessary to regenerate ribulose 1,5-diphosphate. The energy liberated is not coupled to production of ATP. For a certain concentration of CO_2 in the atmosphere, the quantity of CO_2 released by photorespiration plus that from ordinary respiration exactly compensates for the CO_2 absorbed by photosynthesis; this concentration is called the *compensation point,* and its value changes according to the type of plant.

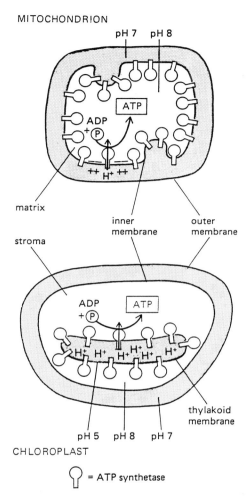

MITOCHONDRION

CHLOROPLAST

= ATP synthetase

FIGURE 9.13 Comparison of proton flows and ATP synthetase orientations in mitochondria and chloroplasts. From Alberts *et al.* (1983), p. 521. Copyright, Garland Publishing, Inc., New York and London.

9.7 LIGHT COLLECTION: RELATIONSHIP BETWEEN THE CHLOROPHYLL ANTENNA AND REACTION CENTER, A CLUE TO EFFECTIVENESS OF PHOTOSYNTHESIS

The energy-collecting structure is called the *antenna*, composed of about 300 pigment molecules (chlorophyll, carotenoids). Light energy is transferred from nearest neighbor to nearest neighbor until a chloro-

LIGHT

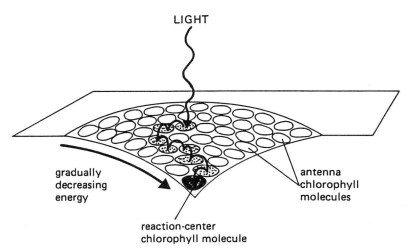

gradually
decreasing
energy

antenna
chlorophyll
molecules

reaction-center
chlorophyll molecule

FIGURE 9.14 The antenna complex in the thylakoid membrane serves as a funnel for transferring an electron excited by light to the reaction center. From Alberts *et al.* (1983), p. 521. Copyright, Garland Publishing Inc., New York and London.

phyll-trap molecule in the reaction center (RC) is photooxidized (Figures 9.14 and 9.15).

9.7.1 Organization of the Photosynthetic Pigments and Transfer of the Excitation Energy

Pigment organization and energy transfer require several steps: (1) collection of energy by a light-harvesting antenna, (2) transformation of singlet excitation energy into electronic energy, (3) formation of electrochemical potentials required for ATP synthesis, and (4) separation of oxidized and reduced chemical products.

The primary charge separation is achieved in a pigment–protein complex, i.e., the RC. Along the antenna, singlet excitations (excitons) may migrate over large distances. In green plants, RC_{II} and RC_{I} work in series to drive electrons from water to $NADP^{+}$. There are 500 antenna chlorophylls (75% Chl a and 25% Chl b) present in different spectroscopic forms, and these can be statistically associated to each electron transport chain. Chl b contributes essentially to the PSII action spectrum whereas Chl a molecules feed PSI. The antenna is organized into photosynthetic units (PSUI or PSUII) containing about 200 Chl per RC (Figures 9.16 and 9.17).

In less than 10^{-13} sec a singlet excitation becomes localized and diffuses in the antenna by successive jumps between molecules, separated at most by a few nanometers. The pairwise transfer rate, L, decreases as R^{-6},

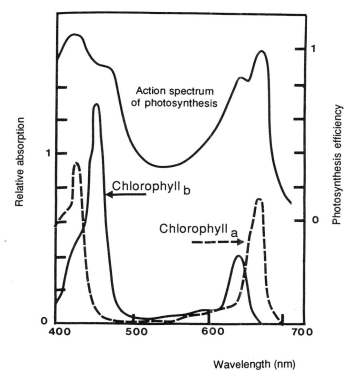

FIGURE 9.15 Action spectrum of photosynthesis superimposed on absorption spectra of chlorophyll *a* and chlorophyll *b*.

with the distance, R, between the donor and acceptor molecules (Förster formula). L also depends on the mutual orientation of these molecules and on the energy gap that may exist between their first singlet excited states. For energy transfer to take place, the fluorescence emission spectrum of the donor molecule must overlap with the absorption spectrum of the acceptor molecule.

Equilibration of the excitation among the different possible states is achieved very quickly within a light-harvesting Chl–protein complex (LHCP). Propagation of excitation from Chl *b* to Chl *a* is "allowed," whereas the reverse propagation is practically forbidden because of the energy gap between the first singlet states.

The functional organization of the antenna can be summarized as follows: (1) four PSUI connected; (2) light-harvesting chlorophyll pigment feeds RC_{II}; (3) inter-PSUI connection; (4) spillover $RC_{II} \rightarrow PSUI$ transfer; and (5) reversible exchange between PSUII and PSUI antennae.

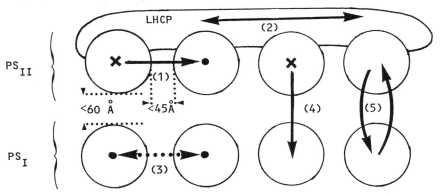

FIGURE 9.16 Functional model of light-harvesting system. LHCP, Light-harvesting chlorophyll–protein complex. Intrasystem transfer and intersystem transfer are shown. (1) Four PSII units connected (the arrow from the X to the dot indicates the directionality of the transfer); (2) LHCP feeds RC_{II}; (3) inter-PSUI connection (the bidirectional arrow indicates that the transfer can take place either way; (4) spillover RC_{II}–PSI transfer (the reverse process, transfer from RC_I to a PSUII, i.e., "uphill transfer," is "forbidden," except for some evidence for step 5); (5) reversible exchange between PSII and PSI antennae. From Paillotin *et al.* (1982), Trends in Photobiology (C. Helene, M. Charlier, T. Montenay-Garestier, and G. Loustriat, Eds.) p. 543. Copyright, Plenum, New York.

9.8 STRUCTURE OF PHOTOSYNTHETIC MEMBRANE

Three kinds of orders are considered in the organization of the phosynthetic membrane: local, intermediate, and long range.

9.8.1 Local Order

Elementary chlorophyll–protein complexes have been isolated. Each building block contains less than 10 Chl molecules (LHCP in Figures 9.16 and 9.17). Resonance Raman, photoselection, and X-ray diffraction data suggest that both positions and orientations of the Chl in these complexes are well defined.

9.8.2 Intermediate Order

Aggregates of the above-mentioned building blocks occur *in vivo* (Figure 9.17). Besides the light-harvesting complex (3 Chl *a* + 3 Chl *b*), which can be obtained in monomeric form, relatively large particles containing either 150 or 40 Chl *a* molecules in association with PSI, and particles of about 100 Chl molecules (probably related to PSUII activity), have been isolated following treatment with mild detergents.

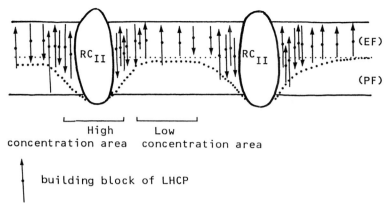

FIGURE 9.17 Schematic organization of photosynthetic unit II (PSUII) building blocks (Chl–protein complexes). The figure is a simplified diagram of the freeze–fracture electron microscopy image of the photosynthetic membrane with its EF and PF faces [the two faces of fractured chloroplast membrane obtained (see line of cleavage ····) by freeze fracture electron microscopy]. The Chl molecules are positionally and orientationally ordered with respect to the membrane plane. This orientational order is preserved at long distances. The centers of gravity of the building blocks are not distributed at random within the photosynthetic membrane; they are condensed into aggregates. The most typical of these aggregates are the EF particles associated with PSII activity. The separation distance between the aggregates must be small to allow energy transfer. The model suggests that the distribution of building blocks, although more important in the vicinity of special structural centers, is rather continuous. The light-harvesting chlorophyll–protein complex (LHCP) plus the core of the PSUII contain about 70% of the Chl molecules. From Paillotin, *et al.* (1982), p. 545. Copyright, Plenum, New York.

9.8.3 Long-Range Order

Linear dichroism measurements on oriented membranes clearly show that the antenna Chl molecules are well oriented with respect to the normal to the membrane plane. This preservation of orientation confers on the photosynthetic membrane some resemblance to a *smectic liquid crystal*. The photosynthetic membranes appear to lie preferentially parallel to each other, stacked in certain regions (*grana*; see Figures 9.1 and 9.2). The stacking phenomenon depends both on the ionic and the LHCP concentration. PSII activity is confined to these stacked regions.

9.9 CHLOROPHYLL FLUORESCENCE

Using Weber's method of "excitation–emission matrix analysis," it has been shown that the fluorescence band in spinach chloroplasts at room temperature originates in two species of chlorophyll *a*. On cooling at $-196°C$, two additional bands appear at 696 and 735 nm, suggesting the participation of four molecular species.

For matrix analysis, the emission must be measured at several wavelengths and excited at several wavelengths, and the data are arranged in rows and columns. The matrix analysis is a statistical method for deciding how many independent emission spectra are included in the sum-spectrum recorded from living cells.

9.9.1 Chlorophyll Fluorescence Lifetimes

Primary photochemical reactions of photosynthesis, beginning with light absorption in the femtosecond time domain, involve a succession of temporal events: exciton migration among the various antenna pigment molecules, exciton trapping at the reaction center chlorophyll (or bacteriochlorophyll), primary charge separation, and stabilization of the charge-separated state. With P680, the photon trap of the special pair, the primary charge separation and recombination in reaction center II can be written as follows:

$$P680 \cdot Pheo + h\nu \longrightarrow {}^1P680^* \cdot Pheo \qquad \text{(excitation process)}$$

$${}^1P680^* \cdot Pheo \longrightarrow P680^+ \cdot Pheo^- \qquad \text{(charge separation)}$$

$$P680^+ \cdot Pheo^- \longrightarrow {}^1P680^* \cdot Pheo \qquad \text{(recombination reaction)}$$

$${}^1P680^* \cdot Pheo \longrightarrow P680 \cdot Pheo + h\nu \qquad \text{(prompt or recombination fluorescence)}$$

The fluorescence decay of open reaction centers (i.e., when both the electron donor P680 and the electron acceptor pheophytin are capable of engaging in charge separation) can be analyzed as a multiexponential decay. The three lifetime components in the open reaction centers participating in the electron transfer chain ending in photosynthetic carbon trapping are ~ 320 psec (33%, fractional intensity), 3.2 nsec (37%), and 27 nsec (30%). On closure of the reaction centers, the fluorescence lifetimes and fractional intensities become 350 psec (50%), 2.2 nsec (41%), and 16 nsec (only 9%). Thus, the most dramatic change is clearly a decrease in lifetime of the slow component, with a concomitant decrease in its fractional intensity.

In open reaction centers a large fraction of components is clustered in the range of 0–300 psec. This is followed by a decay of slow components in the 2- to 20-nsec region. On closure of the reaction center (bottleneck of reduced primary and secondary acceptors in the electron transport chain), a dramatic change occurs: a shift from the broad, almost featureless distribution, to a narrower one with increased fractional contributions from the shorter lifetimes. The most dramatic effect is the disappearance of the long-lived slow (5 to 20 nsec) fluorescence components on closures of the reaction centers. Because the closure phenomenon is interpreted as the reduction of Pheo to Pheo$^-$, the disappearance of these slow components is interpreted to be due to the absence of the recombination reactions:

$$P680^+ \cdot Pheo^- \longrightarrow {}^1P680^* \cdot Pheo$$

$$^1P680^* \cdot Pheo \longrightarrow P680 \cdot Pheo + h\nu \quad \text{(fluorescence)}$$

Thus, the slow components are associated with the recombination reactions, as suggested to explain the variable Chl a fluorescence in PSII. On the other hand, the fast components (0–300 psec) are associated with the prompt fluorescence that may or may not include contributions for energy transfer among the six chromophores (four Chl a and two Pheo molecules) and competition with charge separation.

9.10 THE KAUTSKY PHENOMENON

On illumination, a dark-adapted photosynthetic sample shows time-dependent changes in chlorophyll (Chl a) fluorescence yield; this is known as the Kautsky phenomenon or the OIDP (see legend of Figure 9.18 for the definition of the acronym) transient. The fast OIDP fluorescence transient (Figure 9.18) follows illumination of the dark-adapted sample; O is the original fluorescence level (also called F_o), the so-called constant fluorescence, when all Q_A (primary quinone acceptor, see 9.3.1) is in the oxidized state. It is the minimum fluorescence in PSII. The less photochemistry (i.e., electron movement beyond Q_A ending in photosynthetic carbon trapping) that occurs, such as a bottleneck in the photosynthetic electron transport, with Q_A reduced, the more fluorescence will be observed.

In thylakoid membranes, the measured F_o at room temperature also includes a minor (about 20%) contribution from PSI Chl a fluorescence. In OID the fluorescence intensity F_1 at the first inflection or intermediate peak D (dip) or plateau, was suggested to arise from the interplay of PSII and PSI in intact algal cells, due to the transient changes in $[Q_A^-]$ in the main pathway of photosynthesis. Furthermore, the following and much larger D to P (P for peak) rise reflects the net accumulation of $[Q_A^-]$ as the plastoquinone pool is reduced, and there is a "traffic jam" in the flow of electrons all the way up to the electron acceptors of PSI. *Addition of the electron acceptor methyl viologen, which accepts electrons from PSI, abolishes the D to P rise, but not the OID phase.*

The initial fluorescence rise from F_o to F_1 was identified as the variable fluorescence yield controlled by a type of photosystem II center called

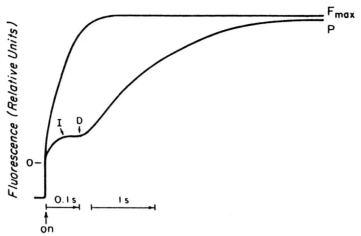

FIGURE 9.18 Chlorophyll *a* fluorescence transient (OIDP transient; see text) of spinach thylakoids. The thylakoids were suspended in a reaction medium containing 0.4 *M* sorbitol, 5 m*M* Mes-KOH (pH 6.5), 20 m*M* KCl, 2 m*M* MgCl$_2$, and 1 μ*M* nigericin. The Chl concentration was 39 μ*M*. The thylakoids were dark-adapted for 10 min before the fluorescence was measured. The original fluorescence level (constant fluorescence, O), the intermediate peak fluorescence (I), the dip or plateau (D), and the fluorescence at the peak (maximum, P) are shown. The fast-rising curve, which approximates the peak level of OIDP (see F$_{max}$), is obtained with 1 μ*M* 3-(3,4-dichlorophenyl)-1,1-dimethylurea (DCMU). When DCMU is added, the forward reaction of electron flow from Q$_A$ to Q$_B$ is known to be totally blocked and the decay of chlorophyll fluorescence is slowed. From Cao and Govindjee (1990), p. 182. With permission of Elsevier Science, Amsterdam, The Netherlands.

PSIIβ. The number of PSII centers capable of active water oxidation is increased by 2,6-dichloro-*p*-benzoquinone (DCBQ), and is decreased when 2,5-dimethyl-*p*-benzoquinone (DMQ) is used as an electron acceptor. These experiments are explained by the hypothesis of the existence of two types of PSII centers, active and inactive, the latter being the ones that are ineffective in electron flow to plastoquinone, and thus inactive in net electron flow and water oxidation. In spinach thylakoids the majority of PSII centers (68%) recover in less than 50 msec. However, 32% of the PSII requires a half-time of 2–3 sec to recover.

DCBQ is able to quench the F_1 observed in OIDPs much more effectively than is DMQ, whereas F_o remains unchanged. Because DBCQ was shown to intercept electrons from the "inactive" PSII centers, the OID phase reflects the reduction of Q$_A$ to Q$_A^-$ in the inactive PSII centers. *DMQ quenches only the DP rise, whereas DCBQ is able to quench the OI phase in addition.*

A heterogeneity among PSII centers has been suggested as an explanation for the above results: fast-rising PSIIα and slow-rising PSIIβ centers. The rate of reduction of electron acceptors in PSII is about three times slower in the PSIIβ centers compared to the PSIIα centers. The PSIIβ centers are supposed to have smaller chlorophyll antennae; electron transport in these centers does not proceed via the two-electron-accumulating plastoquinone. Although PSIIβ centers may be able to oxidize water, PSIIα centers are the dominant source of electrons for reduction of NADP$^+$.

Another heterogeneity is found at the Q$_B$ site (the second quinone site), with impaired Q$_B$ sites not able to transfer electrons efficiently to the plastoquinone pool.

9.11 CHLOROPHYLL FLUORESCENCE OR PHOSPHORESCENCE?

The quantum yield of chlorophyll phosphorescence in the formed photosynthetic apparatus of normal plants is lower by three orders than it is in solutions of monomeric chlorophyll. Confirmation that chlorophyll phosphorescence quenching in leaves is caused by energy transfer from triplet chlorophyll to the carotenoids is obtained from carotenoid-deficient mutants or blockade of carotenoid biosynthesis by herbicides (the pyridazine series).

In the spectra recorded at $-196°C$, the bands of delayed fluorescence (696 and 735–745 nm) and phosphorescence (968–970 nm) of chlorophyll are observed, and also the emission band of nonchlorophyll pigments (760–810 nm). On complete blocking of the synthesis of all colored carotenoids, the values of the quantum yield of delayed fluorescence, of phosphorescence, and of the ratio of first over the second rise 15, 20, and 60 times, respectively. Thus, the absence of carotenes lowers the effectiveness of quenching of the triplet states of chlorophyll in the chloroplasts.

SECTION II *Photosynthetic Bacteria*

9.12 CLASSIFICATION OF PHOTOSYNTHETIC BACTERIA

There are three kinds of photosynthetic bacteria: cyanobacteria, green bacteria, and purple bacteria. The most highly developed are the cyanobacteria; probably the most primitive are the purple bacteria, with the green bacteria somewhere in between. Cyanobacteria are the only prokaryotes that have developed oxygenic photosynthesis. Cyanobacteria, called blue–green algae, are regarded as true bacteria due to their prokaryotic organization. In contrast to cyanobacteria, which have an oxygenic photosynthesis with two light reactions (i.e., PSII and PSI), photosynthesis in purple and green bacteria is driven by only one photoreaction, during which no oxygen is produced. The electron donors are H_2S or organic substances taken from the environment, rather than H_2O. In photosynthetic bacteria, the electrochemical gradient generated is used for ATP synthesis.

9.13 THE DIFFERENT CHLOROPHYLLS AND BACTERIOCHLOROPHYLLS

Chlorophyll is a tetrapyrrole with porphyrin ring structure and a magnesium central atom. Chlorophyll *b* differs from chlorophyll *a* (Figure 9.3) in having a formyl group instead of a methyl group at C-7 (pyrrole ring B). Chl c_1 and Chl c_2 are found in brown algae and Chl *d* is in red algae.

Photoautrophic bacteria possess *bacteriochlorophyll a*, which differs from chlorophyll *a* in two features: it is hydrogenated in ring B and carries an acetyl group instead of a vinyl group on ring A. The green photobacterium, *Chlorobium*, has been found to contain bacteriochlorophylls *c*, *d*, and *e*, with farnesol instead of phytol attached to the pyrrole ring D. Bacteriochlorophyll *b* in *Rhodopseudomonas viridis* probably carries geranylgeraniol instead of phytol.

Chlorophyll *a* absorption maxima are at 430 and 663 nm (shifted to 453 and 642 for *b*). Bacteriochlorophyll *a* has one less double bond than chlorophyll *a* (and two less than protochlorophyll, a precursor in chlorophyll biosynthesis) and an absorption maximum at 760 nm.

9.14 PURPLE BACTERIA

Purple bacteria are classified into sulfur-containing Chromatiaceae and the sulfur-free Rhodospirillacae, which use alcohols and carbonic acids as electron donors. The photosynthetic apparatus is localized in chromatophores resembling thylakoids. The light-harvesting protein complex consists of bacteriochlorophyll proteins $B_{800-850}$ and B_{879}, which enclose the reaction center $BChl_{870}$.

After excitation of the reaction center (P870 is assumed to operate as a dimer, i.e., a bacteriochlorophyll pair), the sequence for electron transport is cyclic, as follows: from P_{870} to the acceptor X (presumably a bacteriophytin) to the secondary acceptors Q_A and Q_B, the quinone pool and several cytochromes and back to P_{870}. Protons are transported through the membrane into the chromatophore interior. The membrane potential of about 200 mV is the main component of the electrochemical gradient necessary for ATP synthesis. *NAD is reduced by reverse electron transport driven by ATP*; the electrons are abstracted from the ubiquinone pool and the electron deficit is compensated by external electron and hydrogen donors such as fumarate.

9.14.1 Trapping of Excitation Energy in Photosynthetic Purple Bacteria

Using steady-state and time-resolved polarized light spectroscopy, it is found that equilibration of the excitation density among different antenna pools is the dominant process before trapping or losses occur. The main antenna pigment protein complex B_{875} contains a minor spectral form B_{896}, absorbing between 890 and 900 nm. B_{896} is the terminal excitation acceptor before transfer to the RC. At 77 K, transfer from B_{875} to B_{896} takes 20 psec, and transfer from B_{896} to the RC takes 40 psec.

RCs are coupled to a large antenna of protein-bound pigment molecules; this transfers the excited state energy with an efficiency of more than 90%. Pigments with a high excited state energy are at the periphery of the PS; those with low excited state energy are at the RC.

The purple bacterium *Rhodobacter sphaeroides* contains the light-harvesting LH2 antenna with major absorption bands at 800 and 850 nm, and the LH1 antenna with a single band at 875 nm. It was generally believed that the LH1 antenna contained from several hundred to 1000 BChl molecules serving at least a few RCs. According to the trap-limited trapping concept, the excitation enters and leaves the RC several times before charge separation takes place. This is in contrast to another model, the diffusion-limited trapping concept. Against an earlier view that excitation transfer in and out of the RC was a fast process, a current view is that it is a rather slow process, neither diffusion limited nor trap limited, but a mixture of both.

At very low temperatures (4 K) in all species of purple bacteria the fluorescence polarization increases on excitation in the red edge of the B_{875} absorption band. The long-wavelength band consists of a major component (at 884 nm) and a minor component (at 896 nm). The polarization of the emission on excitation at 884 nm is low due to extensive energy transfer among identical molecules before the 896-nm pigment is reached. The high polarization on red edge excitation suggests a high ordering of the $BChl_{896}$ pigments.

Pigment-to-pigment transfer rates may easily be of the order of 150 fsec. Recently a third LH2 pigment, $BChl_{870}$, has been identified in a mutant of *Rb. sphaeroides* (lifetime of the excited state, ~ 10 psec). The fast decay of the $BChl_{875}$ excited state (with characteristic absorption changes around 880–890 nm), within 20–30 psec, indicates that in this system the $BChl_{875}$ is not the terminal acceptor of the excitation energy. The radical pair $P870^+$ $Bpheo^-$ is formed after the excitation flash. Measurements above 900 nm show that $BChl_{896}$ (probably the terminal acceptor) has a characteristic decay time of about 150–200 psec.

The specific entry for excitations arriving from the LH1 antenna into the RC, based on distances calculated from the Förster equation, consists of only a few BChl molecules per RC, and these are identified as the *red-shifted pigments, $BChl_{896}$*. At very low temperatures in some species of purple bacteria, the fluorescence associated with active reaction centers increases dramatically, in parallel with a decrease in quantum efficiency of trapping. When the entry pigment is at lower energy than the special pair at low temperature, the trapping will be inhibited and the fluorescence will increase.

Photosynthetic antennae share the universal property that the excitation energy is probably focused by $BChl_{896}$ on the neighborhood of the RC, rather than being distributed equally over all the antenna molecules.

9.14.2 Charge Separation in Photosynthetic Purple Bacteria

In *Rb. sphaeroides*, six tetrapyrrolic subunits are found at the active sites of two structurally similar RCs: a dimeric bacteriochlorophyll "special pair" (P), two "accessory" bacteriochlorophylls (BChls), and two bacteriopheophytins (Bphs), all held in a well-defined but skewed geometry along a C2 axis of symmetry. The BChls are separated from P by center-to-center distances of ~ 1.1 nm and interplane angles of $\sim 70°$. The Bphs are separated by similar distances and angles from the BChls. The electron transport chain consists of the photosensitizer dimer (P), an "accessory" BChl, an intermediate Bph, and a quinone (Q). Charge separation between the excited photosensitizer P* and Bph entities is known to occur on a time scale of 2–4 psec with nearly 100% quantum efficiency in the form of $P^+ - BChl - Bph - Q^-$.

Subpicosecond transient absorption experiments have failed to provide evidence that a $P^+ - BChl^-$ state occurs as a discrete intermediate in the initial $P^* - BChl - Bph \rightarrow P^+ - BChl - Bph^-$ charge separation process.

It has been speculated that the BChl facilitates transfer to the Bph via a "superexchange" mechanism involving a quantum mixing of a virtual $P^+ - BChl^-$ state with the photoexcited dimer, $P^* - Chl$. Mixing low-lying ionic states of the intermediate bacteriochlorophyll BChl with those of the donor P* and acceptor Bph should lead to a greatly increased rate of electron transfer via a superexchange.

9.15 GREEN BACTERIA

Green bacteria contain *chlorosomes*, which are vesicles closely connected with the cytoplasmic membrane, as photosynthetic organelles. Inside the chlorosome, there are rodlike elements that are thought to consist of the *bacteriochlorophyll c–protein light-harvesting complex*. $BChl_{840}$ is the RC pigment that passes electrons to a primary acceptor with a very negative potential. *This acceptor can pass electrons to NAD via ferredoxin.* There is no reverse electron transport involved and no ATP is needed for electron transport to NAD. The electron deficit of P_{840} is compensated eventually by electrons from H_2S or other donors. Molecular sulfur is deposited outside the cell. Chlorobiaceae also may possess a cyclic electron transport.

The chlorosomes structurally resemble the cyanobacterial phycobilisomes. The electron acceptor side (Q) of the photosystem closely resembles that of PSII in higher plants (plastoquinone, PQ), whereas the donor side in bacterial photosynthesis closely resembles that of PSI in higher plants because of the cytochrome–Fe–S complex. Thus, it is possible that both PSII and PSI have developed from a common bacterial photosystem.

9.16 LIGHT-HARVESTING SYSTEM OF CYANOBACTERIA

Light-harvesting pigments (or antennae) that capture light over a long range of the spectrum increase the photosynthetic efficiency of the RC centers and thus allow life in ecological niches under limited and unfavorable conditions. In bodies of water, the light window is between 450 and 600 nm, a spectral range in which chlorophyll does not absorb light efficiently.

The different kinds of light-harvesting systems are (1) membrane-integrated and (2) extramembranous, associated with the surfaces of thylakoids or cytoplasmic membranes. Examples of extramembrane light-harvesting systems are the chlorosomes of green bacteria and the phycobilisomes of cyanobacteria and red algae.

In cyanobacteria the outer surface of the thylakoids is studded with phycobilisomes, with biliproteins as light-harvesting pigments. The four major types of phycobilisomes are (1) bundle-shaped, bound to the inner surface of the cytoplasmic membrane (*Gloeobacter violaceus*, which lacks thylakoids), (2) Hemiellipsoidal (preferably in red algae), (3) hemidiscoidal (cyanobacteria and some red algae), and (4) intermediate between hemidiscoidal and hemiellipsoidal (cyanobacterial *Phormidium* and *Synechococcus* strains).

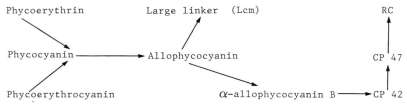

SCHEME 9.1   Energy flow from phycobilisome to PSII. CP, Chlorophyll complexes. The linker polypeptide Lcm is supposed to bind the phycobilisome to the chlorophyll antenna of PSII. CP 42 and CP 47 are chlorophyll antennae with apoproteins 42 and 47. Adapted from Morschel *et al.* (1991), p. 387. Copyright, Plenum, New York.

Phycobilisomes are made of two to three biliproteins and additional linker polypeptides that govern the assembly of phycobilisomes to active light-harvesting complexes. The biliproteins are red phycoerythrins (absorption bands between 500 and 570 nm), blue phycocyanins (absorption maxima at 575 and 590 nm), allophycocyanin and the allophycocyanin B complex (absorption at 650 and 670 nm, respectively).

The spectroscopic properties of the biliproteins are determined by covalently bound linear tetrapyrrole chromophores, the phycobilins, which are in an extended conformation. The bathochromic shift in color from red to blue is generated by increasing the number of double bonds. The spectroscopic properties are further modulated by chromophore–protein interaction and the attachment of chromophores to the protein.

The energy flow from phycobilisome to PSII is shown in Scheme 9.1.

SECTION III   *Bacteriorhodopsin*

9.17 BACTERIORHODOPSIN: DEFINITION AND STRUCTURE

Species from the genus *Halobacterium* are adapted to a habitat with high or even saturated salt concentrations, such as found in the Dead Sea, in the Great Salt Lake in Utah, or in salinas (i.e., natural deposits of common salts or other salts). The cells grow optimally in $3.5-5\ M$ NaCl solution and they lyse in concentrations lower than $3\ M$. The high external NaCl concentration is balanced by an equally high internal KCl concentration. The walls of these cells consist of a single glycoprotein, similar to the surfaces of eukaryotic cells, but lacking the murein layer and lipoproteins of other bacteria. Halobacteria have developed a light energy fixation process without chlorophyll.

With growth at a reduced partial oxygen pressure, patches of purple membrane develop; these consist of bacteriorhodopsin (BR) (molecular weight, 26,000) arranged in a quasi-crystalline hexagonal pattern. BR consists of a chromophoric group, retinal (the aldehyde of vitamin A, i.e., a β-carotene molecule split in half), linked to the lysine (forming a Schiff base) of a protein called *opsin*. With retinal hidden by a hydrophobic pouch inside opsin (see Figure 11.2 in Chapter 11), the absorption maximum is shifted to about 570 nm. BR consists of seven helical domains that span the bacterial membrane perpendicularly.

9.18 PRIMARY PHOTOEVENT, INTERMEDIATES, AND THE BACTERIORHODOPSIN CYCLE

The 13-cis BR isomer absorbing at 548 nm is in equilibrium with the all-trans isomer absorbing at 568 nm. When struck by a photon within 10 psec, BR_{trans} (i.e., the native BR form) absorbing at 568 is changed to BR_{trans} absorbing at 590 nm (Figure 9.19). This reaction is photoreversible. In the forward-moving reaction, BR_{trans}^{590} is followed by BR_{trans} absorbing at

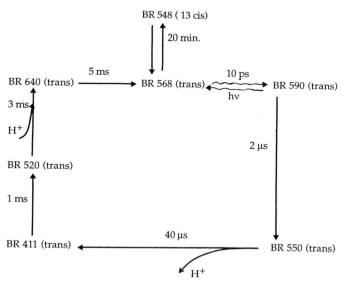

FIGURE 9.19 Light-induced reaction cycle of bacteriorhodopsin (BR) with the absorption maxima of the intermediates. From Hader *et al.* (1987), p. 106. With permission from Pergamon Press, New York.

550 nm and (with a proton release) by BR_{trans} absorbing at 411 nm. The original form is regenerated within a few milliseconds through the sequence BR_{trans}^{412}, BR_{trans}^{520}, and BR_{trans}^{640} with acquisition of a proton, i.e., original BR_{trans}^{568}, in equilibrium with $BR_{13\text{-cis}}^{548}$.

9.19 TRANSDUCTION OF LIGHT ENERGY TO CHEMICAL ENERGY

According to Mitchell's theory, the protons extruded from the cell membrane follow the electrical gradient and reenter the cell through an inward-facing membrane-bound enzyme (ATP synthase). The proton flux drives the ATP synthesis like a water current drives a mill. ATP synthase can also operate in reverse as an ATPase. Thus the cell can convert the proton gradient into ATP and vice versa (to extrude protons or other cations) according to the circumstances. A proton influx can drive an antiport of Na^+ or ATP synthesis. The influx of Na^+ allows a symport influx of amino acids.

There are two mechanisms in *Halobacterium* creating a proton gradient to drive ATP synthesis: (1) a light-dependent process based on bacteriorhodopsin and (2) a dark process, which relies on the electron transport chain. The energy released in this process through breakdown of organic components is stored in the form of a proton gradient and subsequently converted to ATP.

9.20 THE INITIAL 100-PSEC INTERVAL

The initial 100-psec interval of the room-temperature BR photocycle has been examined by picosecond transient absorption (PTA), picosecond time-resolved fluorescence (PTRF), and picosecond time-resolved resonance Raman (PTR) spectroscopy in terms of the molecular properties of the retinal chromophore in the K-590 intermediate, also designated BR-590, Br_{trans}^{590}. The purple membrane of BR has been shown to sustain effective trans-membrane proton pumping. The purple membrane retinal has the same polyene structure as the visual rhodopsin, but (1) the photoinitiated changes in *Halobium* retinal proceed by a photocycle that returns on the millisecond time scale to the original all-trans structure and (2) in the visual system only an enzymatic step recombines the dissociated retinal with the opsin protein.

Retinal stores the large amount of energy (~ 15 kcal/mol) needed to drive ATP synthesis, based in large part on changes in the retinal configuration and conformation.

Sequential events within a span of 100 psec can be studied at a time

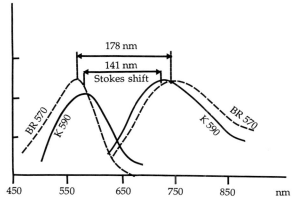

FIGURE 9.20   Absorption spectra of BR-570 and BR-590 (designated K-590) and fluorescence spectra of BR-570 and BR-590 (K-590). Stokes shifts are 178 nm for BR-570 and 141 nm for BR-590. Adapted from Atkinson (1991), p. 550. Copyright, Plenum, New York.

resolution of 10 psec, while maintaining the *in vivo* functional behavior of the BR photocycle. The PTA equipment was similar to that described in Chapter 4 for the femtosecond flash photolysis of hemoglobins.

The 590-nm absorbing intermediate (BR_{trans}^{590}) of the BR photocycle has long been considered to be of major importance in energy storage mechanisms. The fluorescence spectrum of BR-590 has been obtained using the PTRF technique (Figure 9.20). It exhibits a 17-nm shift to the blue in its maximum relative to the fluorescence spectrum of the original species BR-570. BR-590 has a Stokes shift (from emission to absorption maximum) of 141 nm versus 178 nm for BR-570. Thus BR-590 undergoes a smaller change in geometry on electronic state excitation. The quantum yield of fluorescence from BR-590 is approximately twice that of BR-570. Stokes shift and quantum yield characterize the potential energy surfaces of BR-590 as significantly different from that of the BR-570 species, from which it is formed during the initial 10 psec of the photocycle.

PTA and PTRF are measured as *time-dependent* signals relative to the pulsed (5–7 picoseconds) excitation of BR-570 with 565 nm radiation. The probe laser wavelength of 590 nm lies very near the isobestic point of the absorption spectra of the two pigments BR-570 and BR-590. Thus, following the excitation pulse in the first picosecond, the absorption at 590 nm decreases as the pigment BR-570 decreases, until the pigment BR-590 appears, and then starts rising to return to its initial level, when practically all the pigment BR-570 has been converted to BR-590. One of the most important results is derived from the slower rate of PTRF signal increase compared to PTA over the 0- to 40-psec time interval. The simulation of the time-dependent evolution of BR-570 and K-590 is best fitted by a mechanistic model involving another intermediate called J-625. Excellent fits are obtained for PTA data, and partial fits are obtained for PTRF. The proposed linear reaction scheme is BR-570 $\rightarrow$ J-625 $\rightarrow$ K' $\rightarrow$ K-590 (or BR-590), where K' represents vibrationally excited K-590.

The suggestion that K' is populated directly from J-625 and prior to the formation of ground state BR-590 follows reasonable expectations when the potential energy surface describing the photocycle is considered. K' acts as an excited state intermediate in the photocycle; the vibrational relaxation of K' to BR-590 needs to occur over an interval of 6–8 picoseconds to maintain a good simulation fit to PTRF data.

9.21 TIME-RESOLVED ABSORPTION EXPERIMENTS ON THE FEMTOSECOND SCALE

In the halobacterial branch of archaebacteria, a special kind of retinal-based photosynthesis is found with two light-driven pumps: (1) bacteriorhodopsin (BR) as a proton pump and (2) halorhodopsin (HR) as a chloride pump (both occurring in the cell membrane and mediating phototropic growth of halobacterial cells.

In the active form the proteins contain retinal in the all-trans configuration linked via a protonated Schiff base to a lysine residue of the amino acid sequence. After the absorption of a photon, the isomerization to the 13-cis form induces the ion transport, which must proceed via polar side groups of the protein structure. After about 10 msec, ion transport is finished and the chromoprotein has returned to its initial state.

In the very early steps of the photoreaction in BR prior to and during isomerization of the chromophore, extremely rapid (50 fsec) absorption changes reflect the relaxation of high-frequency vibrational modes. The slower kinetics are related to the isomerization of retinal, which starts in the excited electronic state and is terminated after the 500-fsec internal conversion process.

Time-resolved "excite and probe" experiments are performed using 80-fsec pulses. The excitation pulse wavelength is at 620 nm. The time resolution of the system is 50 fsec. Three reactions on the femtosecond scale may occur:

1. The photons incident on BR (i.e., BR_{570}) promote the retinal to the Franck–Condon state $^1BR_1^{**}$ on the $^1BR_1$ potential energy surface. Here a

number of vibrational modes are displaced relative to the $^1BR_1$ equilibrium position.

2. Within 50 psec after light absorption an *equilibration of the high-frequency vibrational modes to the state $^1BR_1^*$ occurs*. During the first reaction the molecule does not have the time to move along the coordinates of *the low-frequency reactive modes*. In the following slower reactive motion of the retinal (180 fsec), part of the isomerization (presumably a rotation by $60-90°$ around the C-13–C-14 double bond) takes place and the retinal arrives at the bottom of the $^1BR_1$ potential energy surface.

3. The system leaves this state via internal conversion with a time constant of 500 fsec.

Two decay pathways are possible: more than 60% of the molecules form the intermediate J-625, while the rest return to the original ground state of BR. J-625 proceeds with a 3-psec time constant to the intermediate K-590, which is stable on the picosecond time scale.

In a molecular system such as the BR photocycle, the vibrational degrees of freedom are extremely sensitive to rapid configurational and conformational changes. Thus vibrational spectra, such as resonance Raman (RR) spectra of photocycle intermediates, can be more useful than absorption and fluorescence data in elucidating the retinal mechanism. To obtain the K-590 spectrum from PTR data measured at a 100-psec delay, the scaled RR spectrum of the other species present, BR-570, must be subtracted (BR-570 comprising 40% of the reaction mixture; K-590, 60%). The first set of changes reflects increased out-of-plane motion in K-590, such as that associated *with twisting of the polyene backbone*, and the second indicates that *the isomeric form of B-570 has been altered*. It is likely that K-590 contains a 13-cis-like isomer, although the complete structural form remains unresolved.

SECTION IV *Artificial Photosynthesis*

9.22 BASIC PRINCIPLES FOR CREATING AN ENTIRELY SYNTHETIC SYSTEM

When the photosynthetic membrane is simplified to the strictly essential, it consists in principle of (1) the phospholipid molecules required to generate a phospholipid membrane, (2) a donor catalyst (e.g., Mn), (3) the

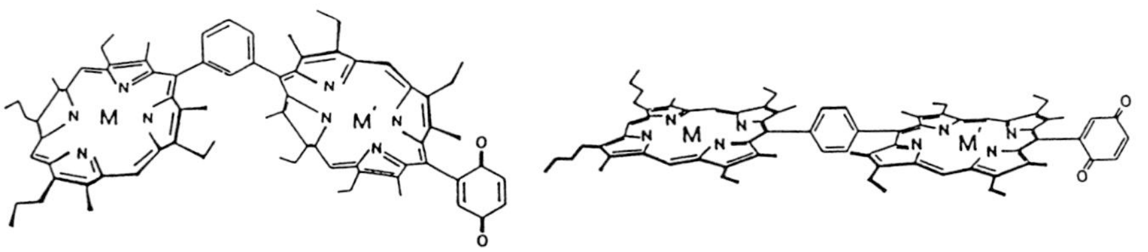

FIGURE 9.21 Selectively metalated, quinone-substituted "gable" (left) and "flat" (right) porphyrin dimer models suitable for studying photosynthetic interchromophore orientation and energetic effects in biomimetic systems. M represents 2H or Zn at the center of each monomer. Adapted from Sessler *et al.* (1991), p. 393. Copyright, Plenum, New York.

two quantum-trapping pigments (PII, PI) on the surface of the membrane, and (4) the iron–sulfur protein on the side of the membrane.

9.23 QUINONE-SUBSTITUTED PORPHYRIN DIMERS AS PHOTOSYNTHETIC MODEL SYSTEMS

A series of photosynthetic models with selectively metalated, quinone-substituted "gable" and "flat" porphyrin dimers has been prepared and characterized (Figure 9.21). An essential feature of these compounds is that they are not "floppy," i.e., they possess a well-defined conformational structure. In both series the flanking methyl substituents at positions 3 and 7 force the porphyrins to adopt a position perpendicular to the bridging phenyl subunits; for the same reason, the quinone is expected to lie perpendicular to the proximal porphyrin. The available X-ray structural data are consistent with these features.

When the quinone-substituted monomers in this family of quinone-substituted porphyrin dimers, are irradiated at the Soret maximum, no detectable fluorescence is seen, suggesting that charge separation is very rapid in these systems. When irradiated in dilute toluene solution with a 350-fsec, 582-nm laser pulse, these monomers show absorption changes that decay completely in 5–15 psec (indicative of the exothermic formation of a charge-separated state, followed by exothermic charge recombination to regenerate the ground state, according to the following equation:

$$\text{MP–Q} \xrightarrow{h\nu} \text{MP*–Q} \xrightarrow{\text{electron transfer}} \text{MP}^+\text{–Q}^- \xrightarrow{\text{charge recombination}} \text{MP–Q}$$

(MP is the unsubstituted porphyrin).

9.24 MIMICKING PHOTOSYNTHETIC ELECTRON AND ENERGY TRANSFER — PHOTODRIVEN CHARGE SEPARATION: FROM DYADS TO TRIADS, TETRADS, AND PENTADS

As suggested by the above example, the key steps in the photosynthetic conversion of light to chemical potential energy include not only photodriven charge separation, but also the prevention of the back reaction (charge recombination). Although charge separation has been achieved in several biomimetic solar energy conversion systems, retarding back reaction has proved difficult. This may be achieved by rapidly moving the electron, the hole, or both away from the site of excitation to more stabilizing environments.

9.24.1 Dyads

The simplest type of covalently linked artificial photosynthetic electron transfer system consists of one donor joined to a single acceptor. P–Q molecules can undergo photoinitiated electron transfer from the porphyrin first excited singlet state with high yield.

P–Q dyad **1** consists of a benzoquinone moiety joined to a tetraaryl-porphyrin via an amide linkage. Excitation of the porphyrin moiety of **1** allows that the porphyrin first excited singlet state can decay by the usual photophysical processes. In model porphyrins that do not bear the quinone moiety, the lifetime of the first excited singlet state is 7.7 nsec. The presence of quinone introduces a new decay pathway: electron transfer to yield a $P^{\ddagger}-Q^{\bar{}}$ charge-separated state. The rate constant for this step may be estimated from the relationship

$$k_2 = 1/\tau - 1/\tau_0$$

where k_2 is the rate constant for the photoinitiated electron transfer, τ is the fluorescence lifetime of the porphyrin in **1**, and τ_0 is the fluorescence lifetime of a model system with the same photophysics as **1**, but which lacks the electron transfer step (i.e., 7.7 nsec). The fluorescence lifetime of **1** was found to be 0.1 nsec. The quantum yield of photoinitiated electron transfer is τk_2, and equals 0.99. The $P^{\ddagger}-Q^{\bar{}}$ state preserves a significant fraction (about 1.4 eV) of the 1.9 eV in the porphyrin first excited singlet state. Thus, the P–Q dyad compound **1** is reasonably successful as a mimic of the gross features of the initial photodriven charge separation step of photosynthesis. However, the very structural and electronic features that ensure rapid photoinitiated electron transfer also favor rapid charge recombination. Thus, the $P^{\ddagger}-Q^{\bar{}}$ state lives at most a few hundred picoseconds in solution.

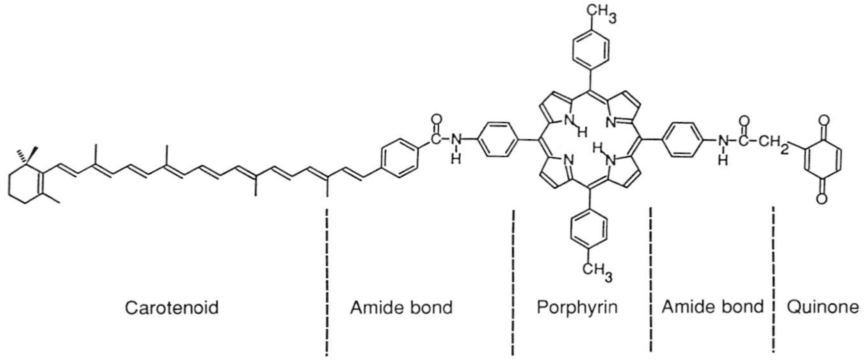

FIGURE 9.22 Molecular structure of triad **2** (carotenoid polyene/porphyrin/quinone). Adapted from Gust and Moore (1991), "Advances in Photochemistry," vol. 16 (D. Volman, G. Hammond, and P. Necker, Eds.), Wiley, New York. p. 9.

9.24.2 Triads

Substantial slowing of the back-reaction has been achieved with a tripartite molecule in which a long-lived photodriven charge-separated state of relatively high potential is formed from an excited singlet state. This molecular triad (compound **2**) consists of a tetraarylporphyrin covalently linked to both a carotenoid electron donor) and a quinone (electron acceptor) (Figures 9.22 and 9.23). Excitation of porphyrin by visible light results within less than 100 psec in a carotene radical cation (car$^+$)/porphyrin–quinone radical anion (Q$^-$) (lifetime in the microsecond time scale, with an energy of more than 1 eV above ground state).

The simplest scheme consistent with the spectroscopic observations is as follows: On excitation (532 nm) of porphyrin, the dominant relaxation pathway will be the quenching of the porphyrin singlet via electron transfer to quinone (P$^+$, 600 nm absorption, and Q$^-$). The decay of the 600-nm absorption and rise of the 950-nm absorption are concurrent with the rise of the car$^+$ and return of P$^+$ to ground state P.

The dramatic shortening of the porphyrin first excited singlet state lifetime suggests that the species with the P$^+$ is formed with a quantum yield approaching unity. The rapid decay of the porphyrin singlet and rise of the car$^+$ rule out the participation of the porphyrin triplet in the primary electron transfer process.

The formation of the final charge-separated state competes with the recombination reaction. The observation of car$^+$ has biological relevance in that the formation of such a radical has been observed in photosystem II.

The redox potential of the charge-separated state is above 1 V, which indicates that a substantial fraction of the 1.8-eV porphyrin lowest excited

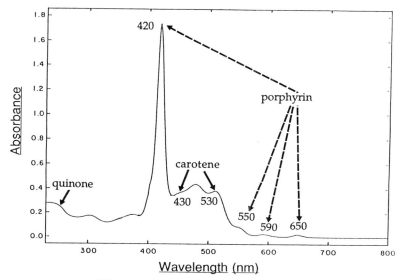

FIGURE 9.23 Absorption spectrum of triad **2**.

singlet state has been conserved. The key to obtaining long lifetimes of the charge-separated state appears to be the interposition of a neutral porphyrin between the widely separated ions, coupled with energetics that are unfavorable for recombination.

Thus, the multistep electron transfer strategy can be successfully applied to artificial photosynthetic systems. The final charge-separated state preserves over 50% of the energy of the initial excited state, and has a lifetime about three orders of magnitude greater than that for the comparable dyad systems.

9.24.3 Tetrads

9.24.3.1 C–P–Q–Q Tetrads

In $C-P-Q_A-Q_B$ the naphthoquinone (Q_A) and benzoquinone (Q_B) moieties are linked by a saturated, rigid bicyclic ring system, and the carotenoid and diquinone species are joined to the porphyrin via amide bonds. The partial double bond character of these linkages restricts conformational mobility. The lifetime of the charge-separated state $C^{\dot+}-P-Q_A-Q_B^{\dot-}$ is 460 nsec.

9.24.3.2 C–P–P–Q Tetrads

In order to investigate electron transfer between tetrapyrrole moieties in model systems, at least two porphyrin or other tetrapyrrole species are

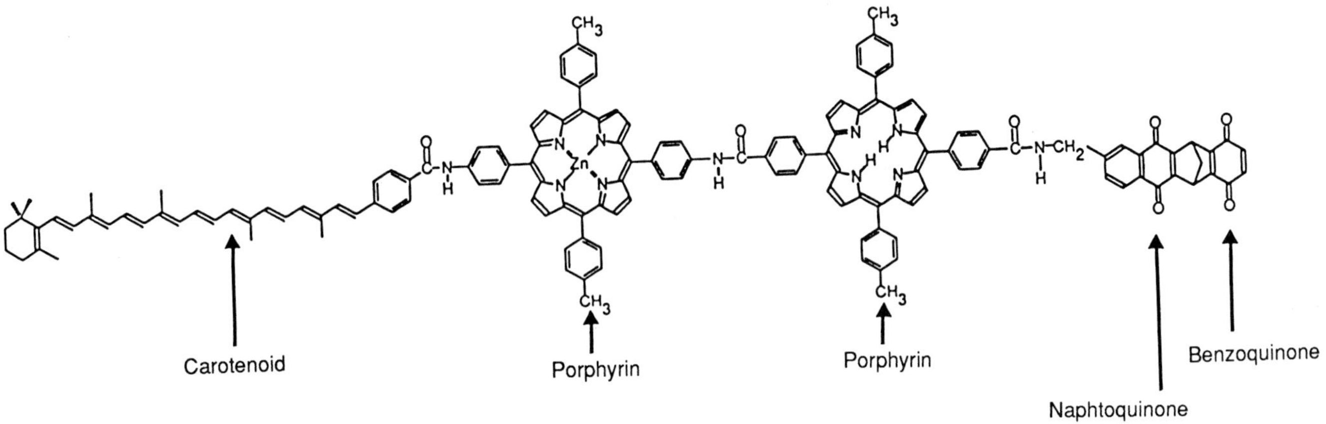

FIGURE 9.24 Molecular structure of a representative pentad. Adapted from Gust and Moore (1991), "Advances in Photochemistry," Vol. 16 (D. Volman, G. Hammond, and P. Necker, Eds.). Wiley, New York.

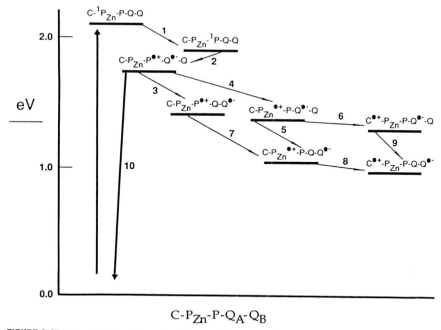

FIGURE 9.25 Photoinduced transient states of the pentad in Figure 9.24. Numbers 1 to 9 are the different transient states resulting from the photoexcitation of the pentad. Number 10 is the return to ground state.

necessary, such as in the C–P–P–Q. Excitation of this tetrad in anisole solution with a 590-nm, 15-nsec laser pulse results in the observation of a transient carotenoid radical cation that decays exponentially with a lifetime of 29 μsec.

The long lifetime of the final $C^{+}_{\cdot}–P–P–Q^{\mp}$ species is the result of the large charge separation between the carotenoid and quinone moieties and the fact that endogonic electron transfer would be needed to place the charges on adjacent chromophores.

9.24.4 Pentads

Phenomena of electron transfer between quinone moieties or two tetrapyrrole moieties are both observed in the natural photosynthetic apparatus. Conceptually, both mechanisms can be fused via construction and study of C–P–P–Q–Q pentad (Figure 9.24). The carotenodiporphyrin portion of the molecule resembles that of C–P–P–Q, but the porphyrin bearing the carotenoid contains a zinc ion. The diquinone is related to that of the C–P–P–Q tetrad, but it is joined to the porphyrin by a different linkage. The final charge-separated state $C^+-P_{Zn}-P-Q_A-Q_{B-}$ (Figure 9.25) is formed with a quantum yield of 0.83 and has a lifetime of 55 μsec.

Future designs of supramolecular systems for charge separation will deal with the problem of interfacing these systems to other chemical reactions and incorporation of such systems into organized assemblies.

SECTION V

9.25 SELECTED READINGS: PHOTOSYNTHETIC PRODUCTION OF HYDROCARBONS

In the chemical transduction of photon energy, some plants are able to carry the carbon reduction (and storage process) beyond the production of carbohydrates, adding carbons and displacing oxygen until hydrocarbons are produced. This, besides its intrinsic interest, is of potential importance as an alternative source of fuel and other products that today are extracted from petroleum. Thus, there exists a possible renewable source for a petroleum substitute.

Natives of the Brazilian rainforest know the species *Copaifera langsdorfii*, the copaiba tree. They drill a 5-cm hole into the 1-m-thick trunk and collect 15–20 liters of the hydrocarbon exudate. The natives use this hydrocarbon for nonenergy purposes. Tests have shown, however, that the liquid can be placed directly in the fuel tank of a diesel-powered car. The hydrocarbons collect in thick capillaries that may extend the full 30-m height of the tree. The yield per tree can be increased by drilling additional holes. An acre of 100 mature trees might be able to produce 25 barrels of fuel per year. This tree can be grown only in the tropics (from just south of São Paulo, Brazil, to just north of Panama). The oil is not identical with diesel fuel. It consists largely of 15 carbon molecules, in a cyclic or ring structure.

The main focus for plant-produced fuel (PPF) lies in the genus *Euphorbia* of the Euphorbiaceae family (best known species, *E. lathyris*, also called gopher plant, gopher wood, gopher weed, mole tree, spring wort, and caper spurge). *Euphorbia lathyris* is very widely dispersed, and has been well known for almost 3 millenia. The name gopher plant comes from a belief that it is repellent to gophers. Pursh, in writing of the plant in 1814, in his *Flora of North America*, says that "It is generally known in America by the name of Mole-Plant, it being supposed that no moles disturb the ground where the plant grows." Darlington states in *Flora Cestrica* (1837) that "this foreigner becomes naturalized in many gardens —having been introduced under a notion that it protected from the incursion of moles."

The name caper spurge comes from the resemblance of its berries to capers (used for seasoning, although they are rather bitter); spurge is related to purge, and the plant is listed in *Materia Medica* as one of the American medicinal plants.

The milky-white latex that oozes out of a broken *E. lathyris* leaf is toxic and irritates the skin. It will raise big blisters if it gets in the mouth, and could blind a person if it gets in the eyes. Another *Euphorbia*, the pencil plant *Euphorbia tirucalli*, actually defended Abyssinia from the Italians in 1938 according to the late Emperor Haile Selassie. The pencil plant grew in great hedges all over the country, and when the Italian cavalry tried to charge through it, their horses balked because the latex got into their eyes and mouths.

For convenience in what follows, we will call *E. lathyris* gopher weed. Almost every species of the genus *Euphorbia* produces some kind of hydrocarbon in the form of latex (an emulsion of oil and water). By the "isoprenoid pathway," the original carbohydrate is broken into two carbon pieces, and then a skeleton is reconstructed that is all hydrocarbon, containing no oxygen. The basic building block has five carbon atoms—i.e., isoprene. The rubber plant, *Hevea*, makes rubber in its latex by stringing together very large numbers of these isoprene units. But euphorbias do not go that far. They do not make two-million-unit long pieces (rubber) but stop at about 32 carbon chains.

It has been reported that we may expect a yield of at least 10 barrels of hydrocarbon material per acre planted with *E. lathyris*, together with enough fermentable sugar to produce about 12 barrels of ethanol.

Fuel from the gopher weed does not come quite so easily as it comes from the copaiba tree. One has first to cut the plants and let them lie in the sun to reduce the water content. Then the whole plant is taken, not just the seeds, and is put into a mill to be ground. This is then boiled with heptane in an extractor. The black, tarry, sticky oil obtained looks like crude oil. This product goes into the refinery (which breaks the product into simpler compounds, usually as a result of heating). In fact, the result is a group of products a little better than what one gets from crude oil, with fewer impurities: no sulfur and no vanadium (both of these poison the cracking catalyst). Thus the product of *E. lathyris* goes through the present petroleum-cracking facilities with ease and yields very useful products, such as ethylene and propylene, which are desirable both as petrochemical gasoline and other fuels. Anything that is made from petroleum can be made out of *Euphorbia*, including plastics.

Two species of *Euphorbia* were used in experimental plantings in southern and northern California: (1) An annual *E. lathyris* (gopher weed) that propagates from seed and comes to the harvesting stage in 6–7 months with 20 inches of rainfall was used. The plant before harvest is 4 feet high and it is cut, crushed, and extracted for its hydrocarbon content. (2) *Euphorbia tirucalli* is perennial and was also planted; it takes 2 to 3 years to come to harvest. With *E. tirucalli* the yields have been of the order of 15 barrels oil equivalent per acre. The product has been designated "green oil."

Both *Euphorbia* species contain in their latex a phorbol ester that is an irritant of the eye and the mucous membranes of the nose. The irritant property can be removed with processing, and causes no problem with machine harvesting of these plants.

Although the *E. lathyris* yield is about 10 barrels of oil per acre with unselected seed and no agronomic experience, the estimates with small agronomic development are over 30 barrels of oil per acre, with 16 inches of water on semiarid desert soil. The cellulosic residues after processing the dried *E. lathyris* can be converted to ethanol, making the entire process energy positive.

The hydrocarbons from *Euphorbia* are primarily a blend of C_{15} compounds (terpene trimers) that, when subjected to catalytic cracking, yield various products identical to those obtained by cracking napththa, a high-quality petroleum fraction that is one of the raw materials of the chemical industry. The material extracted from gopher weed resembles crude oil with the following composition: ethylene, 10%; propylene, 10%; toluene, 20%; xylenes, 15%; olefins, 21%; coke, 5%; alkanes, 10%; and fuel oil, 10%.

Euphorbias grow everywhere in the world; the French grew *Euphorbia resinifera* in Morocco (1938), obtaining 3 tons of hydrocarbon per hectare (about 2.5 acres) with a total planting of 125,000 hectares. In Ethiopia the Italians had plans for using various species of *Euphorbia* (specifically *Euphorbia abyssinica*) to produce a gasoline-like liquid called "vegetable gasoline." There are many species of euphorbias suitable for processing of the oillike materials they contain, which can grow on semiarid land, with a minimum of water, and do not require land that is suitable for food production.

Eucalyptus, a plant that is very common in Australia, California, Spain, and Portugal, also produces oil. It can be cut and regrown every 2 to 3 years. The oil from *Eucalyptus* is a mixture of lower terpenes, whereas the oil from gopher weed is mostly higher terpenes.

For a meaningful exploitation of the hydrocarbon-producing photosynthetic process, hundreds of millions of acres must be cultivated with the above-described plants. Their yield can be improved by selection and it should also be possible to manipulate the plants genetically to increase yield and make more desirable compounds from the standpoint of fuel.

Gopher weed is readily propagated by stem cuttings in coarse sand or in a 1:1 mix of perlite:vermiculite. It is known that cuttings of another *Euphorbia,* i.e., the poinsettia *Euphorbia pulcherrima,* generally root more rapidly and uniformly with auxin treatment.

From the seeds of *E. Lathyris* Linn. a novel bicoumarin has been isolated; it was characterized as the 5,8' dimer of esculetin based on UV, IR, NMR, and mass spectral data.

Seed oil from gopher weed and latex from *Euphorbia ingens* (a tropical, cactuslike tree) contain a cocarcinogen that is structurally different from the active components of croton oil (i.e., phorbol-12,13-diesters) with respect to both ester groups and the parent alcohol.

Through photosynthesis, plants manufacture carbohydrates. In the carbohydrate structure, one atom of oxygen is attached to each atom of carbon. *Euphorbia lathyris* removes the oxygen atoms to create hydrocarbons. Dilution of the seed oil of gopher weed with acetone results in the separation of a crystalline compound, previously named "euphorbiasteroid." This substance is the diacetate–phenylacetate of a diterpene alcohol that is named 6:20-epoxy-lathyrol. Euphorbiasteroid is identical with "substance L_1" obtained from the hydrophilic neutral fraction of the seed oil. Euphorbiasteroid has a molecular formula $C_{32}H_{40}O_8$. As a basis of reference, the chemical formula of glucose is $C_6H_{12}O_6$ (molecular weight 180).

The more reduced a molecule is, the more oxygen it takes to oxidize it, according to the following formula: moles of oxygen per gram (required for full oxidation; that is, yielding CO_2 and H_2O as final products) equal the number of carbons + number of H/4 − number of O/2. Thus, 0.033 mol of oxygen is required to fully oxidize 1 g of glucose. This formula can be used to position euphorbiasteroid on the reduction scale. The moles of oxygen required to fully oxidize euphorbia steroid = 32 + 40/4 − 8/2 = 34. With a molecular weight of 552 for euphorbiasteroid, 34/552 = 0.052 mol oxygen required to fully oxidize 1 g of the steroid, which is in accordance with the fact that, on the scale of reduction, the euphorbia steroid has moved from glucose to the hydrocarbons.

Although some plants, such as *E. lathyris,* go all the way on the reduction path to production of hydrocarbons, other plants are able to take carbon reduction to an intermediate level, beyond the carbohydrates, but not as far as hydrocarbons. One typical example is a plant related to the euphorbias, *Ricinus commonis,* which is known for its pharmacological product castor oil. The composition of castor oil is ricinoleic acid (87%), oleic acid (7%), linoleic acid (3%), palmitic acid (2%), stearic acid (about 1%), and dihydrostearic acid (trace amounts). The chemical formula of ricinoleic acid is $C_{18}H_{12}O_3$ (molecular weight 298). As indicated above, 0.033 mol of oxygen is required to fully oxidize 1 g of glucose versus 0.083 mol for ricinoleic acid. Therefore, on the scale of reduction, ricinoleic acid, the product of *R. commonis*, has moved farther then glucose toward the hydrocarbons.

9.26 THE GRAETZEL CELL: A SOLAR CELL BASED ON PHOTOSYNTHESIS AND PHOTOGRAPHY

A research group in Lausanne, Switzerland has combined spectral sensitization and optical absorption enhancement to produce a 7–10% efficient solar cell. The cell, which is called a *nanocrystalline dye sensitized solar cell*, or *Graetzel* cell (Figure 9.26), resembles natural photosynthesis in two respects: 1) it uses an organic dye to absorb light and produce a flow of electrons, and 2) it uses multiple layers to enhance both the absorption and collection efficiencies.

To create the Graetzel cell, a solution of nanometer-size particles of titanium dioxide (TiO_2) is deposited directly on conductive (SnO_2) glass. The film is heated to form a porous, high surface red TiO_2 structure which resembles a thin sponge. The glass plate is dipped in a solution of a dye such as Ruthenium bis-pyridyl [Ru(bpy)] or chlorophyll derivative. A single layer of dye molecules attaches to each particle of TiO_2 via carboxylic groups and acts as the primary absorber of sunlight. To form the final cell, a drop of liquid electrolyte containing iodide is placed on

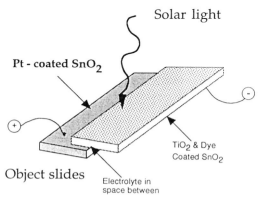

FIGURE 9.26   Schematic representation of requirements for building a Graetzel cell based on photosynthesis and photography.

the film. A counter electrode of conductive glass (SnO_2) coated with a thin layer of platinum or carbon is placed on top. The "sandwich" obtained [*i.e.* TiO_2 and dye coated SnO_2 (negative pole) and platinum or carbon coated SnO_2 with electrolyte in between] is illuminated through the TiO_2 side.

Thus, in a nanocrystalline cell a rough TiO_2 substrate is used for a thin layer of dye in order to increase the light absorption while allowing for efficient charge collection. Since the dye layer is so thin, the excited electrons produced from light absorption can be injected into the TiO_2 with near unity efficiency via sensitization. The TiO_2 functions in a similar way as the silver halide in photography except, that instead of forming an image, the injected electrons produce electricity. The many interconnected particles of the porous membranes which can absorb 90% of visible light are analogous to the stacked thylakoid membrane in chloroplasts.

When light strikes, at first dye electrons move from ground state S to excited state S*, just as for chlorophyll in photosynthesis. The dye functions as a photochemical "pump." The TiO_2 serves as the electron acceptor (like the pheophytin or quinone in photosynthesis), and the iodine in the electrolyte solution serves as the electron donor by neutralizing and restoring the photochemically semioxidized dyes with production of iodine or triiodide.

With light striking the dye the competing processes are: (1) luminescence from the dye, (2) non-radiative recombination through surface defects on the TiO_2, and (3) back reaction of the triiodide to iodide at any exposed surface of the TiO_2 or SnO_2.

With the excited state of the sensitizer dye effectively quenched by the TiO_2, the luminescence component is quite weak.

Bibliography

Adoph, W., Hecker, L., Balmain, A., Lhomme, M. F., Nakatani, Y., Ourisson, G., Ponsinet, G., Pryce, R. J., Santhannakrishnan, T. S., Matyukina, L. G., and Saltikova, I. A. (1970). Euphorbiasteroid. *Tetrahedron Lett.* **25**, 2241–2244.

Alberts, B., Bray, D., Lewis, J., Raff, M., Roberts, K., and Watson, J. D., eds. (1983). "Molecular Biology of the Cell." Garland Publ., New York and London.

Atkinson, G. H. (1991). Picosecond time-resolved spectroscopy of the initial events in bacteriorhodopsin photocycle. *In* "Photobiology: The Science and Its Applications" (E. Riklis, ed.), pp. 547–559. Plenum, New York.

Berkaloff, A., Bourguet, J., Fayard, P., and Lacroix, J. C. (1981). "Biologie et Physiologie Cellulaires," Vol. III, pp. 26–36, 40–41, 45–50, 102–105. Hermann, Paris.

Berkaloff, A., Bourguet, J., Fayard, P., and Lacroix, J. C. (1982). Chloroplastes. *In* "Biologie et Physiologic Cellulaires," Vol. III, Chapter 11, p. 3. Hermann, Paris.

Calvin, M. (1982). Bioconversion of solar energy. *In* "Trends in Photobiology" (C. Hélène, M. Charlier, T. Montenay-Garestier, and G. Laustriat, eds.), pp. 645–659. Plenum, New York and London.

Cao, J., and Govindjee (1988). Bicarbonate effect on electron flow in a cyanobacterium *Synechocystis* PCC 6803. *Photosynth. Res.* **19**, 277–285.

Cao, J., and Govindjee (1990). Chlorophyll *a* fluorescence transient as an indicator of active Photosystem II in thylakoid membranes. *Biochim. Biophys. Acta* **1015**, 180–188.

Chen, R. F., Govindjee, Papageorgiou, G., Rabinowitch, E., and Wehry, E. L. (1973). Chlorophyll fluorescence and photosynthesis. *In* "Practical Fluorescence: Theory, Methods and Techniques" (G. G. Guilbault, ed.), pp. 543–573. Dekker, New York.

Dutta, P. K., Banerjee, D., and Dutta, N. L. (1973). Isoeuphorbetin, a novel coumarin from *Euphorbia lathyris* Linn. *Indian J. Chem.* **11**, 831–832.

Fisher, A. (1982). Auto fuel into plants. *Pop. Sci.* **182**, 84–66.

Fruit Grower (1978). **98**, 12.

Govindjee (1991). A unique role of carbon dioxide in photosystem II. *In* "Global Climactic Changes and Photosynthesis" (Y. Abrol, ed.), pp. 349–369. Oxford Univ. Press, London.

Govindjee, and Coleman, W. J. (1991). How plants make oxygen. *Sci. Am.* **262**, 50–58.

Govindjee, and Yang, L. (1966). Structure of the red fluorescence band in chloroplasts. *J. Gen. Physiol.* **49**, 763–780.

Govindjee, van de Ven, N. M., Preston, C., Seibert, M., and Gratton, E. (1990). Chlorophyll *a* fluorescence lifetime distribution in open and closed photosystem II reaction center preparations. *Biochim. Biophys. Acta* **1015**, 173–179.

Graan, T., and Ort, D. R. (1986). Detection of oxygen evolving photosystem II centers inactive in blastoquinone reduction. *Biochim. Biophys. Acta* **852**, 320–330.

Gust, D., and Moore, T. A. (1991). Mimicking photosynthetic electron and energy transfer. *Adv. Photochem.* **16**, 1–64.

Hader, D.-P., and Tevini, M. (1987). "General Photobiology," pp. 103–111, 144–148. Pergamon, Oxford.

Hecker, E. (1968). Cocarcinogenic principles from the seed oil of *Croton tiglium* and from other Euphorbiaceae. *Cancer Res.* **28**, 2338–2348.

Horticultural Division, Bull. **61**, 331–332.

Indianapolis News (1981). Monday, June 22.

Joliot, P. (1969). How plants make oxygen? *Sci. Am.* **262**, 50–58 (in Govindjee and Coleman, 1991).

Kobayashi, T., Ohtani, H., Tsuda, M., Ogasawara, K., Koshihara, S., Ichimura, K., Hara, R., and Terauchi, M. (1991). Primary processes in sensory rhodopsin and retinochrome. *In* "Photobiology: The Science and Its Applications" (E. Riklis, ed.), pp. 561–570. Plenum, New York and London.

Kovalev, Y. V., Krasnovskii, A. A., Jr., Lehoczki, E. and Maroti, J. (1981). Intensification of phosphorescence and delayed fluorescence of chlorophyll in barley leaves on selective suppression of biosynthesis of carotenes. *Biophysics* **26**, 910–912.

Kurlhara, K., Tundo, P., and Fendler, J. H. (1983). Aspects of artificial photosynthesis. Photosensitized electron transfer and charge separation in redox active surfactant aggregates. *J. Phys. Chem.* **87**, 3777–3782.

Maugh, T. H., II (1979). Unlike money, diesel fuel grows on trees. *Science* **206**, 436.

Melis, A. (1985). Functional properties of photosystem II in a spinach chloroplase. *Biochim. Biophys. Acta* **808**, 334–342 (in Cao and Govindjee, 1988).

Moore, T. A., Gust, D., Mathis, P., Mialocq, J.-C., Chachaty, C., Bensasson, R., Land, E. J., Doizi, D., Liddell, P. A., Lehman, W. R., Nemeth, G. A., and Moore, A. L. (1984). Photodriven charge separation in a carotenoporphyrin–quinone triad. *Nature (London)* **307**, 630–632.

Morschel, E., Schatz, G.-H., and Lange, W. (1991). The supramolecular structure of the light-harvesting system of cyanobacteria and red algae. *In* Photobiology: The Science and Its Applications" (E. Riklis, ed.), pp. 379–389. Plenum, New York and London.

Nanba, O., and Satoh, K. (1987). Isolation of a photosystem II reaction center consisting of D-1 and D-2 polypeptides and cytochrome b-559. *Proc. Natl. Acad. Sci. U.S.A.* **84**, 109–112, (in Wasielewski *et al.*, 1989, p. 524).

O'Regan, B., and Grätzel, M. (1991). A low-cost high-efficiency solar cell based on dye-sensitized colloidal TiO_2 films. *Nature* **353**, 737–739.

Paillotin, G., Vermeglio, A., and Breton, J. (1982). Organization of the photosynthetic pigments and transfer of the excitation energy. *In* "Trends in Photobiology" (C. Helene, M. Charlier, T. Montenay-Garestier, and G. Laustriat, eds.), pp. 539–547. Plenum, New York.

Preece, J. E., and Wollbrink, E. B. (1983). Vegetative propagation of *Euphorbia lathyris* by stem and leaf bud cuttings. *HortScience* **18**, 193–194.

Rabinowitch, E., and Govindjee (1965). The role of chlorophyll in photosynthesis. *Sci. Am.* **213**, 74–83.

Sessler, J. L., Johnson, M. J., Creager, S., Fettinger, J., Ibers, J. A., Rodriguez, J., Kirmaier, C., and Holten, D. (1991). Quinone substituted porphyrin dimers: New photosynthetic model systems. *In* Photobiology: The Science and Its Applications" (E. Riklis, ed.), pp. 391–399. Plenum, New York and London.

Stryer, L. (1975). "Biochemistry." Freeman, San Francisco.

Tabushi, I., and Kugimiya, S.-I. (1985). Phase-transfer-aided FMNH production using artificial photosynthesis cells of bacterial type. Cell system optimization by the concept of flux conjugation. *J. Am. Chem. Soc.* **107**, 1859–1863.

Tecnicos del College of Agriculture, University of Arizona (1981). Planta que produce petroleo. Adapted from "Progressive Agriculture in Arizona," Vol. 31, No. 4. Agricultural Experimental Station, Tucson, AZ.

Terry, S. Lowly gopher weed contains oil harvest. *Christian Science Monitor.*

Van Grondelle, R., Bergström, H., and Sunström, V. (1991). Trapping of excitation energy in photosynthetic purple bacteria. *In* "Photobiology: The Science and Its Applications" (E. Riklis, ed.), pp. 371–378. Plenum, New York.

Wasiliewski, M. R. (1992). Photoinduced electron transfer in supramolecular systems for artificial photosynthesis. *Chem. Rev.* **92**, 435–461.

Wasielewski, M. R., Johnson, D. G., Seibert, M., and Govindjee (1989). Determination of the primary charge separation rate in isolated photosystem II reaction centers with 500-fs resolution. *Proc. Natl. Acad. Sci. U.S.A.* **86**, 524–528.

Wolfe, S. L. (1972). "Biology of the Cell." Wadsworth, Belmont, CA.

Xu, C., Auger, J., and Govindjee (1990). Chlorophyll *a* fluorescence measurements in isolated spinach thylakoids obtained by using single-laser based cytofluorometry. *Cytometry* **11**, 349–358.

Zinth, W., and Oesterhelt, D. (1991). The primary photochemical process in bacteriorhodopsin. *In* "Photobiology: The Science and Its Application" (E. Riklis, ed.), pp. 531–535. Plenum, New York.

10

Photobioregulatory Mechanisms

Photomorphogenesis

10.1 LIGHT AS A SOURCE OF INFORMATION

Living organisms can utilize the energy of light in two ways: light can be used as a source of energy to facilitate a process, such as photosynthesis, that otherwise would be thermodynamically unfeasible; alternatively, light can be used as a source of information, as in photomorphogenesis.

There are five properties of light that can serve as a source of information: (1) quantity, (2) quality (e.g., color), (3) spatial asymmetry, (4) periodicity, and (5) polarization.

Photoreceptor chromophores are of different types, e.g., in order of increasing wavelength and decreasing energy of absorbed light, (1) solar UV radiation absorbers (tryptophan, DNA, RNA, tRNA), (2) blue light

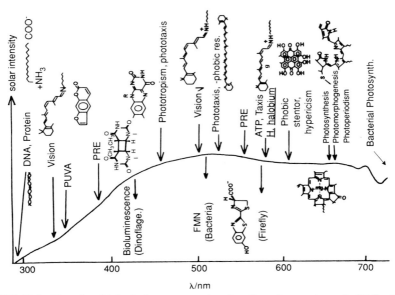

FIGURE 10.1   The photobiological spectrum, showing the solar spectrum (solid line) and photoresponses at the wavelength maximum of photosensitivity (arrows). Arrows inside the line representing the solar spectrum represent light-emitting processes. The structures of possible photoreceptor chromophores for each photoresponse are indicated. From Song *et al.* (1991). p. 24.

absorbers (flavins, carotenoids), (3) green light absorbers (rhodopsin, bound carotenoids), (4) yellow–orange light absorbers (phycocyanins, bacteriorhodopsins), and (5) red light absorbers (stentorin, chlorophylls, phytochrome). Figure 10.1 shows the solar spectrum and the photoresponses at the wavelength maximum of photosensitivity for different chromophores.

10.2 THE PIGMENTS OF PHOTOMORPHOGENESIS

Pigments that may play a role in photomorphogenesis, using light as the source of information, include flavins and flavoproteins, carotenoids and carotenoproteins, biliproteins, cytochromes, and chlorophylls. Phytochrome is used in the photomorphogenetic system about which most is known.

10.2.1 Red and Far-Red Absorbing Phytochrome

Phytochrome is a light receptor protein, the best known photoreceptor for light-dependent development and differentiation in higher plants.

The phytochrome nomenclature is described in detail by Quail *et al.* (1993). The historical account of phytochrome research is given by Sage (1992). The most recent monograph on phytochrome is by Koornneef and Kendrick (1994). *Biologically inactive red-absorbing phytochrome (P_r) is synthesized in the dark, and on irradiation with red light it is converted into an active, far-red-absorbing form (P_{fr}) (Figure 10.2).*

Far-red light suppresses the germination of light-sensitive lettuce seed and reverses the promotive effect of red light. Such antagonistic effects of red and far-red light are repeatedly reversible through as many as 100 cycles. The photoreversibility of other photomorphogenetic responses reinforces the notion that a single photoreceptor in two photointerconvertible forms is responsible; one form (P_r) should absorb red light most strongly, and the other (P_{fr}) should absorb far-red light best. P_{fr} is known to (1) promote germination of light-sensitive seed, (2) induce flowering of long-daylength plants, (3) inhibit potassium uptake, (4) induce anthocyanin synthesis, (5) induce gene transcription, and (6) modulate biolectric potentials. Two possibilities exist: either P_r is active, preventing the display of photomorphogenetic responses, or P_{fr} is active, promoting the display of photomorphogenetic responses.

The photoconversion of less than 0.1% of P_r to P_{fr} can lead to a significant biological response. Although this photoconversion represents an insignificant portion of the relative P_r level (i.e., a change from 100 to 99.9%), it represents a relatively large increase in P_{fr} from nothing to a detectable value. Thus it is likely that a plant responds to the appearance of P_{fr}, not to the disappearance of P_r.

Both red and far-red light are effective at low fluences, with reciprocal relationships between fluence rate and time of exposure, such that the outcome reflects the product of the two parameters.

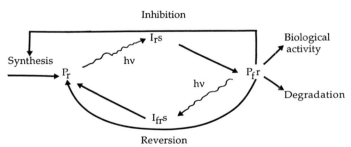

FIGURE 10.2 The phytochrome system. Synthesis of phytochrome, which occurs as P_r, is inhibited by P_{fr}. The photoconversion pathways between P_r and P_{fr} involve several intermediates (I_rs and I_{fr}s). P_{fr}, which yield biological activity by an as yet unknown mechanism, disappear from the cell via both thermal reversion to P_r and exhibit enhanced proteolytic degradation of their protein moiety, a process that is referred to as *destruction*. From Pratt and Cordonnier (1989), p. 286. With permission of Plenum, New York and London.

10.2.2 Function of Phytochrome as a Photosensory System

Following the initial excitation of either P_r or P_{fr} (Figure 10.2), the first intermediate decays to the opposite form of phytochrome via different sets of thermally unstable intermediates (I_rs and I_{fr}s). The slowest reactions are those leading to P_r, which occur on a millisecond time scale. Appreciable levels of phytochrome can be present in intermediate form at steady state under bright illumination (under certain circumstances a phototransformation intermediate may have biological activity).

Except in the far-red region, where P_r absorbs negligibly, the two forms of phytochromes have overlapping spectra. Under conditions of constant illumination, phytochrome will be cycling continuously between its two forms. The proportion that is present as P_{fr} is strongly wavelength dependent.

10.2.3 A "Cycling Process"

The system seems to "count" the number of times a molecule cycles between the P_r and P_{fr} forms. The greater the irradiance, the more rapid the interconversion between the two forms. If the response is proportional to the rate of interconversion rather than simply to the concentration of P_{fr}, the magnitude of the response will be irradiance dependent. The synergistic effects for two wavelength mixtures may be due equally to an increased rate of photoconversion.

If all the light and dark processes are included in a mathematical model showing the interdependence of these processes, the equilibrium value for total phytochrome is irradiance dependent. The steady state P_{fr} level is found to be proportional to the reciprocal of the irradiance of the far-red light. A threshold value exists below which the phytochrome no longer decays out of the system, and the rate constants vary only slightly from one biological system to another.

10.2.4 The Action of High Irradiance

For continuous high irradiance, the response magnitude is not simply proportional to the amount of P_{fr} initially present, but rather to both the level of P_{fr} maintained over a period of time and the irradiance value. These processes are irradiance dependent, reciprocity is not obeyed, and it is difficult to demonstrate that they are photoreversible.

The time course for anthocyanin pigment accumulation in mustard seedlings was followed under continuous red (675 mW m^{-2}) or far-red (3.5 W m^{-2}) light as a function of irradiance. The far-red light maintains a nearly constant level of P_{fr} over the 12-hr irradiation, but increasing the irradiance 1000-fold increases the response size 4-fold. Increasing the red irradiance 100-fold increases the response size about 2-fold.

Action spectra for such high irradiance responses do not coincide with the absorption spectra of either P_r or P_{fr}. These action spectra have a maximum near 720 nm. Simultaneous irradiation at 658 and 766 nm is synergistic, but the effectiveness of 717 nm is greatly decreased when mixed with 758 nm.

10.2.5 Nonphotochemical Processes

Phytochrome is involved in four important nonphotochemical processes: (1) *de novo* synthesis of P_r, (2) destruction, preferentially of P_{fr}, (3) nonphotochemical reversion of P_{fr} back to P_r, and (4) a yet unknown reaction or reactions leading to transduction of the primary photoprocess into biological activity and gene expression.

10.3 CHEMICAL STRUCTURE OF PHYTOCHROME

Phytochrome is now understood to be a dimer of two monomers, each bearing a linear tetrapyrrole chromophore joined to the protein moiety via the thioether linkage (Figure 10.3). It is the variable interaction between the P_r and P_{fr} forms of the protein moiety with the two forms of chromophore that gives rise to their different absorption spectra (Figure 10.4).

The overall shape of the tetrapyrrole π electron network remains approximately invariant in the P_r to P_{fr} transition. A possible mechanism is isomerization at the 15,16 carbon-to-carbon double bond (C-15 methine

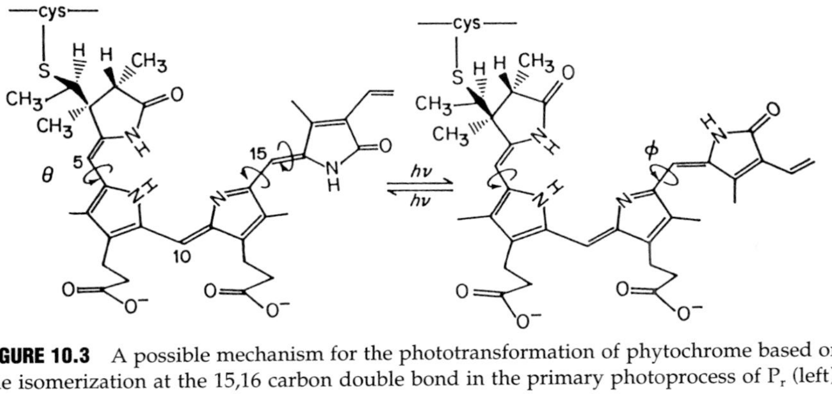

FIGURE 10.3 A possible mechanism for the phototransformation of phytochrome based on the isomerization at the 15,16 carbon double bond in the primary photoprocess of P_r (left). From Song (1988), p. 48.

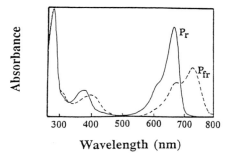

FIGURE 10.4   Absorption spectra of P_r and P_{fr} forms.

bridge) in the primary photoprocess. It is possible that the free rotation of the pyrrole ring D (see Figure 10.5) is accompanied by a conformational change of the protein pocket to allow such movement. According to the proposed model for phototransformation, the P_{fr} chromophore of phytochrome is more exposed than the P_r chromophore.

Major discrepancies arise in attempts to quantitatively assess the accessibility of the chromophore and to predict the secondary structure of phytochrome. Intact phytochrome has been shown to be a dimer. Small-angle X-ray scattering, as well as electron microscope studies, have yielded a model of the phytochrome dimer that entails a distinct separation of the nonchromophore and chromophore domains.

The phytochrome chromophore lies in a hydrophobic pocket and becomes exposed in the P_{fr} form (Figure 10.5). From spectroscopic as well as chemical accessibility studies, it is apparent that the N-terminal sequence interacts with the hydrophobic chromophore and presumably the chromophore environment (hydrophobic pocket). This induces in the N-terminus chain a photoreversible α-helical folding that can be suppressed by monoclonal antibody binding near the amino-terminus sequences. It has been observed that the specific binding of Zn ions preferentially bleaches the P_{fr} form and suppresses the photoreversible conformational change. The Zn accessibility test supports the proposed model for phototransformation of phytochrome, in which the P_{fr} chromophore is exposed to the environment and is accessible to chemical agents, whereas the P_r chromophore lies within a hydrophobic crevice and is therefore less accessible. Crystallographic studies of the zinc–biliverdin complex (structurally very similar to the phytochrome chromophore–zinc complex) demonstrate that the four nitrogen atoms of the tetrapyrrole bind to one zinc ion, resulting in a slightly distorted configuration around the zinc atom.

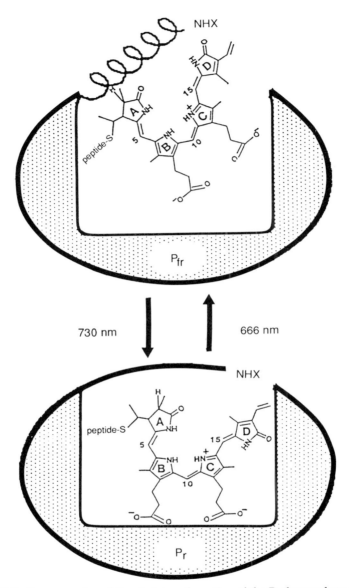

FIGURE 10.5 Representation of the phototransformation of the P_r chromophore to the P_{fr} chromophore. The overall semiextended conformation of the tetrapyrrolic chromophore is retained during the transformation, with conformation of the exocyclic dihedral angle at ring D by a chromophore–apoprotein interaction. The protonation of ring C is based on work by Fodor *et al.* (1988, 1989). The model also incorporates the postulated orientation and the more exposed nature of the P_{fr} chromophore, as well as the increase of the α-helical folding of the N terminus. X = $COCH_3$, as determined by Grimm *et al.* (1988). The postulated slight out-of-plane twisting of ring D for the P_{fr} chromophore is not shown. From Rospendowski *et al.* (1989), *FEBS Lett.* **258**, p. 2. With permission of Elsevier Science Publishers, Amsterdam, The Netherlands.

10.3.1 Subunit Structure of Phytochrome: Domains and Quaternary (Dimer) Structure

The phytochrome sequence may be divided into structural domains according to correlation and structural prediction results. There are 11 domains (A–K).

Loss of spectroscopically crucial chromophore–protein interaction and/or insertion of the divalent cation in the hydrophobic chromophore pocket is apparently responsible for the large change in secondary structure occurring on Zn^{2+} chelation. This suggests that the chromophore stabilizes α-helical structure in the phytochrome. Much more "precise" primary and secondary structures are required for β-sheet formation than for α-helix formation. The possibility exists that the apoprotein possesses a distinct structure rich in β-sheets. It may be that this conformation accommodates insertion of the chromophore. The insertion of the chromophore would then facilitate the concomitant formation of the hydrophobic chromophore pocket present in the native protein. The chromophore and nonchromophore domains are connected by a hinge region.

10.3.2 Surface-Enhanced Resonance Raman Scattering Spectroscopy

Surface enhanced resonance Raman scattering spectroscopy (SERRS) is employed for the study of the P_r and P_{fr} forms of the protein, in the form of citrate-reduced silver colloid suspensions, to preserve the native structure of the two forms and to allow the use of low sample concentrations.

The SERRS spectral differences observed in the high- and low-wavenumber region, between the two forms of phytochrome excited at 406.7 nm, are consistent with those observed at 413.1 nm excitation. The wavenumbers and relative band intensities of modes in the middle- to low-wavenumber region are expected to be more susceptible to the influence of the protein, compared to the high-wavenumber region. Furthermore, the spectra obtained for *the same phytochrome species in the low-frequency region* are substantially different for 406.7 nm as compared to 413.1-nm excitation.

The emergence of new bands and substantial relative intensity differences in the SERRS spectra of a single form of phytochrome, with a change of only 7 nm in the excitation wavelength, reflects the complex nature of the electronic structures of P_r and P_{fr}, i.e., the existence of at least two near-degenerate electronic states. The observation of relative intensity rather than frequency differences between SERRS of the two forms indicates that a subtle, protein-controlled structural variation is responsible for the spectroscopic difference.

It is postulated that following the primary photoisomerization the overall semiextended conformation of the tetrapyrrolic chromophore is conserved. A slight out-of-plane twisting may occur in the case of P_{fr}. This could be responsible for the finely tuned light sensing of the phytochrome. The structural association between the different components may be important in the dimeric model of phytochrome action. In this model, a receptor would be able to distinguish between a P_r–P_{fr} and a P_{fr}–P_r dimer. Hinge region and/or nonchromophore domain movement is necessarily involved in one monomer's effect on the other monomer. Thus, there is some separation of the two major domains by a flexible "hinge." Given this model, it seems reasonable that the chromophore domain may be modified while the dimer region maintains native structure. There are five types of phytochrome (phy A, phy B, phy C, phy D, and phy E) as described by Quail (1991) and Furaya (1993); of these phy A is for etiolated plants and phy B for green plants.

10.4 PROTEIN KINASE ACTIVITY OF PHYTOCHROME

The highly purified phytochrome possesses polycation-dependent protein kinase activity. Reversible phosphorylation is involved in a variety of metabolic regulations in cells. The kinase activity is separable from phytochrome by repeated gel filtrations. This indicates that phytochrome has no intrinsic kinase activity. The kinase of the phytochrome preparation is not specific for intact phytochrome; it degrades 114-kDa phytochrome as well as other proteins when phosphorylated. There are indications that the protein kinase may be a contaminant, although the intrinsic nature of the kinase activity cannot be completely ruled out at this time.

10.5 MUTANTS IN PHYTOCHROME RESEARCH

Phytochrome mutants include photoreceptor mutants, transduction chain mutants, and response mutants. As an example, there are long-hypocotyl *Arabidopsis* mutants that appear to be defective with respect to blue light-absorbing pigment, although response in the red light and far-red spectral region is retained.

In the tomato, during the selection of gibberellin (GA) mutants, a mutant was isolated that requires GA for germination. Spectroscopic and immunological studies have resulted in the *au* mutant in tomato cell cultures, being the best characterized example of a photoreceptor mutant, having a phytochrome content of 5% of that in the wild type. *The au mutant demonstrates that the light-labile phytochrome in which it is deficient is functional in the control of gene expression.*

10.6 COACTION BETWEEN PHYTOCHROME AND BLUE/UV RECEPTORS

Three different sensor pigments occur in higher plants: (1) phytochrome (a photochromic photoreceptor, the photochromic property being expressed in the red-absorbing to far-red-absorbing change; (2) cryptochrome (operating in the blue/UVA spectral range); and (3) the UVB photoreceptor.

A number of case studies have contributed to a unifying model of coaction between these pigments. Phytochrome (P_{fr}) can in some cases act on growth, development, and gene expression without any requirement for blue/UV light [i.e., light-mediated synthesis of anthocyanin in the mustard (*Sinapis alba* L.) seedling cotyledons, where red light can fully replace white light]. For anthocyanin synthesis in the mesocotyl of the milo (*Sorghum vulgare* Pers.) seedling, red and far-red light alone have no effect; dichromatic illumination is required with blue/UV light needed to establish responsiveness toward the effector P_{fr}. Blue and UV light are equally effective and far more effective than red light in causing responsiveness to P_{fr} for induction of plastidic glyceraldehyde-3-phosphate dehydrogenase in the primary leaf of the milo seedling.

Studies have confirmed that sesame as well as mustard seedlings respond very sensitively to unilaterally oriented blue light, although, with regard to straight growth in omnilateral light, both seedlings are totally "blue light blind" at these low fluence rates. In other words, a strong phototropic response can be elicited by unilateral blue light, which has no effect on straight growth if applied omnilaterally. On the other hand, if the same seedlings are kept in red light for some period, their rate of phototropic response toward unilateral blue is much higher than is the case for dark-grown seedlings.

Comparing photomorphogenesis and phototropism, it can be stated that phytochrome (P_{fr}) is the effector proper in bringing about photomorphogenesis in higher plants, whereas cryptochrome and UVB photoreceptor (together with phytochrome) determine the plants' responsiveness toward P_{fr}. Phototropism, on the other hand, can be elicited only by blue/UV light. In this case it is the phytochrome that modulates the rate of the response.

SECTION II *Regulatory Mechanisms of Photoperiods and the Pineal Gland*

Circannual rythms in physiology are particularly apparent in animals that inhabit progressively higher latitudes, where the extremes of daylength, temperature and food availability are especially obvious. The 'choice' of using photoperiodic information as opposed to seasonal temperature swings to regulate the waxing and waning of certain functions is predicated on the fact that daylength is a much more accurate predictor of season than is ambient temperature. Photoperiod-sensitive mammals depend on their retinas to provide the brain with information about seasonally changing daylengths. It is, however, the pineal gland, an end organ of the visual system, which translates the photoperiodic message into a chemical signal which serves as a *messenger* to every organ in the body.

10.7 ANATOMICAL CONNECTION BETWEEN RETINOHYPOTHALAMIC TRACT, SUPRACHIASMATIC NUCLEI, AND PINEALOCYTES

Because the mammalian pineal gland is incapable of direct photoreception, the eyes communicate with the pineal by a neuronal route that includes *the retinohypothalamic tract, the suprachiasmatic nuclei (SCN), and*

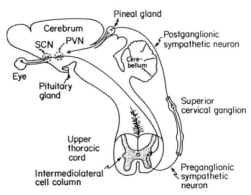

FIGURE 10.6 Neural pathways that connect the eyes to the pineal gland in mammals, including humans. The fibers connecting the eyes to the suprachiasmatic nuclei (SCN) of the hypothalamus comprise the retinohypothalamic tract. Besides fibers in the central nervous system, neurons in the central nervous system and pre- and postganglionic sympathetic fibers also participate in relaying light/dark information from the eyes to the pineal gland. PVN, Paraventricular nuclei. From Reiter (1981), reproduced in Reiter and Richardson (1992), p. 2284, Fig. 1.

the pre- and postganglionic fibers of the peripheral sympathetic nervous system (Figure 10.6). The SCN contain an endogenous circadian clock that generates a rhythm of pineal melatonin synthesis with a periodicity slightly greater than 24 hr. Restricting the cycle to precisely 24 hr is the function of the prevailing light/dark environment. Not only must the pineal signal, encoded in the circadian production of melatonin, provide information about absolute daylength or nightlength, it must also "tell" the responsive organs if the days are getting shorter (autumn) or longer (spring). In this way a given organ can begin its physiological adjustment well in advance of the time the change in function will be required. In many mammals, circannual fluctuations in reproductive potential are inextricably linked to photoperiodic information provided by the circadian changes in circulating melatonin levels. The pineal gland is the site at which light/dark information is translated, or transduced, into a chemical messenger.

10.8 CONTROL OF MELATONIN SYNTHESIS AND SECRETION

The melatonin synthetic and secretory activities are determined by the interaction of the postganglionic sympathetic neurons with the pinealocyte. The release of norepinephrine (NE) is strictly associated with darkness, when the SCN are relieved of the inhibitory signal from the eyes due to the interaction of photons with the retinas. NE release onto the pinealocytes, i.e., the melatonin-producing cells, occurs exclusively during darkness. The postsynaptic transduction mechanisms governing the nighttime production of melatonin following the release of NE involve both β- and α-adrenergic receptors on the pinealocyte membrane. These receptors exhibit 24-hr rhythms in their density; such fluctuations are generally believed to be determined by NE, which, following its release, causes the desensitization or internalization of the receptors. The second messenger regulated by the β and α receptors is cyclic adenosine 3',5'-monophosphate (cAMP), which is required for the rise in nocturnal melatonin production. Levels of a second messenger cGMP, also rise in the rat as a consequence of adrenergic receptor activation; its functional significance, however, remains unknown. The β-adrenergic receptors are linked via a stimulatory guanine-binding protein (Gs protein) to adenylate cyclase, which results in large increases of cAMP. Activation of β-adrenergic receptors alone induces up to a 10-fold increase in cAMP; stimulation of α receptors alone is without effect. When combined stimulation of both β and α receptors by NE occurs, the rat pineal response in terms of cAMP is greatly potentiated, up to 100-fold. Thus, β receptor activation is a requirement for cAMP accumulation, with α receptor stimulation amplifying the response. The marked interactive effect of β-

and α-adrenergic stimulation of the second messenger cAMP does not carry over into a similar large augmentation of the amount of melatonin formed; augmentation is only 15%.

The rapidity with which these processes are accomplished varies greatly among species. In the pineal gland of the Syrian hamster, the nocturnal rise in cAMP and in the synthesis of the mRNA for the rate-limiting enzyme in melatonin production, N-acetyltransferase (NAT), seem to precede, by several hours, the maximal expression of NAT activity and the rise in intracellular melatonin. The rather sluggish intracellular mechanisms in the hamster contrast with those in other species, where NE stimulation leads to rapid increases in pineal cAMP closely followed by NAT mRNA transcription and induction of NAT activity.

In the rat pineal gland the activity of NAT increases 50- to 100-fold each night; in other species nocturnal NAT activity only doubles. The magnitude of the NAT rise seems to have little to do with the actual melatonin concentrations in the blood, in that all species so far examined have comparable nighttime blood melatonin levels.

10.9 PATHWAY OF MELATONIN SYNTHESIS

The pathway of melatonin synthesis (Figures 10.7 and 10.8), starts with the amino acid tryptophan, via 5-hydroxytryptamine (5HT, or serotonin); 5HT, besides being converted to melatonin, is also metabolized via another pathway to the metabolite 5-hydroxyindole acetic acid (5HIAA). This compound serves as a precursor for other substances produced by the pineal gland; their function remains unknown. The product of the N-acetylation of serotonin by NAT within the pinealocyte is N-acetylserotonin (NAS). Once formed, NAS is quickly converted to melatonin by the action of the cytosolic enzyme hydroxyindole-O-methyltransferase (HIOMT). Monoclonal antibodies to bovine HIOMT, which shows a high degree of structural similarity with HIOMT from other species, have allowed the immunocytochemical localization of the enzyme. The melatonin metabolites are 6-hydroxymelatonin conjugates formed in the blood and N-acetyl-5-methoxy-kynurenamine formed in the SCN. The 6-hydroxymelatonin conjugates are excreted in the urine.

Three patterns of nocturnal melatonin formation have been described: (1) a discrete peak in the late dark phase, (2) a peak near the mid-dark phase, and (3) a prolonged peak during the majority of the dark phase. As the duration of the daily dark period increases, the nocturnal duration of elevated melatonin is proportionally increased.

Of considerable interest is the observation that when animals are acutely exposed to light at night, melatonin levels are markedly inhibited

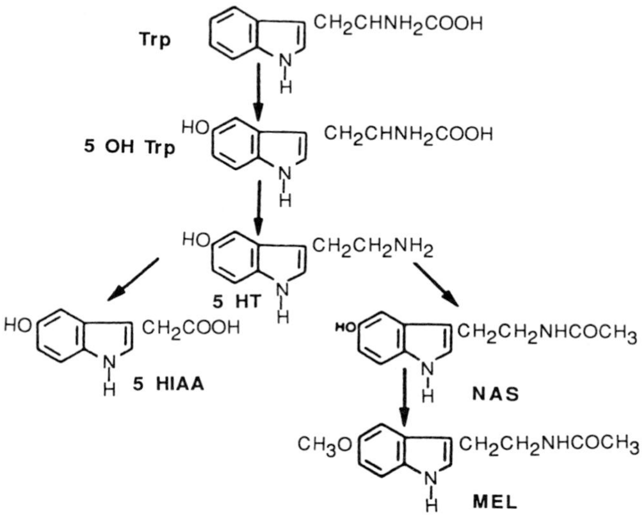

FIGURE 10.7 Pathway for the production of melatonin (MEL), an important pineal hormone, from the amino acid tryptophan (Trp) via serotonin (5HT). The conversion of 5HT to melatonin involves two enzymes, N-acetyltransferse, which converts 5HT to N-acetylserotonin (NAS) and hydroxyindole-O-methyltransferase, which converts NAS to melatonin. The activity of N-acetyltransferase increases at night in the pineal gland and determines the amount of melatonin produced; as a result, pineal and blood levels of melatonin also increase at night. Besides being metabolized to melatonin, 5HT can be converted to 5-hydroxyindole acetic acid (5HIAA) by the enzyme monoamine oxidase. HIAA serves as a precursor for other substances produced in the pineal gland, whose function in terms of organismal physiology remains unidentified. The intermediate factor in the conversion of Trp to 5HT is 5-hydroxytryptophan (5OHTrp). From Reiter and Richardson (1992), p. 2284.

even though the duration of light is brief (1 sec to 1 min). The brightness of light required either to prevent or suppress pineal melatonin synthesis and secretion during the night is highly species specific. The Syrian hamster pineal gland is very sensitive to light inhibition (e.g., irradiance of 0.1 μW/cm^2); the rat pineal gland is sensitive to even lower light intensities. Both are nocturnally active species. In diurnal rodents (e.g., ground squirrel, chipmunk) the pineal gland seems somewhat less sensitive to inhibition by light. The pineal gland of the human, in terms of sensitivity to light, falls between the extremes of nocturnal and diurnal animals. Excessive light intensity is "toxic" to the gland, leading to a total cessation of melatonin production. The differential responses between species may not in fact relate directly to the pineal gland, but may be due to varying sensitivities of other neurons to light, for example, the SCN,

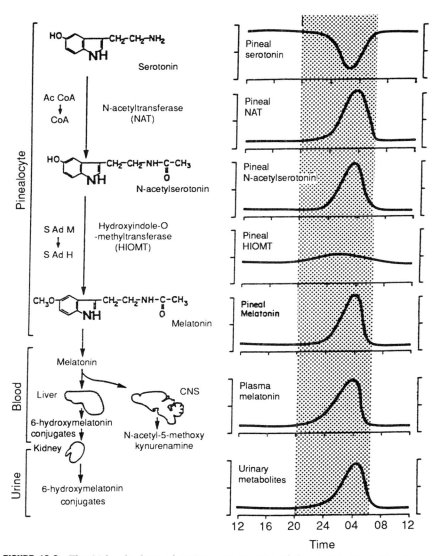

FIGURE 10.8 The 24-hr rhythms of various constituents of the pineal gland, blood, and urine. The shaded area represents the daily period of darkness. When the activity of the rate-limiting enzyme (NAT) in melatonin synthesis is elevated, it is typical for pineal and blood levels of the hormone also to be high. From Reiter (1992), p. 231. Ac CoA, acetylcoenzyme A; S Ad M, S-adenosyl-methionine; CNS, central nervous system. With permission of Academic Press, Orlando, Florida.

which are key structures in relaying photoperiodic information from the eye to the pineal gland.

Earlier studies using rats suggested that pineal HIOMT activity was most sensitive to green wavelengths of light. A subsequent study using hamsters suggested that blue wavelengths were primarily responsible for light-induced suppression of indolamine metabolism in the gland. In humans, both blue and green wavelengths have been implicated in the control of the amplitude and the phasing of the circadian production of melatonin.

One of the most frequently reported observations after the exposure of animals to extremely low-frequency (ELF) electric and magnetic fields includes alteration of the circadian melatonin rhythm. All life forms are exposed to the natural geomagnetic field (field strength approximately 50 μtesla, depending on latitude) as well as to the natural geoelectric field (field strength of about 100 V/m on the Earth's surface during clear weather conditions and approaching several thousand volts per meter under the clouds of a thunderstorm). Superimposed on these relatively static fields there is also a natural background of electric and magnetic "noise" at extremely low and ultralow frequency, as a consequence of worldwide thunderstorms. Besides these natural electric and magnetic fields, there are human-introduced electric and magnetic fields in the ELF range (50–60 Hz). Magnetic fields of 60 Hz in the immediate vicinity of household appliances can reach field strengths of 10^{-5} tesla, and man-made magnetic field strengths can reach several tenths of milliteslas in occupational environments. The electric field strength under high-tension transmission lines may be 7 kV/m at ground level. Power frequency electric fields inside modern homes are in the range of 10–64 V/m. A severe attenuation of the melatonin circadian rhythm has been observed in rats exposed to ELF electric fields in the range of 2–40 kV/m for 3 weeks. The field strength threshold for this response was estimated to be between 0.2 and 2 kV/m. It was also found that the induced electrical currents that accompany the rapid pulsing of the magnetic field depressed pineal melatonin production.

It seems that the inhibition of pineal melatonin synthesis by magnetic field perturbations involves complex mechanisms, with parameters such as eddy currents, field strength, duration of exposure, and time interval between darkness onset and exposure onset. During the night in the absence of light, the SCN of the hypothalamus become neurally active in reference to the pineal gland; as a result, action potentials in the postganglionic sympathetic fibers innervating the pineal gland release NE. If visible light is introduced at night, the retina transmits this information to the SCN via the retinohypothalamic tracts; this stimulus inhibits the electrical activity of the SCN in reference to the pineal gland and melatonin synthesis drops. It seems, however, that the electromagnetic field effects on the pineal gland are not mediated through the retina. Even when organ-cultured pineal glands are stimulated to produce melatonin by use of the β-adrenergic receptor agonist, isoproterenol, the melatonin response is attenuated in the presence of pulsed static magnetic fields. The field exposure may accomplish this by causing the internalization of the β-adrenergic receptors. Despite such *in vitro* observations on the apparent direct sensitivity of the rat pineal gland to magnetic fields, the possibility that the retina is involved in *in vivo* situations cannot be excluded. Magnetic field fluctuations can induce visual sensations of light referred to as *magnetophosphenes*, which may, via the usual retina/ pineal connections, inhibit melatonin formation. Yet the strengths of the magnetic fields used in pineal studies are well below those required to produce *retinal phosphenes*. Even less known is how ELF electric fields alter pineal indolamine metabolism. These produce *electrophosphenes*, but at much higher field strengths than those used in the studies wherein the suppression of pineal melatonin synthesis was seen. Rather than an effect on the retina, ELF fields could change the firing rate of postganglionic sympathetic neurons that enter the pineal gland.

The UVA wavelengths (320–400 nm), which reportedly do not penetrate to the level of retinal photoreceptors, suppress the ability of the pineal gland to synthesize its hormonal product. UVA light presented during the dark period changes the physiology of animals so treated in a way suggesting an interruption of circadian melatonin production. In Syrian hamsters, monochromatic UVA light at 360 nm suppresses pineal melatonin levels at night in a dose-related manner. The mechanism is unknown. The Harderian glands (porphyrin-containing organs in the orbital cavity of some animals; these organs may absorb UVA energy and reemit red wavelengths) are not related to the UVA suppression of melatonin production.

SECTION III *Photomovement*

10.10 STIMULUS PERCEPTION AND TRANSDUCTION

Photosensory transduction, known as *photomovement*, refers to any light-induced motility or behavioral response involving the spatial displacement of all or part of an organism. The various examples of photomovement have an adaptive advantage for the organism such that the organism is optimally positioned in light or darkness.

10.11 FUNDAMENTAL PHOTOMOVEMENT TERMS

Any light-induced change in an organism's motor activity, resulting in an alteration of the movement or orientation of the organism, is called *photoresponse.* The three major classes of photoresponses of motile organisms are photokinesis, photophobic responses, and phototaxis. Each of these may lead to exactly the same photomovement, because the term refers not to the final response but rather to the mechanism for achieving that response.

A photomovement that is restricted to a cellular organelle is termed a photodinesis (e.g., an acceleration of cytoplasmic streaming or induced movement of resting organelles in response to light stimuli).

The orientation to light of a nonmotile organism is termed *phototropism* (e.g., growth-mediated directional response of a sessile organism or organ to an asymmetric external light stimulus, such as growth of canary grass seedlings toward a lighted window).

The bending of an organ relative to the electric vector of a polarized light stimulus is *polarotropism,* and is usually excluded from the definition of phototropism in the strict sense.

10.11.1 Photokinesis

Photokinesis defines a change in the velocity of an organism depending on light. The steady-state rate of activity of the organism is affected by the steady-state of light stimulus. If the activity parameter measured is swimming velocity, photokinesis may be called *photoorthokinesis.* If the activity parameter represents the frequency of directional change, photokinesis may be referred to as *photoklinokinesis.* A positive photokinesis refers to an activity rate that is higher in the presence of the stimulus light than in its absence.

10.11.2 Photophobic Responses

Photophobic responses refer to a transient alteration in activity, e.g., directional change or swimming velocity of an organism elicited by a change in the stimulus light intensity. The photophobic response most commonly observed is a stop response, typically followed by a change in movement direction.

A photophobic response may be due either to a step-up stimulus (a sudden increase in light intensity) or a step-down stimulus (a sudden decrease in light intensity).

10.11.3 Phototaxis

Phototaxis is an oriented movement of an organism toward or away from the source of the light stimulus, referred to as positive or negative phototaxis, respectively. The direction of the taxis often depends on the stimulus intensity, with many organisms showing positive phototaxis at low light intensity and negative phototaxis at high light intensity.

Transverse taxis or *diaphotaxis* is when the organism moves perpendicular to the axis of the light stimulus (occurs at an intermediate light intensity).

Photoaccumulation or *photodispersal* is the end result of photokinesis, photophobic response, or phototaxis, corresponding to an organism's accumulating in or dispersing from a region of higher fluence, respectively. Photodispersal is a behavioral consequence of positive photokinesis, a step-up photophobic response, or negative phototaxis. Photoaccumulation is a behavioral consequence of negative photokinesis, step-down photophobic response, or positive phototaxis.

10.12 METHODS FOR THE STUDY OF PHOTOMOVEMENT

10.12.1 Individual Cell Methods

Actinic light (i.e., light of a proper wavelength and fluence rate to produce a response) is provided from the side of a culture chamber, and the directional movement of the cells is then photographed or video-recorded using a nonactinic monitoring beam. To quantitate the phototaxis of a population of cells or organisms, one can measure the direction of movement of each organism with respect to direction of the stimulus light. The mean direction of a population is between 0 and 360°, and directedness ranges from 0.0 for a completely random distribution to 1.0 for a distribution in which every individual is oriented in exactly the same direction.

Stentor coeruleus shows negative phototaxis. To establish the phototactic response, it is necessary to distinguish between a true phototactic orientation and a photophobic response of the organism. A *Stentor* cell might exhibit movement away from a light source, with direction maintained through a series of step-up photophobic responses (movement in the "wrong" direction would elicit a step-up photophobic response, permitting a course correction). On the other hand, a *Stentor* cell with a capacity for phototaxis will have a mechanism for measuring the direction of the light, and maintaining its course relative to that light direction. Thus, to distinguish between photophobic and phototactic responses, one must be able to discriminate between *course correction* in case of "wrong" direction movement and *course maintenance,* through a mechanism for measuring the direction of light.

A focused light beam can be used for such a purpose, *using a convex lens such that the focal point is within the culture chamber, with the light provided from the side of the chamber.* In the absence of an absorbing medium, the fluence rate is highest at the focal point of the lens. A cell possessing only a step-up photophobic response, and lacking a negative phototactic response, cannot actually measure the direction of light, but a cell with negative phototactic response can do so, and can move through the high fluence focal point further away from the side of the chamber through which the actinic light is entering (the *Stentor* cells can do this).

Chlamydomonas moving toward one light source can be subjected to a second light source of the *same intensity* applied from a direction normal to the direction of the first source at the time the first source is turned off. When the first source is turned off, and the second is on, if the cell changes its direction smoothly toward the second light source, then the response is a true positive phototaxis (because direction has been sensed, while there has been no change in fluence).

10.13 PHOTOSENSORY TRANSDUCTION MECHANISMS

The steps of photosensory transduction are photon stimulus, perception (excited photoreceptor), signal generation, signal processing, amplification and motile response. The generation of signal usually begins by an ultrafast primary reaction of the excited photoreceptor pigment. This reaction must be fast enough to compete effectively with the other relaxation processes of the excited state of the pigment molecules. The primary reactions are more likely to originate from the singlet state S_1 than the triplet state T_1 (efficiency is limited by the slow rate of intersystem crossing). Bimolecular processes, however, require the longer lifetime of the triplet state.

10.14 PHOTOMOVEMENT OF WHOLE CELLS

Because of various errors in the action spectrum measurements and optical biases *in vivo* (light scattering, reflection, refraction, screening pigments, polarization), the action spectrum usually does not quantitatively match the absorption spectrum of the photoreceptor pigment *in vitro.* Nevertheless, action spectroscopy permits pinpointing a few or, optimally, one candidate for the physiological photoreceptor pigment.

10.14.1 Cyanobacteria

In *Phormidium uncinatum,* the action spectrum for photokinesis exhibits maxima in the red and blue regions, and it is very similar to the absorption spectrum of Chl *a.* However, the photokinesis action spectrum shows relatively high efficiency beyond 700 nm, whereas Chl *a* absorption drops. There are three possible explanations for this discrepancy: (1) the efficiency of far-red wavelength results are exaggerated, because of the screening effect of other bulk pigments below 700 nm; (2) the chlorophyll involved in photokinesis may be of a special type; and (3) the receptor pigment may be other than chlorophyll.

For the photophobic response of *P. uncinatum,* the primary photoreceptor could be photosystems I and II, with Chl *b,* phycocyanobilin, and phycoerythrobilin contributing to the action spectrum in the 500- to 650-nm region.

The phototaxis action spectrum of *P. uncinatum,* with maxima in the near-UV, blue, and green wavelengths, extends well beyond 550 nm, where flavins do not absorb. It is possible that phototaxis in *P. uncinatum* uses a pigment complex (primary photoreceptor pigment and antenna pigments) and if the concentration of photoreceptor is negligibly low compared to that of the antenna pigments, its absorbance would contribute little to the action spectrum. Chl *a* could be such a primary photoreceptor.

In the cyanobacterium *Anabaena,* positive phototaxis is probably mediated by C-phycocyanin, and negative phototaxis by chlorophyll *a.*

10.14.1.1 The Primary Photoreaction

In cyanobacteria the primary photoreaction is most likely to be photoionization and charge separation in the reaction center, which leads to photophosphorylation of ADP. The ATP formed accelerates the motile activity of cyanobacteria in the photokinetic response. For the phototactic and photophobic responses, the action spectra are quite different.

In the absence of light stimulus, the step-down photophobic response of *Phormidium* can be simulated by an abrupt pH change. The light-induced membrane potential changes are similar to the action spectrum of the photophobic responses.

Cyanobacteria move using a gliding mechanism. The contractile microfilaments presumably undergo a significant conformational transformation on binding Ca^{2+} ions.

10.14.2 *Halobacterium halobium*

10.14.2.1 Photoresponses

The wavelength-dependent photomovement of *Halobacterium halobium* has been called *color sensing*. A step-up stimulus in the blue region results in photodispersal, and in the yellow–green regions, in photoaccumulation.

10.14.2.2 Receptor Pigments for Photomovement

The membranes of *H. halobium* are purple, and rhodopsin-like pigments are present, two of which probably mediate the two opposing photoresponses (i.e., photodispersal and photoaccumulation). In addition to bacteriorhodopsin and halorhodopsin, which are involved in proton and chloride ion transport, a *slow-cycling rhodopsin* (sR_{587}) is probably the photoreceptor pigment for the attractant response of the bacterium. The blue/UV photoreceptor pigment for the repellent response is S_{373}, a photocycling intermediate of sR_{587}. Less excitation of sR_{587} will mean less production of S373. sR_{587} and S_{373} can be considered as the photoreceptor pigments of *H. halobium*.

Through a *one-photon* cycle, attractant signal transduction results in (1) photon strike, (2) a configurational isomerization of the retinylic chromophore (sR_{587} converts to S_{600}, which within 20 μsec converts to S_{373}, which within 800 msec converts to sR_{587}), (3) a conformational change in the sR protein embedded in the membrane, and (4) modulation of an ionic flux as a processible form of the signal.

Through a two-photon cycle, (1) a photon strikes sR_{587}, a repellent signal transduction results; (2) a second photon strikes S_{373} (a conversion product of S_{600}, itself originating from sR_{587}); (3) a new photoelectron leads to conversion of S_{373} to S_{b510}; (4) the signal generated from the photoreaction of S_{373} may lead to a conformational change in the sR protein, which could modulate an ionic flux; and (5) S_{b510} is converted to sR_{587} within 80 msec.

Thus the same sR protein has a dual role, mediating both the attractant (one-photon cycle) and repellent (two-photon cycle) photoresponse.

The flagella motors in prokaryotes (as in *Halobacterium*) appear driven by a proton-motive force (PMF). The signal generated by the photoreceptors possibly results in a change in PMF amplified through Ca^{2+} ion flux. This process ultimately triggers a change in the direction of the flagella, from clockwise to counterclockwise or vice versa.

The photoreceptor pigment in its excited or metastable state may interact with a methyl-accepting protein (Map), which acts as the site of signal integration. Attractant light causes methylation and repellent light causes demethylation of this protein. It is suggested that MaP lowers the level of a hypothetical regulator, which serves as the flagellar rotor switch.

Bacteria employ one or more flagella as the motor apparatus. Each flagellum is driven by a rotary motor called the M ring. PMFs are the driving forces for the rotary motor. Protons flow through the interstice between the M ring and the S ring (a stator), which powers the rotary motion of the motor apparatus. The photoresponse is either a suppression or an induction of a reversal in the flagellar rotary motor. It is not known whether Ca^{2+} alone or Ca^{2+} plus protons are directly involved in the possible conformational changes of the proteins of the rotary motor of *H. halobium.*

10.14.3 *Chlamydomonas*

10.14.3.1 Photoresponses

Chlamydomonas rheinhardtii, a green flagellate, exhibits step-up photophobic responses and a negative phototactic response. These cells disperse from lighted areas.

10.14.3.2 Receptor Pigments for Photomovement

Chlamydomonas shows a maximum positive phototactic response to light of 500–510 nm and 440 nm (action spectrum for phototaxis). The differences between the flavin spectrum and the action spectrum of *Chlamydomonas* are considerable. The photoreceptor pigment may be a flavin other than FMN and FAD, which show a broad spectrum around 450 nm, whereas deazaflavins and hydroxyflavins show absorption maxima on both sides of 450 nm. Although flavins are not ruled out, the current thesis strongly favors rhodopsin in *Chlamydomonas* (Foster *et al.,* 1983; Hagemann, 1992, 1993).

A quarter-wavelength reflector may radically alter the light absorbed by the photoreceptor pigment such that the action spectrum differs dramatically from the absorption spectrum. A quarter-wavelength reflector is any structure composed of layers of alternating refractive indices with thickness appproximately one-quarter of the incident light wavelength. Light reflected from such a structure is constructive (reinforced light intensity) for a particular wavelength, and destructive (decreased light intensity) for other wavelengths.

The function of the eye spot (stigma) of *Chlamydomonas* has been defined according to two models: (1) In the *shading* model, as the cell rotates along its axis, with periodic shading of the photoreceptor by the stigma acting as an optically dense body, the cell effectively measures light direction through two measurements with the same photoreceptor pigment. (2) In the *quarter-wavelength* model, the stigma serves as a directional antenna, which is rotated somewhat like a radar antenna as the cell rotates while swimming. In this case *the granular layers of the stigma can provide alternating layers of different refractive index media, with the thickness or the spacing between the layers in the 120-nm range (quarter-wavelength interval for 480 nm,* the maximum in the *Chlamydomonas* action spectrum). With the photoreceptor pigment located in the plasma membrane in front of the granular rows of the stigma, the stacked layers of stigma granules can in fact act as a reflector. Thus, the photoreceptor apparatus can be excited both by the incident and the reflected light beams; with the direction of the latter dependent on the direction former, the mechanism would be highly directional.

In the shading model, the action spectrum must be a function of absorption by the photoreceptor pigment and the shading pigment. In the quarter-wavelength model, the action spectrum must be a function of absorption by the photoreceptor, reflectance of the sigma granular layers, and absorption by those layers.

The phototaxis action spectrum of *Chlamydomonas* resembles the action spectrum of carotenoid more closely than that of a flavin. A nontypical flavin, or a flavin chromophore with substantial screening, however can account for the spectrum recorded. The current view favors the rhodopsin theory in connection with findings in *carotenoidless chlamydomonas* mutants.

The photodispersal movement can be restored in a *carotenoidless* mutant by exogenously added retinal isomers. The best evidence at this time indicates that a retinal is the photoreceptor pigment for phototaxis of *Chlamydomonas*.

The present scheme for the primary photoreaction of *Chlamydomonas* is (1) photon strike and signal perception by the photoreceptor, (2) primary reaction (conformation change?), (3) signal generation, (4) depolarization of the membrane, (5) signal amplification (Ca^{2+} influx), and (6) conformational change in the axoneme, resulting in flagellar reversal.

10.14.4 *Stentor coeruleus*

The protozoan *Stentor coeruleus* possesses numerous cilia around the membranelles of the oral (or adoral zone). Locomotion consists of a successive series of propulsive stroke–recovery motions with a characteristic beat frequency and amplitude. The forward swimming motion involves a clockwise rotation of the *Stentor* cell body. Reversal of the ciliary beating, while pointing the membranelles forward, causes the organism to stop or swim backward, in conjunction with a coordinated motion of the body cilia.

This organism exhibits a step-up photophobic response and a negative phototactic response.

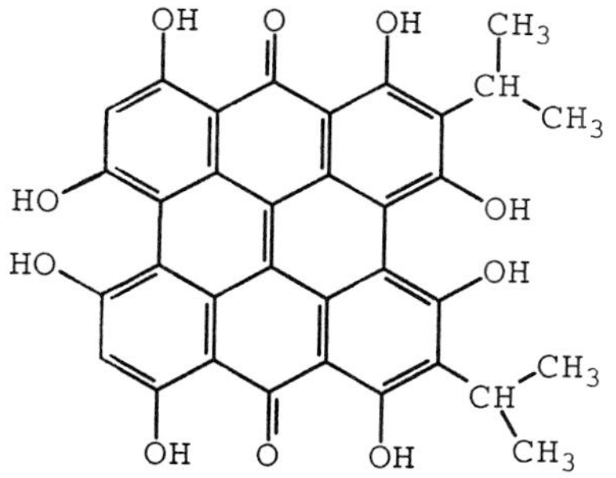

FIGURE 10.9 The correct form of stentorin, with the isopropyl groups on the right.

10.14.4.1 Receptor Pigments

The action spectra for both responses match the absorption spectrum of the pigment stentorin, a hypericin-like chromophore, located in pigment granules, presumably in the ectoplasm and cell membrane. Photosensory-responsive *Stentor* species protect themselves from photodynamic damage by their own endogenous sensitizer, stentorin. From the two structures proposed in Tao *et al.* (1993) the one with the isopropyl groups on the right is the correct one (see Figure 10.9).

The stentorin photoreceptor pigments are either embedded in the membrane or entrapped within the pigment granule as a distinct vesicle. The peculiar property of blue–green. *S. coeruleus* exhibiting photosensory response and of fluorescent *Stentor* spp. exhibiting photodynamic responses can be explained in terms of the function and topography of the photoreceptor pigment.

10.14.4.2 Primary Photoreaction and Stimulus–Response Coupling

The primary photoreaction comprises photon strike and signal perception; photoresponse occurs only at $pH_{out} > 6$, which means that proton release from excited pigment granules must be sufficient to generate an internal pH (pH_{in}) < 6. Further, there is pH activation of depolarization of calcium-sensitive Ca^{2+} channels and Ca^{2+} influx and signal amplification; light-induced influx of Ca^{2+} triggers a conformational change by the axoneme, resulting in a reversal of ciliary stroke.

Triphenylmethylphosphonium (TPMP), which is active on the proton-motive force, the protonophores carbonylcyanide-*p*-trifluoromethoxy-phenylhydrazine (FCCP) and carbonylcyanine-*m*-chlorophenylhydrazone (CCCP), the calcium ionophore calimycin, and ruthenium red all produce a strong inhibitory action on the photophobic response of *Sten-*

tor, but a weak action on organism motility. Other calcium blockers [methoxyverapamil, verapamil, lanthanum, and poly(L)lysine] specifically inhibit both the step-up photophobic and phototactic responses in *Stentor.*

D_2O substantially sensitizes the cell to a step-up photophobic response. The enhanced response in heavy water may be due to the higher steady state concentration of singlet oxygen in D_2O compared to H_2O (see Chapter 4, Section 4.4.2). Heavy water enhances the step-up photophobic response and inhibits the negative phototactic response. Ions of lanthanum, which are competitive with calcium, strongly inhibit the photophobic response but do not affect the phototactic response. Inhibition of the phototactic response by lanthanum is partially restored by heavy water.

10.14.5 *Blepharisma*

The pigment *blepharismin* in the red-colored ciliate *Blepharisma* is apparently the photoreceptor for the photodynamic sensitization of the organism. It exhibits many of the spectroscopic and chromatographic properties of stentorin. Its proposed chemical structure differs from that of hypericin only by the positions of two hydroxyl groups. This strongly red fluorescent pigment does not appear to be tightly bound to a protein or membrane. There are two forms of blepharismin: the red pigment with λ_{max} 572, 535, and 445 nm and the blue pigment with λ_{max} 592 and 548 nm.

The action spectrum of the step-up photophobic response in *Blepharisma japonicum* shows a maximum at 400 nm and no response to red light, which is active for photodynamic response. Thus, the photoreceptor for photophobic response may not be the same as the receptor for photodynamic action. According to Gualtieri (1989), however, Euglena's photoreceptor is rhodopsin.

10.14.6 *Euglena*

The photoreceptors of the flagellate *Euglena* are usually composed of primary (paraflagellar swelling) and secondary (stigma) photopigments.

The action spectrum of *Euglena* for the step-up phobic response shows maxima at 365, 412, 450, and 480 nm. The action spectrum for the step-down phobic response is structureless, with maxima at 375 and 480 nm (similar to the action spectrum for positive phototaxis).

The action spectrum of *Euglena gracilis* for photokinesis resembles the action spectrum of photosynthetic antenna pigments, including chlorophyll *b*. It is apparent that the photoreceptor for *Euglena* photokinesis is a photosynthetic pigment, rather than a flavin. The stigma granules of *Euglena* mainly contain carotenoids. If only a small fraction of the photosynthetic pigment (chlorophyll) is specifically located in primary photoreceptor organelles, the action spectrum of the system will match the action spectrum of carotenoids.

The dichroic arrangement of the photoreceptor molecules is also suggested by the differential photoresponses of E. gracilis *to blue light polarized parallel and perpendicular to the long axis of the cell.* Near-UV light polarized parallel to the long axis of the cell is more effective. This suggests that the transition dipoles involved in the blue and near-UV absorption bands of the photoreceptor are essentially orthogonal and that the chromophores are in bound form. Using a holographic method, the paraflagellar body (PFB) was found to have a rodlike lamellar structure, with each rod arranged in a helical strand.

The action dichroism of *E. gracilis* is difficult to accommodate in terms of flavin alone as the photoreceptor chromophore. It is more likely that dichroically oriented carotenoid pigments in the stigma contribute to the action dichroism, because the angle between the two transition dipoles is almost 90°. The flavinic nature of the photoreceptor for the phototactic response of *Euglena* has been reasonably well established on the basis of action spectra and other spectroscopic evidence.

Euglena gracilis shows also a positive phototactic response at a very low light intensity, indicating a highly efficient primary photoprocess and a large amplification in the transduction reaction chain. The fluorescence quantum yield of flavin in the PFB is 0.01 and its lifetime 0.19 nsec. The nature of the fast primary photoprocess *is unlikely to involve the triplet state of flavin*, because intersystem crossing in flavins is at least an order of magnitude slower than the decay rate estimated from luminescence quantum yield and lifetime.

One possible mechanism involves photoreduction of a membrane-bound flavin, followed by reoxidation of the reduced flavin by a *b*-type cytochrome [*light-induced absorbance change (LIAC)*]. The hypothetical model would be (1) vectorial proton release, (2) activation of ATPase, and (3) opening/release of Ca^{2+} channels and bound Ca^{2+}.

10.15 PHOTODINESIS

Photodinesis is the photomovement of cell organelles.

10.15.1 Photoreceptor Pigments for Photodinesis

Some algae, such as *Mougeotia*, exhibit a maximum sensitivity of chloroplast movement in the blue region at higher fluence rates, in the red region at low fluence rates. The far-red/red reversibility of chloroplast movement in response to polarized light (polarotropism) suggests that the photoreceptor pigment is a phytochrome.

10.15.2 Polarotropism of *Mougeotia* Chloroplasts

Each cell of a trichome of the filamentous alga *Mougeotia* contains one flat chloroplast. The rectangular plate-shaped chloroplast turns its flat face toward relatively low-intensity actinic red light of 660 nm within minutes of irradiation ("face presentation"). At high fluences the chloroplast rotates such that the edge faces the actinic beam ("profile orientation"). In darkness the *Mougeotia* chloroplasts remain in the orientation established by the previous exposure to light. *It is assumed that phytochrome serves as the red light receptor.* There is a distinct possibility that phytochrome in *Mougeotia* is membrane or organelle bound, based on the fact that chloroplast rotation is polarotropically controlled.

In response to red light polarized along the short axis (i.e., perpendicular to the long axis) of the chloroplast, as P_{fr} builds up at the front and back of the cell, this organelle rotates 90°, with its edges turning toward the low P_{fr} concentration (its edges now facing the actinic beam). Its transition moment perpendicular to the chloroplast membrane, P_{fr}, can only be reversed to P_r by far-red light polarized parallel to the long axis of the cell (perpendicularly polarized light is no longer effective).

There are several mechanisms by which the inversion of dichroism may occur, assuming that the transition dipole of P_r lies parallel to the cell surface and the transition dipole of P_{fr} is perpendicular:

1. The direction of the transition dipole changes by 90° on transformation from P_r to P_{fr}.
2. The whole protein rotates by about 90° on phototransformation, with the chromophore transformation dipoles of P_r and P_{fr} remaining fixed with respect to the chromophore binding site/crevice.
3. The chromophore moves and reorients relative to the binding site, with the protein moiety fixed.
4. Extensive conformational change of the protein moiety occurs.

In the absence of a gross conformational/geometric isomerization mechanism, mechanism 3, with rotation of the chomophore, seems the most plausible, although mechanism 2 is also possible.

Under certain conditions, e.g., at high fluence, blue light also elicits chloroplast movement in *Mougeotia*. Ca^{2+} influx can be mediated by red and blue light by both *Mougeotia* and *E. gracilis*. Interestingly, phytochrome and flavin are both capable of releasing protons upon light reception. It is possible that local acidification in the plasmalemma depolarizes the membranes and opens a Ca^{2+} channel in the *Mougeotia* cell. Alternatively, acidification may cause a shift in the binding equilibrium of receptor-bound Ca^{2+} ions in an internal sequestrating vesicle.

The blue light receptor acts in large part independently of phytochrome in the choroplast movement of *Mougeotia*. However, under conditions such as weak actinic light intensity, the two photoreceptors interact at least indirectly, as can be seen from the differential effects of blue and red light on the membrane viscosity. Interestingly, "endogenous" flavins bind to phytochrome and energy transfer from the bound flavin to the P_r chromophore occurs preferentially (but not to the P_{fr} chromophore), resulting in the blue light-induced $P_r \rightarrow P_{fr}$ phototransformation.

Bibliography

Ahmad, M., and Cashmore, A. R. (1993). HY4 gene of *A. thaliana* encodes a protein with characteristics of a blue-light photoreceptor. *Nature* **366**, 162–166.

Batschelet, E. (1965). "Statistical Methods for Analysis of Problems in Animal Orientation and Navigation." Am. Inst. Biol. Sci., Washington, DC.

Batschelet, E. (1972). Statistical methods in the analysis of problems in animal orientation and certain biological rhythm. *In* "Symposium on Animal Orientation and Navigation" (S. R. Galles, K. Schmidt-Koenig, G. J. Jacob, and R. F. Belleville, eds.), pp. 61–91. NASA, Washington, DC.

Borthwick, H. A. (1972). The biological significance of phytochrome. *In* "Phytochrome" (K. Mitrakos and W. Shropshire, Jr., eds.), pp. 3–23. Academic Press, New York, London.

Butler, W. (1980). Remembrances of phytochrome twenty years ago. *In* "Photoreceptors and Plant Development" (J. A. DeGreef, ed.), pp. 3–7. Antwerpen Univ. Press, Belgium.

Flint, L. H., and McAllister, E. D. (1935). Wavelengths of radiation in the visible spectrum inhibiting the germination of light-sensitive lettuce seed. *Smithson. Misc. Collect.* **94**, 1–11.

Fodor, S. P. A., Lagarias, J. C., and Mathies, R. A. (1988). Resonance Raman spectra of the Pr. form of phytochrome. *Photochem. Photobiol.* **48**, 129–136.

Fodor, S. P. A., Lagarias, J. C., and Mathies, R. A. (1989). Studies of the chromophores in the P_r and P_{fr} forms of phytochrome from resonance Raman spectrum. *Photochem. Photobiol.* **49s**, 63s.

Grimm, R., Kellermann, J., Schafter, W., and Rudiger, W. (1988). The amino-terminal structure of oat phytochrome. *FEBS Lett.* **234**, 497–499.

Gualtieri, P., Barsanti, L., and Passarelli, V. (1989). Absorption spectrum of a single isolated paraflagellar swelling of *Euglena gracilis*. *Biochim. Biophys. Acta* **993**, 293–296.

Gualtieri, P. (1991). Microspectroscopy of photoreceptor pigments in flagellated algae. *Crit. Rev. Plant Sci.* **9**, 475–495.

Gualtieri, P., Pelosi, P., Passarelli, V., and Barsanti, L. (1992). Identification of a rhodopsin photoreceptor in *Euglena gracilis*. *Biochim. Biophys. Acta* **1117**, 55–59.

Gualtieri, P. (1993). A biological point of view on photoreception (no-imaging vision) in algae. *J. Photochem. Photobiol.* (*B*) *Biol.* **18**, 95–100.

Hendricks, S. B. (1964). Photochemical aspects of plant periodicty. *Photophysiology* **1**, 305–331.

Henry, A. S., and Nielsen, P. J. (1980). *J. Protozool.* **27**, A32–A33 (in Song, 1983, p. 40).

Kendrick, R. E., Adamse, P., Lopez-Juez, E., Koornneef, M., Peters, J. L., and Wesselius, J. C. (1991). The significance of mutants in phytochrome research. *In* "Photobiology: The Science and Its Application" (E. Riklis, ed.), pp. 437–443. Plenum, New York and London.

Koornneef, M., and Kendrick, R. E. (1994). Photomorphogenetic mutants of higher plants. *In* "Photomorphogenesis in Plants" (R. E. Kendrick and G. H. M. Kronenberg, Eds.) 2nd ed., pp. 601–628. Kluwer Academic Publishers, Dordrecht, The Netherlands.

Land, M. F. (1972). The physics and biology of animal reflectors. *Prog. Biophys.* **24**, 77–105.

Matsuoka, T. (1983). Distribution of photoreceptors inducing ciliary reversal and swimming acceleration in *Blepharisma japonicum. J. Exp. Zool.* **225**, 337–340.

Matsuoka, T. (1983). Negative phototaxis in *Blepharisma japonicum. J. Protozool.* **30**, 409–414.

Matsuoka, T., Mamiya, R., and Taneda, K. (1990). Temperature-sensitive response in *Blepharisma. J. Protozool.* **37**, 323–328.

Matsuoka, T., Murakami, Y., Furukohri, T., Ishida, M., and Taneda, K. (1992). Photoreceptor pigment in *Blepharisma*: H^+ release from red pigment. *Photochem. Photobiol.* **56**, 399–402.

Meischke, D. (1936). Uber den Einfluss der Strahlung auf Licht-und Dunkelkeimer. *Jahrb. Wiss. Bot.* **83**, 359–405.

Moeller, K. M. (1962). On the nature of Stentorin. *C. R. Trav. Lab. Carlesberg, Ser. Chim.* **32**, 472–497.

Mohr, H., and Drumm-Herrel, H. (1991). Mode of action between phytochrome and blue/UV photoreceptors. *In* "Photobiology: The Science and Its Applications" (E. Riklis, ed.), pp. 445–453. Plenum, New York and London.

Nagano, K., Sugimoto, T., and Suzuki, H. (1978). *J. Phys. Soc. Jpn.* **45**, 236–243 (in Song, 1983).

Niess, D., Reisser, W., Wiesner, W. (1981). The role of endo symbiotic algae in photoaccumulation of green *Paramecium bursaria. Planta* **152**, 268–271, (in Song, 1983, p. 40).

Pado, R. (1972). Spectra activity of light and phototaxis in *Paramecium busaria. Acta Protozool.* **11**, 387–393 (in Song, 1983, p. 40).

Parker, W., Romanowski, M., and Song, P.-S. (1991). Conformation and its functional implications in phytochrome. *In* "Phytochrome Properties and Biological Actions" (E. Thomas, and C. B. Johnson, eds.), pp. 85–108. Springer-Verlag, Berlin.

Pratt, L. H., and Cordonnier, M.-M. (1989). Photomorphogenesis. *In* "The Science of Photobiology" (K. C. Smith, ed.), 2nd ed., pp. 273–304. Plenum, New York and London.

Quail, P. H. (1991). Phytochrome: Light-activated molecular switch that regulates plant gene expression. *Annu. Rev. Gen.* **25**, 389–409.

Quail, P. H. (1992). Phytochrome genes and their expression. *In* "Photomorphogenesis in Plants" (R. E. Kendrick and G. H. M. Kronenberg, Eds.), pp. 71–104. Kluwer Academic Publishers, Dordrecht, The Netherlands.

Reiter, R. J. (1981). The mammalian pineal gland: structure and function. *Am. J. Anat.* **162**, 287–313.

Reiter, R. J. (1991). Melatonin: The chemical expression of darkness. *Mol. Cell. Endocrinol.* **79**, C153–C158.

Reiter, R. J. (1992). Alterations of the circadian melatonin rhythm by the electromagnetic spectrum: A study in environmental toxicology. *Regul. Toxicol. Pharmacol.* **15**, 226–244.

Reiter, R. J., and Richardson, B. A. (1992). Magnetic field effects on pineal indoleamine metabolism and possible biological consequences. *FASEB J.* **6**, 2283–2287.

Rospendowski, B. N., Farrens, D. L., Cotton, T. M., and Song, P.-S. (1989). Surface enhanced resonance Raman scattering (serrs) as a probe of the structural differences between the P_r and P_{fr} forms of phytochrome. *FEBS Lett.* **258**, 1–4.

Rudiger, W. (1991). Molecular properties of phytochrome. *In* "Photobiology: The Science and Its Applications" (E. Riklis, ed.), pp. 423–434. Plenum, New York and London.

Sage, L. C. (1992). *In* "Pigments of the Imagination: A History of Phytochrome Research." Academic Press, N. Y.

Shropshire, W., Jr. (1989). Photomorphogenesis. *In* "The Science of Photobiology" (K. C. Smith, ed.), 2nd ed., pp. 281–312. Plenum, New York and London.

Sommer, D., and Song, P.-S. (1990). Chromophore topography and secondary structure of 124-kilodalton *Avena* phytochrome probed by Zn^{2+}-induced chromophore modification. *Biochemistry* **29**, 1943–1948.

Song, P.-S. (1981). Photosensor transduction in *Stentor coeruleus* and related organisms. *Biochim. Biophys. Acta* **639**, 1–29.

Song, P.-S. (1983). Protozoan and related photoreceptors: Molecular aspects. *Annu. Rev. Biophys. Bioeng.* **12**, 35–68.

Song, P.-S. (1988). The molecular topography of phytochrome: Chromophore and apoprotein. *J. Photobiol. Photochem. B. Biol.* **2**, 43–57.

Song, P.-S., and Poff, K. L. (1989). Photomovement. *In* "The Science of Photobiology" (K. C. Smith, ed.), 2nd ed., pp. 305–346. Plenum, New York and London.

Song, P.-S., Suzuki, S., Kim, I.-D., and Kim, J. H. (1991). Properties and evolution of photoreceptors. *In* "Photoreceptor Evolution and Function" (M. Holmes, ed.), pp. 21–63. Academic Press, San Diego.

Wong, Y. S., Chen, H. C., Walsch, D. A., and Lagarias, J. C. (1985). Phosphorylation of *Avena* phytochrome *in vitro* as a probe of light-induced conformation changes. *J. Biol. Chem.* **261**, 12089–12097.

PART III

LIGHT
AND
HEALTH

CHAPTER 11

Molecular Mechanism of Visual Transduction

11.1 THE PRIMARY PHOTOEVENT IN VISION

When a rod cell of the retina absorbs light as a primary photoevent, an isomerization of the photoreceptor chromophore 11-*cis*-retinal to 11-*trans*-retinal occurs within picoseconds. The cis–trans isomerization process was discovered in 1957 by George Wald and Ruth Hubbard of Harvard University. Chromophore photon absorption leads to isomerization about half of the time. In contrast, spontaneous photoisomerization in the dark takes place roughly once in 1000 years.

The photoreceptor cells that make it possible to form black and white images in dim light, i.e., the rod cells, are exquisitely sensitive detectors. When a photon strikes the retina, the rhodopsin molecule (the macromolecule holding the retinal) that is struck reports the event with high efficiency, while the millions of other rhodopsin molecules in the cell remain silent. The photoreceptor of rod cells, rhodopsin, located in the rod disks as a transmembrane protein, has two components: (1) *cis*-reti-

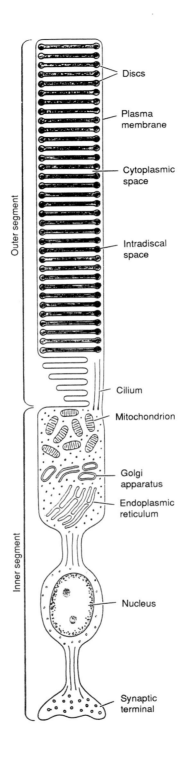

Outer segment

Discs

Plasma
membrane

Cytoplasmic
space

Intradiscal
space

Cilium

Mitochondrion

Golgi
apparatus

Endoplasmic
reticulum

Inner segment

Nucleus

Synaptic
terminal

nal, an organic molecule derived from vitamin A, and (2) opsin, a protein that has the capacity to act as an enzyme.

Opsin is a single polypeptide chain of 348 linked amino acids. The sequence has been worked out in the laboratories of Yuri A. Ovchinikov of the M. M. Shemyakin Institute of Bioorganic Chemistry in Moscow and by Paul A. Hargrave at Southern Illinois University. Opsin has the form of seven helices (α-helices) arranged vertically in the disk membrane of the rod cell and connected by short nonhelical segments. Attached to one α-helix is a single molecule of 11-*cis*-retinal, which lies near the center of the membrane, its long axis aligned in the plane of the membrane.

Retinal is nested at the center of a complex and highly structured protein environment, responsible for "tuning" retinal by influencing the spectrum of radiation it can absorb. Whereas in solution the maximum absorption of retinal is in the near-UV, within the rhodopsin molecule the maximum is shifted to 500 nm in the green region. Functionally this shift matches the absorption spectrum of rhodopsin with the light reaching the eye.

When a photon is absorbed by *cis*-retinal, the energy of light straightens the bend in the retinal carbon chain, a bend that is due to the presence of hydrogen atoms attached to the C-11 and C-12 on the same side of the chain. In the transduction process that follows, a cascade of reactions results in a nerve signal.

11.2 THE ROD CELLS

A cascade of molecular reactions amplifies the minute piece of information corresponding to the excitation of a rod cell by a single photon into a signal that is useful to the nervous system. Because the degree of amplification varies with the background illumination of the rod, rod cells are able to function efficiently over a wide range of background illumination. The rod cell (Figure 11.1) is a long, thin structure, divided into two parts. The outer segment contains most of the molecular apparatus for detecting light and initiating a nerve impulse. The inner seg-

FIGURE 11.1 The rod cell is divided into two parts that have specialized functions. The apparatus for detecting light is in the outer segment, which holds a stack of some 2000 disks derived from the plasma membrane. The inner segment contains organelles for making specialized molecules required in photoreception. When light strikes the disk, the primary photoprocess takes place at the level of the rhodopsin molecule localized in the disk as a transmembrane molecule (see Fig. 11.2). The signal is sent by a chain of reactions to the plasma membrane. It travels through the plasma membrane to the synaptic terminal, from which it is sent to other retinal cells. Courtesy of L. Stryer (1987), Scientific American **257**, p. 44. Copyright George V. Kelvin, Science Graphics, Great Neck, New York.

ment is specialized for generating energy and renewing the molecules needed in the outer segment. It also includes a synaptic terminal that provides the basis for communication with other cells.

The outer segment of the rod is a narrow tube filled with a stack of some 2000 tiny disks. Both the tube and the disks are made of bilayer membrane. The disk membranes contain the protein molecules that absorb light and initiate the excitation response (Figure 11.2). The outer membrane serves to convert a chemical signal into an electrical one.

The rod plasma membrane is selectively permeable to ions, which carry a net electric charge. As a result, in the resting state the inside of the cell is about 40 mV negative with respect to the outside. In 1970 Tsuneo Tomita of Keio University, as well as William A. Hahins and Shuko Yoshikami at the National Health Institutes, Bethesda, Maryland, showed that following illumination, varying with the strength of the stimulus and the background illumination, the potential difference can increase up to -80 mV. The increase in potential difference is known as *hyperpolariza-*

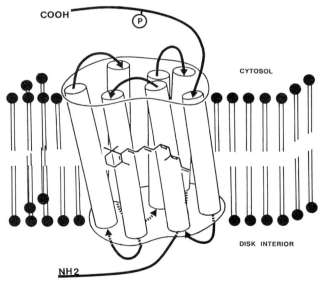

FIGURE 11.2 Rhodopsin molecule embedded in the membrane of the disk receives a photon and initiates the excitatory cascade. Like the plasma membrane from which it is derived, the disk membrane is basically a lipid bilayer. Rhodopsin has two components: 11-*cis* retinal and opsin. Opsin is a protein that has the form of seven helical structures, threaded through the membrane, connected by short linear segments. 11-*cis*-Retinal lies near the center of the membrane, attached to one helix. The absorption of a photon by retinal alters its shape and activates rhodopsin. This model for the structure of rhodopsin was proposed by Edward Dratz at the University of California at Santa Cruz and Paul Hargrave of Southern Illinois University.

tion. This *hyperpolarization of the plasma membrane generates the neural signal.* Hyperpolarization is then passed along the rod outer membrane to the synaptic terminal at the other end of the cell, where the nerve impulse arises. The signal must be carried first from the disk membrane to the rod outer membrane.

The sensitivity of a rod cell is continuously adjusted according to the background light level, enabling it to detect incremental stimuli over a 10^5-fold range of ambient light intensity.

11.3 SIGNAL TRANSDUCTION BETWEEN THE DISK MEMBRANE AND THE ROD OUTER MEMBRANE

What is the transmitter that carries the excitation signal between the disk membrane and the rod outer membrane? The light-induced release of calcium ions sequestered in the disk membrane in the dark seemed at first an attractive hypothesis. The light-induced decrease of the calcium level is important for both recovery and adaptation. *Calcium ion is the remembrance of photons past.* Recoverin, a cytoplasmic protein, may be the calcium sensor in adaptation as well as recovery. Recent work has shown that *the actual transmitter is not calcium ion but rather cyclic guanosine monophosphate.*

When the ring formed by the joining of the 3',5' carbons of cGMP through a phosphate group is intact, this molecule is capable of keeping the rod outer membrane sodium channels open. When the ring is cleaved by cyclic guanosine monophosphate phosphodiesterase (PDE), the sodium channels close spontaneously. Denis A. Baylor showed that the absorption of a single photon blocks the influx of millions of sodium ions in the plasma membrane.

The cascade that follows the primary photoevent provides the link between the disk membrane and the outer membrane. The cascade has been worked out in molecular detail. The photoisomerization of the 11-*cis*-retinal chromophore of rhodopsin (R) to the all-*trans* form generates the excited rhodopsin R*. A key intermediate is a protein called transducin (T), or G protein (Figure 11.3). Excitation is mediated by the formation of the active form of transducin ($T_\alpha \cdot GTP$). This complex, which is an activator of cyclic GMP PDE, is formed by interaction of R* (the activator of T) with $T \cdot GDP$, which leads to the complex $R^* \cdot T \cdot GDP$, followed by GDP/GTP conversion through reaction with a molecule of GTP. R* and two transducin subunits ($T_{\beta,\gamma}$) subsequently dissociate from $R^* \cdot T \cdot GTP$, and the resulting $T_\alpha \cdot GTP$ goes on to activate a potent phosphodiesterase.

Rapid removal of cyclic GMP by the switching on of phosphodiesterase closes cation-specific ion channels in the plasma membrane of the rod

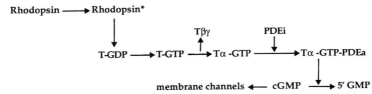

FIGURE 11.3 Scheme of molecular processes involved in transduction of signal from photoexcited rhodopsin to cleavage of cyclic GMP, the transmitter that carries the signal between the disk membrane and the rod outer membrane. Note the three-level cascade of events, which suggests mechanisms of hormonelike signal transduction. PDEi, inactive PDE; PDEa, activated PDE.

outer segment. In essence, light triggers a nerve signal by activating a cascade. The cascade continues by activation of PDE, removal of cGMP, and closing of sodium and calcium channels (which are kept open in the dark by cGMP) (Figure 11.4).

The capacity of rods to detect single photons and to operate over a wide range of background light levels is conferred by multiple stages of positive and negative feedback that are coordinated and precisely timed. The plasma membrane of the rod outer segment is in essence a *cGMP electrode. The sodium channel responds to the instantaneous levels of cGMP, which it samples each millisecond.*

There are various feedback mechanisms involved in termination of the excitation and restoration of the initial dark state prior to a photon striking rhodopsin: (1) deactivation of transducin, (2) restoration of 11-*cis*-retinal, and (3) stimulation of guanylate cyclase.

The reopening within 2 sec of channels that close within a second of a dim light flash is due to restoration of cGMP levels. This requires activation of guanylate cyclase and inhibition of PDE. Deactivation of PDE is brought about by hydrolysis of $T_\alpha \cdot GTP$ to $T_\alpha \cdot GDP$ through the intrinsic GTPase activity of transducin. Deactivation of transducin and hence PDE are necessary but not sufficient for restoration of the dark state.

Photoexcited R (i.e., R*) must also be quenched. The back isomerization of all-*trans*-retinal to *cis*-retinal *takes many minutes. A much faster shutoff is achieved by the phosphorylation of multiple serines and threonines in the carboxy-terminal region of R* by rhodopsin kinase.* A 68-kDa cytosolic protein, called *arrestin,* then caps multiply phosphorylated R* to prevent it from interacting with transducin. Rhodopsin is regenerated several minutes later by insertion of 11-*cis*-retinal, release of arrestin, and removal of the phosphate of the COOH terminus by protein phosphatase A.

Restoration of the dark state also requires stimulation of guanylate cyclase. There is a negative feedback loop between cGMP and cytosolic calcium. Following illumination, the influx of Ca^{2+} through the cGMP-gated channel ceases, but its export by the $Ca^{2+}/Na^+,K^+$ exchanger

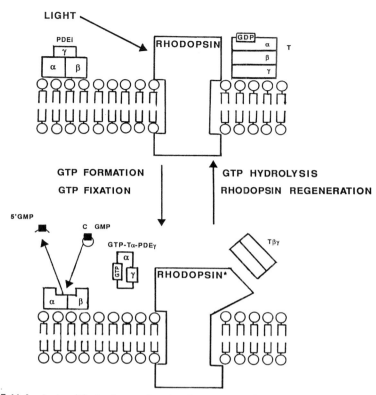

FIGURE 11.4 A simplified scheme of molecular events and structural changes involved in the visual cascade from formation of excited rhodopsin by light, to GTP formation, T · GTP fixation on PDE, activation of PDE, GTP hydrolysis, and rhodopsin regeneration. In this figure PDEγ is the inactive form of PDE.

continues, until Ca^{2+} drops markedly from 500 to 50 nM, which stimulates guanylate cyclase. This enzyme, like PDE, consists of *separable regulatory and catalytic subunits*. In its calcium-free form, a calcium-sensitive stimulatory cytoplasmic protein called *recoverin* binds to guanylate cyclase and stimulates it.

11.4 MULTIDISCIPLINARY STUDY OF SENSORY TRANSDUCTION IN ROD VISION

Many different disciplines—biochemistry, biophysics, molecular genetics, and electrophysiology—have come together, each enriching the other in revealing the molecular mechanisms of vision and providing glimpses of how this sensory transduction came into being.

The several steps intervening between the excitation of rhodopsin and the enzymatic cleavage of cGMP have been described above. Excited rhodopsin activates an enzyme called transducin (i.e., G protein), an activator of PDE.

In the summer of 1978, Paul A. Liebman of the University of Pennsylvania reported his finding that a single photon could trigger the activation of hundreds of phosphodiesterase molecules per second in preparations of rods outer segments. Earlier work carried out in the presence of ATP had shown much less amplification. Liebman observed that considerably more amplification could be obtained with GTP. GTP is a nucleotide closely related to the noncyclic form of GMP. Instead of having one phosphate attached to its 5′ carbon, however, it has a chain of three phosphates bound to each other by phosphodiester linkages. *The energy stored in these bonds provides for many cellular functions.* The splitting off of one phosphate converts GTP into GDP and liberates considerable energy to drive chemical reactions that are otherwise energetically unfavorable. Liebman's key finding was that this process seemed to be at work in the activation of the phosphodiesterase, where GTP is an essential cofactor.

Miller and his co-worker Grant Nicol succeeded in injecting cGMP into the outer segment of intact rod cells. cGMP quickly reduced the potential difference across the plasma membrane, and sharply increased the delay between the arrival of a light pulse and the hyperpolarization of the membrane. Thus, cGMP kept open sodium channels, until degraded by light-activated PDE.

These findings led to the idea that a protein capable of binding GTP (i.e., the nucleotide found effective by Liebman, in amplification of PDE) might be a significant intermediate in activation. The following sequential studies by Stryer led to the isolation of transducin by Hermann Kuhn at the Institute for Neurobiology at the University of Julich (Germany):

1. Incubation of radioactively labeled GTP with rod outer segments followed by washing of the preparation over a filter (which retained membrane fragments or molecules as large as proteins but allowed small molecules such as GTP to pass) revealed a substantial amount of radioactivity bound to the membrane (but as GDP, not as GTP).

2. This suggested that there was, in the rod membrane, a protein capable of binding GTP and converting it to GDP (at the time Walter Godchaux III and William F. Zimmerman of Amherst College (Amherst, MA) reported finding just such a protein in the rod membrane).

3. The release of GDP from the membranes when they were illuminated was strongly enhanced by the presence of GTP in the surrounding solution. It seemed that one part of the activation process involved an

exchange of GTP for GDP in the membrane. The release of GDP depended on the substitution of a molecule of GTP for the GDP.

4. It was found that the excitation of one rhodopsin molecule led to the binding of about 500 molecules of a GTP analog (selected because it resists being reduced to GDP and therefore enabled the isolation of the exchange step). The discovery of this *amplification* pointed toward an explanation of the overall amplification characterizing the excitatory cascade.

5. The above study led to the proposal that, in the excitation cascade, there is a protein intermediate (i.e., transducin, a G protein) that can exist in two states. In one state the protein binds GTP, in the other it binds GDP. The substitution of GTP for GDP is the signal for the protein activation. That exchange is triggered by rhodopsin; in turn, it serves to activate PDE.

Two conclusions followed from the isolation of transducin: it was inferred that transducin could be converted to its GTP form in the absence of PDE, and it also followed that PDE could be activated in the absence of photoexcited rhodopsin.

There were trials to verify these two conclusions: (1) A synthetic membrane from which PDE was absent was assembled, and purified transducin in the GDP form was introduced into the artificial membrane. When rhodopsin was activated by light, each rhodopsin molecule catalyzed the uptake of 71 molecules of a GTP analog to the membrane. Thus it appeared that each molecule of rhodopsin was activating transducin by catalyzing the exchange of GTP for GDP in many transducin molecules (amplification). (2) In purifying the active form, the transducin–GTP complex, it was found that while in the inactive form (i.e., with GDP) transducin is complete [with all three of its subunits (α, β, γ) joined]; in the active form two subunits came apart. GTP was bound to the α subunit. *It is this subunit that activates PDE also in the absence of rhodopsin.*

PDE consists of three chains, α, β, and γ. Trypsin digestion of the γ subunit removes the inhibitory constraint of PDE. The rate of destruction of the γ unit by trypsin closely resembles the rate of activation of PDE in the excitatory cascade. Marc Chabre and colleagues at the Center for Nuclear Studies at Grenoble (France) found that transducin in the GTP form can bind to the γ subunit of PDE, unleashing the catalytic activity of the enzyme, which can hydrolyze 4200 molecules of cGMP per second.

If the organism is to see more than once, this cycle must be turned off. The α subunit of transducin has a built-in chemical timer (an intrinsic phosphorylase) that terminates that activated stage by converting the bound GTP to GDP. As GTP is cleaved to GDP, the α unit of transducin releases the inhibitory γ unit of PDE. Transducin is restored to its preactivation form by rejoining of the α unit and the $\beta-\gamma$ unit. Rhodopsin is returned to the deactivated state by a kinase that attaches multiple phosphate groups to amino acids at one end of the opsin chain (Figure 11.2). Rhodopsin then forms a complex with a protein called *arrestin*, which blocks the binding of transducin and restores the system to the dark state.

To study the action of cGMP on the sodium channels of the outer rod membrane, Fesenko and co-workers at the Institute of Biological Physics, Moscow, employed a micropipette to pull off a small patch of the rod cell plasma membrane. The patch adhered tightly to the end of the pipette, with the site that would normally be inside facing out. Exposure of the membrane to various solutions in order to test their effect on sodium conductance unambiguously showed that the channels were directly opened by cGMP. *Thus the overall information flow is from rhodopsin to transducin to PDE and then to cGMP.*

Liebman's group and that of Benjamin Kaupp at the Neurobiology Institute at the University of Osnabruck (Germany) have purified cGMP-gated sodium channels and reconstituted their function in model membranes. In addition, a feedback loop that restores cGMP to its preillumination level is required, otherwise the rod could fire only a few times before exhausting its own capacity.

The genetic information on transducin and three G proteins shows that the proteins include some regions that have been conserved quite stringently as well as some that have diverged widely. Each protein contains three binding sites: one for guanyl nucleotides, one for the activated receptor (rhodopsin or a hormone–receptor complex), and one for the effector protein (PDE or adenylate cyclase). Because its crucial function in the activation cascade, the binding site for GTP or GDP is the most highly conserved among the various proteins. Cyclic GMP and cytosolic calcium concentrations are set by a feedback loop. Ca^{2+} enters through a cGMP-gated channel, and this influx is matched by efflux through a Na^+/K^+, Ca^{2+} exchanger. *Photoexcitation blocks Ca^{2+} influx, but not its efflux, so that there is Ca^{2+} depletion within 0.5 sec of light onset, leading to membrane hyperpolarization.* Guanylate cyclase activity increases when Ca^{2+} concentration is lowered to less than 100 nM. This activation is cooperative and mediated by *recoverin*, a protein that can be detached from the guanylate cyclase catalytic moiety in low-ionic-strength buffer. A study with the affinity-purified antibody to isolated recoverin specifically recognized a 26-kDa protein in retina and pineal homogenates. Fluorescence spectroscopy confirmed that recoverin binds Ca^{2+}. The tryptophan residue fluorescence decreased and the emission spectrum shifted to the red when Ca^{2+} concentration was increased from less than 10 nM to 1.4 μM.

Raising the concentration of free Mg^{2+} did not alter the fluorescence properties of recoverin, which specifically binds to Ca^{2+}.

The affinity of recoverin for Ca^{2+} and its presence in rods and cones led Stryer's group to investigate its relation to the putative Ca^{2+}-sensitive regulator of guanylate cyclase. Addition of purified recoverin to rod outer segment membranes stripped off the endogenous activator by washing in low-ionic-strength buffer, activated guanylate cyclase at low Ca^{2+} concentrations, but at high Ca^{2+} concentrations the cyclase returned to its basal activity. The activity of the enzyme increased nearly fourfold when Ca^{2+} was lowered from 450 to 40 nM. The Hill coefficient for activation of cyclase by recoverin is about 3, showing that the effect is highly cooperative, as reported for the rod outer segment. Adding 50 μg of recoverin polyclonal antibody to native rod outer segment membranes (about 100 μg of rhodopsin) completely blocked guanylate cyclase activation at low Ca^{2+} concentration without interfering with basal cyclase activity at high Ca^{2+} concentration.

Recoverin differs from Ca^{2+}-dependent activators such as calmodulin and troponin C, which require Ca^{2+} to activate their targets. Rather, recoverin is a Ca^{2+}-sensitive activator that must be liberated from Ca^{2+} before it activates its target. Polans and co-workers have identified antibodies to recoverin in sera from patients with retinas that have degenerated as a result of cancer-associated retinopathy. Recoverin and a homologous protein, visinin, are suggestively Ca^{2+}-sensitive regulators that, through cooperative interactions, act as switches at submicromolar Ca^{2+} levels.

11.5 THE STRYER SCHEME OF MOLECULAR MECHANISM OF VISUAL TRANSDUCTION (AS REVISED BY LIEBMAN)

Liebman, at the University of Pennsylvania, has summarized the sequence (Figure 11.5) of events that follows the picosecond kinetics associated with the conversion of *cis*-retinal (Figure 11.2) to *trans*-retinal. This corresponds to the "visual transduction cascade" (Figure 11.3), which involves the following steps:

1. The *cis*- to *trans*-retinal isomerization that follows the absorption of a single photon of light by rhodopsin R is accompanied by the conversion of R to the activated form R*, also known as metarhodopsin-2. R* undergoes lateral diffusion on the photoreceptor disk membrane surface, where it serially encounters several hundred G protein (transducin) molecules complexed with GDP at the rate of $3 \times 10^3/\sec^{-1}$. The G protein

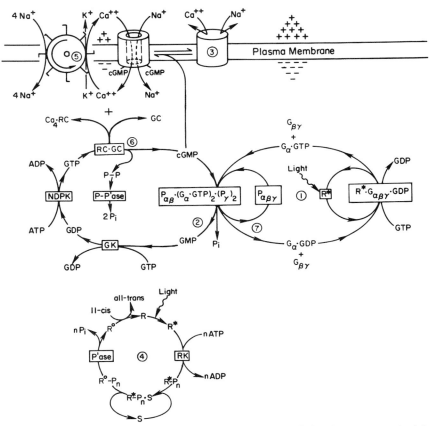

FIGURE 11.5 Molecular mechanism of visual transduction [after Stryer, as revised by Liebman (see text)]. G_α corresponds to T_α (transducin$_\alpha$). $P_{\alpha,\beta}$ corresponds to PDE with only α and β subunits, which is the active form. In this figure $P_{\alpha,\beta,\gamma}$ is the inactive form of PDE. The numbers in the figure refer to steps 1, 2, 3, and so on. Courtesy of P. Liebman, School of Medicine, University of Pennsylvania, Philadelphia, PA.

has α, β, and γ subunits. R* catalyzes GTP/GDP exchange on each G protein. The complex $G_\alpha \cdot$ GTP separates from the β and γ subunits of the G protein. $G_\alpha \cdot$ GTP binds pairwise to the heterotetrameric inactive PDE, i.e., $P_{\alpha,\beta,\gamma,\gamma}$. Cooperative activation of the two catalytic sites on α and β is caused by removal of the two γ subunits.

2. The activated phosphodiesterase PDE (or $P_{\alpha,\beta}$) rapidly destroys surrounding cytoplasmic cGMP, which causes net cGMP loss from the cooperative binding sites holding the plasma cation channels open in the dark depolarized cell.

3. This results in the closing of the cation channels and, with the cation

influx so prevented, in the hyperpolarization of the cell. Concomitantly, Ca^{2+} influx is stopped on the resting retina, Ca^{2+} enters in darkness to block guanylate cyclase activity from forming cGMP while PDE is quiescent).

4. At about this time, R* becomes quenched by serial phosphorylation. This reaction is catalyzed by rhodopsin kinase, RK, to form R^*-P_n (i.e., R* with n phosphate groups). Depending on the light level, the number n of phosphates binding to R* can be as high as 9. R^*-P_n then binds to S, which is also known as the 48-kDa protein, S-antigen, S-arrestin, or simply arrestin. The resulting noncovalent complex $R^*-P_n \cdot S$ sterically occludes further G protein activation at step 1 and prevents access of the type 2A protein phosphatase (P'ase), to the phosphates of R^*-P_n. This lasts until some minutes later, to the time when R has thermally decayed to R^0. The n phosphates are subsequently removed and R_0 is regenerated by conversion to R after about 30 min by entering into reaction with a new protagonist molecule, 11-*cis*-retinal.

5. Recovery from hyperpolarization begins as cytoplasmic Ca^{2+} is rapidly removed by a $Na^+/K^+,Ca^{2+}$ antiporter. As cytoplasmic Ca^{2+} concentration falls, bound Ca^{2+} dissociates from the cooperative binding sites on recoverin (RC).

6. Recoverin in the Ca^{2+}-free state can activate guanylate cyclase, GC, to synthesize cGMP.

7. The GTP bound to G_α is hydrolyzed by intrinsic GTPase activity and the active state of PDE (i.e., $PDE_{\alpha,\beta}$) decays by two P_γ subunit recombinations so that there is no further loss of cGMP. Thus both PDE and G return to their dark quiescent state.

Other coupled reactions that return all the cellular nucleotides to their initial condition are guanylate kinase (GK), nucleoside diphosphokinase (NPDK), and pyrophosphatase (P-Pase).

11.6 RODS AND CONES

Photoreceptor cells are found with two different morphologies: rods and cones. Rods predominate in the peripheral retina whereas cones are packed in the fovea. Rods are responsible for dim light vision. Cones work at much brighter light and are responsible for color perception. Rods and cones share the same photoactive material, e.g., rhodopsin. Much less is known about the sequence of events in the cone cell struck by a photon. Transducin, PDE, and the cGMP-controlled channel in the cones are like their counterparts in the rods. However, microspectrophotometry has demonstrated that, in contrast to rods, cone rhodopsin has its absorption maximum in the blue, green, and red regions. The difference in the absorption spectrum of each category of cones is due to minute changes in the microenvironment of the retinal at the binding site with opsin. These spectral properties of retinal in cone rhodopsin support the involvement of cones in the trichromatic theory of color vision.

Bibliography

Dizhoor, A. M., Ray, S., Kumar, S., Niemi, G., Spencer, M., Brolley, D., Walsh, K. A., Philipov, P. P., Hurley, J. B., and Stryer, L. (1991). Recoverin: a calcium sensitive activator of retinal rod guanylate cyclase. *Science* **251,** 915–918.

Stryer, L. (1987). The molecules of visual excitation. *Sci. Am.* **257,** 42–50.

Stryer, L. (1991). Visual excitation and recovery. *J. Biol. Chem.* **266,** 10711–10714.

12

Biological Effects of Solar Ultraviolet Radiation

The most important aid for the recognition of the healing power and avoidance of the destructive effects of natural and artificial ultraviolet radiation in both science and everyday life remains still the human intelligence.

Reinhard Breit

12.1 UVB EFFECTS

12.1.1 Introduction

The UV absorption spectra of major epidermal chromophores (trypto-phan, tyrosine, DNA, and urocanic acid) provide a glimpse of primary targets of UVB effects (Figure 12.1). In the case of urocanic acid, the action spectrum is almost superposable with the action spectrum of immuno-suppression.

The combined action spectra of UVB–UVA irradiation for lethality, mutagenesis, and pyrimidine dimer formation (Figure 12.2) show an abrupt break at about 330 nm. The UVB effects on DNA are described in

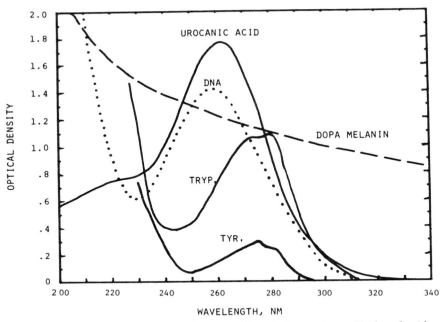

FIGURE 12.1   UV absorption spectra of major epidermal chromophores. The broad epidermal absorption band near 275 nm is the result of absorption by proteins, urocanic acid, nucleic acids, and other aromatic chromophores. TRYP, Tryptophan; TYR, tyrosine. From Anderson and Parrish (1982), p. 165. With permission of Plenum, New York.

Chapter 5. The action spectra are discussed in Chapter 7 on environmental photobiology. DNA repair processes are discussed in the next section.

12.1.2 UV Radiation Survival Curves

On a logarithmic scale, straight-line survival curves (Figure 12.3) can be characterized by a single parameter: the F_{37}, i.e., the dose or fluence to yield a survival of 37% ($1/e$). For a shouldered survival curve three parameters apply: (1) the extrapolation number n is one measure of the width of the shoulder, and is the point where the extrapolation of the linear portion of the survival curve crosses the ordinate; (2) F_q, the quasi-shouldered dose, is also a measure of the width of the shoulder and is the dose at which the extrapolation line intersects the 100% survival curve; (3) F_0 is based on the same concept as F_{37}, but is determined on the linear portion of the survival curve.

The F parameters are similar to those previously developed for absorbed ionizing radiation doses.

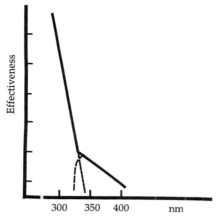

FIGURE 12.2 Combined spectr ɩ for lethality, mutagenesis, and pyrimidine dimer formation. From F. Urbach (1992), p. 2. With permission from Valdenmar Publishing Company, Oveland Park, KS.

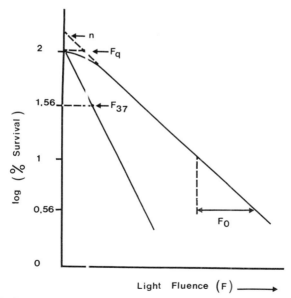

FIGURE 12.3 Radiation fluence curves; the various parameters are discussed in Section 12.1.2.

Escherichia coli cells grown to logarithmic phase in rich medium and plated on rich medium are 1.5 times more resistant to UV radiation. Thus, these cells will have larger F_0 and/or F_{37} than are cells grown in and plated on minimal medium: they show medium-dependent resistance (MDR).

12.2 DNA REPAIR

12.2.1 Introduction

Escherichia coli cells in a nutrient-free environment for about 4.5 hr post-UV irradiation exhibit a much higher survival than if they are plated on nutrient agar immediately. This is called the "liquid-holding recovery" (LHR). The requirements for LHR are (1) proficiency in nucleotide excision repair and (2) at least partial deficiency in postreplication repair (*recA* mutants completely deficient in postreplication repair show high LHR).

The *postreplication repair process,* which results *in DNA daughter strand gaps,* requires a complete growth medium with nutrients, whereas nucleotide excision repair occurs in buffer devoid of nutrients. Avoidance of postreplication repair by cells in stationary phase, with nutrients in the growth medium exhausted, results in higher survival.

The repair-deficient double mutant *uvrA6 recA13* is much more radiation-sensitive than either single mutant, suggesting that the two genes act on different pathways of DNA repair: the *uvrA* mutation blocks excision repair, and the *recA* mutation blocks postreplication repair and inducible processes.

The discovery of *excision repair* in *E. coli* came from the observation that the UV-induced lesions (i.e., thymine dimers) were selectively cut out in the wild strain, but not in the radiation-sensitive strain.

12.2.2 Photoreactivation

The damaged part of a DNA molecule can be restored *in situ,* without breaking its sugar phosphate backbone, by photoreactivation (PR). The enzymatic splitting of cyclobutane-type pyrimidine dimers is mediated by intense blue light. The PR enzyme is DNA photolyase. In the dark, the enzyme binds tightly to a cyclobutane-type dimer to form an enzyme–substrate complex. The absorption of light between 300 and 450 nm activates this complex. The pyrimidine dimer is converted to monomeric pyrimidines, and the enzyme is released.

The action spectra for PR in *Euglena gracilis* and *Saccharomyces cerevisiae* are very similar to the action spectrum for PR of yeast cells. The occurrence of PR is confirmed in a broad variety of species, from Cyanophyta to Ungulata (domestic beef). In the marsupial *Monodelphis domestica*, effective PR suppresses the capacity of UV radiation to induce erythema. In primates, PR prevention or suppression of UV-induced skin pathology appears to be low or nonexistent.

Escherichia coli DNA photolyase binds to DNA containing *cis,syn*-pyrimidine dimers and catalyzes the cycloreversion of the dimers by photoinduced electron transfer. The enzyme has two chromophores, 5,10-methyltetrahydrofolate (MTHF) and flavin adenine dinucleotide (FAD). Evidence for electron transfer between photoexcited flavin in the photolyase active site and a pyrimidine dimer is provided by the occurrence of a pyrimidine dimer radical after a picosecond laser flash. The flavin radical anion thereby formed undergoes rapid conversion. Fluorescence polarization shows that MTHF is relatively mobile whereas the dihydro form of flavin ($FADH_2$) is rotationally restrained during excitation energy transfer. There is a close match between the photoreactivation enzyme–$FADH_2$ complex (E–$FADH_2$) absorbance and the absolute action spectrum for pyrimidine dimer splitting.

12.2.3 Excision Repair

The first step in nucleotide excision repair (Figure 12.4) in *E. coli* is recognition by the uvrABC nuclease of the cyclobutylpyrimidine dimer, through the distortion in the DNA helix. This complex enzyme, coded for by the *uvrA*, *uvrB*, and *uvrC* genes, *produces breaks in the DNA chain on the 5' and 3' sides of a pyrimidine dimer*. This is the *incision* function of the endonuclease, followed by exonuclease action to remove the DNA fragment containing the lesion.

The removal (excision) of the lesion-containing fragment in *E. coli* includes the following steps: (1) the *uvrD* gene product, helicase II, appears to cause the release of the UvrC protein through protein–protein interaction rather than helicase activity and (2) DNA polymerase I then displaces the DNA fragment with the lesion and fills the gap (about 12 nucleotides long) using the opposite DNA strand as the template.

Finally, the break in the repaired DNA strand is filled by DNA ligase. Thus the enzymatic sequence is UvrABC nuclease, helicase II, DNA polymerase I, and DNA ligase.

The major pathway is the *polA* pathway, which produces short patches (20–30 nucleotides). The minor repair pathway, the *recA* pathway, produces long patches (200–1500 nucleotides; only in replicated portions of the chromosome).

Lesions can be produced in both the replicated and nonreplicated regions of the chromosome. In the *recA*-dependent repair, an intrachro-

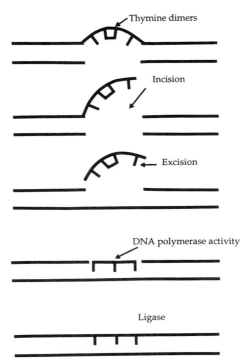

FIGURE 12.4   A model for the major pathway of nucleotide excision repair. An endonucle-ase recognizes the lesion (i.e., a cyclobutyl-pyrimidine dimer) and cuts on both sides of the lesion, producing a gapped structure that is about 12 nucleotides long. Repair replication fills this gap, using the opposite strand of DNA as the template. Finally, the break in the repaired DNA strand is sealed by DNA ligase. Adapted from Smith (1989), p. 121; modified from Sancar and Rupp (1983).

mosomal recombination process leaves a gap in the homologous sister duplex that can be filled by long-patch repair replication using the parental strand opposite the gap as a template.

Deficiency in excision repair seems to cause the induction of fatal skin cancer in xeroderma pigmentosum (XP) patients after exposure to sunlight. If an individual is deficient in "error-free" excision repair, then the remaining "error-prone" system takes over in the repair of DNA lesions, leading to mutagenesis, as a possible early step in some mechanisms of carcinogenesis.

12.2.4 Inducible SOS Response

UV radiation mutagenesis is presumed to be due to error-prone SOS repair. The SOS process is under the direction of the *recA* and *lexA* genes

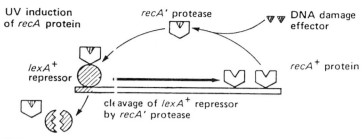

FIGURE 12.5 Regulation of the *recA* gene *in vivo*. The regulatory features of this model are described in the text. The broad arrow represents increased transcription of the *recA* gene. Redrawn from R. McEntee (198?), p. 193. Plenum, New York and London.

(Figure 12.5); it can be induced by (1) UV radiation and (2) other DNA-damaging agents (ionizing radiations) and chemicals (mitomycin).

A plausible candidate for the SOS-inducing "signal" is the sudden increase in the amount of single-stranded DNA that occurs in a cell due to distortion of the DNA helix or stalling of DNA replication at the site of the lesion. Single-stranded DNA activates *in vitro* the protease function of the RecA protein, the product of the *recA* gene. The activated RecA protein catalyzes the cleavage of the LexA protein (the product of the *lexA* gene), which is the repressor of the SOS genes. Thus the SOS genes become derepressed. Other inducible SOS genes include (1) *uvrA, uvrB*, and *uvrD*, which are involved in excision repair, (2) *recN* and *ruv*, which are involved in recombination, and (3) a number of *din* (damage-inducible) genes, some of which are still uncharacterized.

A large set of genes is activated in the SOS process to optimize the cell's chance for survival, but at the cost of mutagenesis, because the SOS system is more error prone than other repair systems. The SOS process is terminated when decrease of the damaged DNA ends RecA protease function (i.e., activation) and the decrease to a low level of LexA protein depresses SOS genes, including the *recA* and *lexA* genes. It is not totally resolved whether mammalian cells show an SOS response.

12.2.5 Postreplication Repair

DNA synthesis proceeds up to and then skips past the lesions in the parental strands, leaving gaps in the daughter strands. Filling of the gaps in the daughter strands takes place by a recombinational process (Figure 12.6) that also requires a functional *recA* gene. In mammalian cells, very few dimers are detected in daughter DNA strands after postreplication repair. However, in *E. coli* about half of the genetic transfers result in the covalent joining of daughter strand DNA to parent DNA with pyrimi-

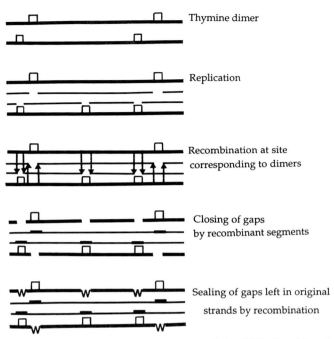

FIGURE 12.6   Model for the postreplication repair of the DNA daughter strand. *Boxes* indicate photochemical lesions produced in the two strands of DNA. DNA synthesis proceeds up to and then skips past the lesions of the parental strands, leaving gaps in the daughter strands. Filling of the gaps in the daughter strands proceeds with material from the parental strands by a recombinational process. Gaps in the parental strands are closed by repair replication. Adapted from Smith (1989), p. 125; modified from Smith (1971). With permission of Plenum, New York.

dine dimers, thus requiring several cycles of DNA repair before all the dimers are "diluted" out.

12.2.6 Impact of DNA Repair in Photomedicine

12.2.6.1 Epidemiology of UV-DNA Repair Capacity

A UV-damaged expression vector plasmid with the encoded reporter chloramphenicol acetyltransferase *(cat)* is transfected into peripheral blood T lymphocytes from subjects. The host repair capacity is measured by reporter *cat* expression. Young basal cell carcinoma (BCC) cases repair DNA damage poorly. With age, the differences between BCC cases and controls disappear. The normal decline in DNA repair with increasing age may account for the increasing risk of skin cancer; BCC in the young may represent precocious aging. Patients with reduced DNA repair capacity and overexposure to sunlight have an estimated risk of BCC 5-fold greater than controls in male subjects, and 10-fold greater in female subjects.

12.2.6.2 Photorepair of Pyrimidine Dimers and Suppression of UV-Induced Pathologic Lesions

Post-UV exposure to photoreactivating light (PRL; 320–500 nm) resulted in the reversal of UV-induced pyrimidine dimers in cutaneous and corneal cells of a South American opossum, *M. domestica*. Photorepair is accompanied by suppression of a number of pathologic changes in the acutely and chronically irradiated skin and cornea. Post-UV (but not pre-UV) PRL suppressed (1) erythema, edema, and desquamation of the skin, (2) formation of sunburn cells and hyperplasia, (3) loss of the so-called ATPase Langerhans cells from exposed skin, and (4) loss of contact hypersensitivity response to hapten applied directly to UV-irradiated or -protected sites on UV-irradiated opossums. The PRL also delayed UV-induced opacities, vascularization, and tumors of the cornea, nonmelanoma, and melanoma skin tumors.

12.3 UVA EFFECTS

12.3.1 Action Spectrum of UVA

The action spectrum (Figure 12.2) for human erythema and pigmentation shows a distinct change of slope at about 335 nm, suggesting that UVA and UVB effects are the result of different mechanisms. The action spectra for cell killing, mutagenesis, and DNA dimer formation are made of a UVB component that closely correlates with the DNA event spectrum, and a UVA component that does not. Chronic irradiation with UVA can cause significant damage to connective tissue (especially photoelastosis). Human exposure risk is similar in elastosis and photocarcinogenesis. There are differences in the biological effectiveness over small-wavelength ranges in the UVA–UVB band.

In albino hairless mice exposed to UVA, the slope of tumor induction time for the UVA-induced tumors is significantly less steep than for UVB (280–320 nm). UVA (320–340 nm), sometimes called UVAII, produces the same pigmentation response as exposure to UVB; UVA (340–400 nm), i.e., UVAI, produces non-UVB effects.

The photocarcinogenesis action spectrum for mouse skin falls below (i.e., less effectiveness) the erythema spectrum for UVA. Except at sun angles below 60° or at the winter solstice, UVA does not contribute more than 25% of the total erythema effectiveness of solar radiation. Very long times of exposure are necessary to produce UVA erythema even with very intense solar simulators.

The data on UV tumorigenesis in Skh:HR1 mice show that UVA at wavelengths longer than 340 nm is tumorigenic, although with an efficiency 10,000 times smaller than radiation at 290–300 nm. The poorly matching results obtained for the mutagenic action spectrum, corrected for epidermal thickness, underline the point that UVA carcinogenesis may be envisaged as an intricate dynamical and multistep process, with no straightforward interpretation. The concept of photoadditivity has been demonstrated for erythema and can be used for photocarcinogenesis. Because the sun is a polychromatic source, the obvious way to study UVA radiation is by polychromatic irradiation. In looking for some sort of bulk ecological effect, polychromatic spectra with bandpass filters are required. In looking for a mechanism, a narrower band or monochromatic approach (i.e., analytical action spectrum) is preferable.

The action spectrum for pigmentation is more toward the longer wavelength part of the UVA spectrum (UVAI) than is the erythema spectrum. Even though they may be equally effective against sunlight-induced erythema, sunscreens absorbing primarily UVAII or UVAI are different in terms of immediate pigmentation.

Lasers emitting in the UVA include flash-lamp pumped-dye lasers (340 nm), nitrogen lasers (CW or P; 337 nm), helium–cadmium lasers (CW), argon lasers (CW; 350 nm), and krypton lasers (CW; 350 nm).

12.3.2 Effects of Pollutants on UV Radiation

A typical smoggy day in Los Angeles can reduce the erythema radiance by 15–20%. Particulates in the air have more effect than do gaseous pollutants. At the Dead Sea, people can sit exposed to the sun for hours without getting burned, because the UVA radiation is higher than normal and the UVB is lower. This is likely due to the fact that the Dead Sea is 1200 ft below sea level, and to the haze from the very high rate of evaporation due to high temperatures (ranging from 40 to 50°C).

12.3.3 The Biological Effect of UVA

12.3.3.1 Single-Strand DNA Breaks

Because of their relatively low energy (a few electron volts) and because critical biological targets (nucleic acids, proteins) have low or no extinction coefficients at these wavelengths, UVA photons, previously considered innocuous, are weakly mutagenic in cultured cellular systems. They produce single-strand breaks (SSBs), DNA-to-protein crosslinks (DPCs), and double-strand breaks (DSBs). Compared to F_{37}

(fluence necessary for ≤7% survival), UVA photons are 1–2 orders of magnitude more efficient for induction of these lesions than are γ-rays, X-rays, or UVC photons. Different kinetics in the induction of SSBs at 365 and 405 nm in mammalian cell targets may relate to different sensitizing pathways or levels of endogenous antioxidants (reduced glutathione). The mutant Chinese hamster ovary (CHO) line EM9, which is deficient in repair of ionizing radiation-induced SSBs, is also deficient in repairing UVA- and visible-light-induced SSBs.

SSBs induced by UVA (365 m) are reduced by 90% if the cells are irradiated anoxically, due to the role of hydroxyl radical ($\cdot$OH) and singlet oxygen in these processes. SSBs are not increased or decreased in cells lacking or overproducing catalase.

12.3.3.2 Photoreactivation–Photokill Effects

On one hand, UVA radiation can drive photoreactivation of DNA and prevent inactivation of a UV-irradiated organism; on the other hand UVA has a clear photokill effect.

The action spectra for pyrimidine dimer formation (corrected for transmission through 60 μm of Caucasian epidermis) and for killing and transformation of mammalian cells in culture (not shielded by overlying layers) agree at wavelengths below 330 nm. At longer wavelengths the skin action spectrum for dimer formation falls below that for killing of mammalian cells, which may involve non-DNA damage. There are two caveats: excision repair and photorepair (in the range 360–400 nm) are quite rapid, and both repairs during irradiation might significantly reduce the measured dimer frequency (dependent on duration of irradiation).

Radiation in the photoreactivation range (300–600 nm) produces (1) photochemical pyrimidine dimer reversal, (2) pyrimidine dimer formation with wavelengths as long as 365 nm, but not 385 or 405 nm, and (3) pyrimidine-6,4-pyrimidone dimer (see Chapter 5) photoisomerization to a Dewar isomer on irradiation in the 300- to 380-nm range.

For straightforward measurements of photoreactivation without complications of dimer induction or reversal or photoconversion of other lesions, photoreactivating light should include no radiation shorter than about 380 nm, or longer than the long wavelength edge of the actual spectrum for photorepair of the PR enzyme of interest.

12.3.3.3 Formation of Active Oxygen Intermediates

The absorption of UVA by many cellular compounds can lead to generation of active oxygen intermediates: H_2O_2 by photochemical degradation of tryptophan photolysis products, and both H_2O_2 and superoxide anion by irradiation of NADH and NADPH. *In vivo*, iron complexes react with H_2O_2 to generate the highly reactive hydroxyl radical in a superoxide-driven Fenton reaction. Furthermore, cells contain chromophores such as flavins, quinones, and porphyrins, which may generate singlet oxygen via a type II photodynamic process.

Several types of DNA damage, including single-stranded breaks and DNA–protein cross-links, are dependent on the generation of active oxygen intermediates. UVA also leads to peroxidation of membrane lipids and oxygen-dependent photoinactivation of many enzymes. Active oxygen intermediates also participate in the acute response of skin to UVA. Free radical intermediates may also be implicated in UV-induced carcinogenesis. Endogenous glutathione (unique hydrogen donor for glutathione peroxidase) plays a critical role in protection against the cytotoxic effects of UV radiation.

In bacteria, a UVA-inducible phenomenon independent of the *recA* gene product involves the appearance of several new proteins that can repair a major fraction of UVA-induced lethal damage.

12.3.3.4 Induction of Stress Proteins

In eukaryotes, inducible defense systems can consist of heme oxygenase, for elimination of porphyrins, i.e., endogenous photosensitizers generating singlet oxygen.

The induction of the heme oxygenase gene is a general response to oxidant stress in cultured human and mammalian cells. The response is characteristic of UVA radiation and high levels of expression are induced by sublethal fluences. The response is entirely dependent on the redox state of the cell; depletion of glutathione leads to increased gene expression.

12.3.3.5 Cytoplasmic Targets of UVA

The cellular targets of activated oxygen species (i.e., oxidizing equivalents such as peroxides, superoxide anion, and hydroxyl radical) formed by UVA radiation are multiple. The hydroxyl radicals react most effectively with all amino acids. The oxidation by activated oxygen species of key residues of protein can (1) provoke their inactivation in the case of enzymes or membrane proteins, and (2) induce structural modifications sufficient to destabilize the microarchitecture of the cell in cases when cytoskeletal proteins are affected.

The membranes, rich in polyethylenic fatty acids, are another important target of activated oxygen species. The peroxidation of polyethylenic fatty acids (LH) is triggered in an initial phase by a free radical (for instance, hydroxyl radical) or during type I photodynamic reactions leading to the formation of the radical $L\cdot$ ($LH + OH\cdot \rightarrow L\cdot + H_2O$). The rearrangement of the $L\cdot$ radical and the fixation of molecular oxygen lead to the formation of the peroxyl radical LOO· (L· + O_2 → LOO·). The latter, by extracting in turn one atom of hydrogen, forms the hydroperoxide LOOH and regenerates the radical L· (LOO· + LH → LOOH + L·). This cycle which can theoretically renew itself *ad infinitum* (propagation phase), is interrupted by the recombination reactions of radical species (termination phase).

The chemistry of lipid peroxidation is in fact much more complex, with, among others, reactions of cyclization (formation of endoperoxides) and of fragmentation, and it depends considerably on the fatty acid concerned. The LOOH hydroperoxides also constitute a reservoir of oxygenated radical species. A propagation phase can only exist if these hydroperoxides are decomposed ("activated") by metallic ions (see Chapter 5). The peroxidation of the membrane lipids will alter the structure and thereby the function of membranes (cytoplasmic membrane or organelle membranes), with resulting modifications in transmembrane potentials, ionic fluxes, and transmembrane transport, and also inactivation of receptors, deregulation of messenger systems, etc. Within the membrane structure these peroxidations can, on the other hand, lead to the modification of the protein residues or, further, to lipid–protein bridges.

It is important to note that some of the degradation products of lipid peroxides or fragmentation products during the peroxidation, particularly those of an aldehyde nature, described as mutagenic substances, can interact with DNA. They can also interact with free NH_2 groups to yield Schiff bases. It is suspected that such reactions participate in the formation of polymerized complexes that are the origin of the lipofuscins (blue-fluorescing pigments) encountered in the course of degenerative processes.

12.3.3.6 Induction of Anuclear Organelle Damage

Several cell types were used to study the effects of UVAI in induction of anuclear organelle damage. These include distortion of the plasma membrane, vacuolization of the endoplasmic reticulum, nuclear–cytoplasmic separation, and Heinz bodies in erythrocytes. Heinz body formation indicates cytoskeletal damage and oxidative stress. Addition of reduced glutathione (GSH) prior to irradiation protects from cell lysis in a concentration-dependent manner.

From 20 to 30% of the UVA dose used in unshielded red blood cell studies can reach the cells *in vivo*, which may result in oxidative stress. These estimates have been questioned, as clinicians have never reported a hemolytic patient after a day at the beach. However, in dermatology, after chronic irradiation with UVAI, extensive membrane damage is seen in the vessels of the dermis, and damage to mitochondrial membranes in all of the cells of dermis. Oxidative damage has been shown in conversion of oxyhemoglobin to methemoglobin in the skin of volunteers.

12.3.4 Pigmentation by UVA

For the sake of brevity, information on the biochemistry and photobiology of melanogenesis is presented in Section 12.5. The following material deals with the more general aspects of skin pigmentation induction by UV radiation.

Immediate pigment darkening (IPD) can be used to evaluate the protection factor (PF) of any sunscreen. The *in vivo* transmission of the human epidermis at 313, 365, and 436 nm is, respectively, 9.5, 19, and 34% of the radiation reaching the surface. More than 80% of the UVA radiation is transmitted by the stratum corneum, producing several effects, including IPD. PF-UVA has been estimated for different physical and chemical absorbers [buthylmethoxydibenzoylmethane (DBM, benzophenone)].

Tanning by UVA radiation involves two separate processes: immediate pigment darkening and delayed tanning (DT), also called melanogenesis. DT involves (1) a two- to three-fold increase of Dopa-positive melanocytes, (2) a doubled volume of melanocyte cell body, (3) an increased arborization of dendrites with melanosomes within, (4) ultramorphological signs of secretion in the melanocytes, (5) melanosomes in all stages of development, and (6) signs of transfer of melanosomes to keratinocytes and redistribution within, leading to lasting pigmentation.

DT can be stimulated by suberythemogenic doses of UVA. IPD and DT can be quantified by colorimetry, which cannot distinguish between changes of color due to high-dose UVA from pigmentation elicited by UVB. However, immediately after irradiation with low doses of UVA (9 J/cm^2), IPD differs significantly from DT (brown pigment turns black, in agreement with the assumption of an oxygen-dependent chemical alteration of preexisting pigment).

In contrast to UVA, UVB does not produce an IPD at a dose smaller than required to produce an intense erythema reaction. The skin reaction to UVB involves an erythema that reaches a maximum at 8 hr and decreases thereafter.

UVA-induced hyperpigmentation (persistant pigmentation, UVA) is spectrally different from native epidermal melanin pigmentation (EMP) and UVB-induced pigmentation. Furthermore, UVA produces significant photochemical alterations by introduction of a pigment that, while appearing to be photoprotective in the visible region, actually offers less protection than the native pigment in the UVA region.

The big difference between UVA and UVB is that the system induced by UVA damage is very fast, but *UVB neomelanogenesis* takes place between 24 and sometimes 36 to 48 hr. The whole basal layer in the epidermis is full of melanosomes, which is the difference between UVB- and UVA-induced tanning. *There is hyperplasia associated with UVB, but very little with UVA.*

12.3.5 Photoaging of Skin

Visible skin sagging is induced most efficiently near 340 nm, and wavelengths up to 380 nm are significant contributors. UVA-induced changes affect elastic and collagen fibers and increase dermal cellularity. Typical antiinflammatory agents are effective in protection.

Pathak *et al.* have tested the free radical theory involving generation of reactive oxygen species (singlet oxygen and superoxide anion) in the pathogenesis of photoaging. Kohen *et al.* were able to produce a kind of "accelerated photoaging" recognized by the accumulation of Schiff bases (lipofuscin-like pigment), with UVA irradiation of porphyrin-sensitized cultured murine L fibroblasts.

UVA exposure for 32 weeks resulted in a highly significant increase in the ratio of type III collagen in hairless albino mice. UVA effects on secondary messengers, epidermal growth factor (EGF), Ca^{2+}, inositol phosphate, etc., deserve consideration. The UVA erythema effect on blood vessels may be indirect via mediators. Thus, UVA is a potent inducer of cytokines, tumor necrosis factor (TNF-α), and interleukin 1 (IL-1α and IL-1β), which can induce changes in the levels of type III collagen in human dermal fibroblasts *in vitro*. The action spectrum for photoaging in the mouse is quite similar to that for mouse skin cancer and human erythema, which points to the possibility of common chromophore(s) for all endpoints.

A stiffening of the skin as a result of increased cross-linking, accompanied by reduced collagen synthesis, points to the need for broad-spectrum protection.

12.3.6 Effect of UVA on Immune Function

UVA causes morphological alterations in Langerhans cells similar to those induced by UVB. In shaved C3H mice 24 hr post-UVA, the number of ATPase+, Ia+, and Thy+ dendritic epidermal cells is reduced by 50%, and the remaining cells exhibit morphological alterations. No impairments are detected in the induction of contact hypersensitivity or in the antigen-presenting activity of draining lymph node cells.

Exposure of murine keratinocytes to UVA *in vitro* causes the release of immunosuppressive cytokines. The effect on natural killer (NK) cell activity may be a direct effect of UVA on NK cells rather than on some mediator. As to the sense of well-being experienced by people when they go on solarium beds it may be due to the possible release of opioids. The decrease in antibodies and the elimination of rash are the most impressive effects.

12.3.6.1 Isomerization of Urocanic Acid

Irradiation with broadband sources emitting in the UVA region causes significant isomerization of urocanic acid (UCA) from trans to cis. The mechanism by which UVA radiation causes isomerization of UCA is unclear, because absorption by this molecule decreases precipitously at wavelengths longer than 320 nm. UVA radiation is photobiologically active by mechanisms that usually differ from those that occur with UVB. Most notably, UVA is well known for its photosensitizing reactions. A reasonable hypothesis is that the isomerization of UCA after exposure to UVA is catalyzed by an interaction between UVA and an endogenous photosensitizer in the skin. *In vitro* isomerization of UCA is readily triplet sensitized, which favors the endogenous sensitizer hypothesis.

The hypothesis that UCA is the photoreceptor for UV-induced suppression of contact and delayed-type hypersensitivity responses is supported by the action spectrum of UVB-induced suppression (peak effectiveness at 270 nm and decrease of effectiveness to 3% of maximum at 320 nm). UVA does not suppress contact hypersensitivity, which raises different possibilities: (1) isomerization of UCA may not be involved in immunosuppression—given the plethora of information supporting the role of *cis*-UCA as an initiator of immunosuppression, this is highly unlikely; (2) possibly on interaction with UVA—either the *cis*-UCA molecule remains bound, incapable of delivering the signals for immune suppression, or UVA inhibits the delivery/formation of such signals.

12.4 THE CELL DEFENSE SYSTEM AGAINST PHOTOOXIDATIVE DAMAGE

12.4.1 Definition of Photooxidative Stress

In Chapters 4 and 5 it was shown that many photochemical reactions of biological interest are due to direct or photosensitized oxidation reactions in which excess electronic energy is converted into chemical energy via activation of molecular oxygen. Metabolic pathways normally produce "activated oxygen species." Examples are the mitochondrial electron transfer chain in enzymatic reactions involving mono- or dioxygenases or activation of polyunsaturated fatty acids to form prostaglandin intermediates. By virtue of chemical reaction kinetics, there is some possibility for oxyradicals formed during normal physiological processes to escape from their beneficial migration site and to migrate toward other cellular sites or organelles where they can be very harmful. However, aerobic life has been maintained because an armamentarium of defense mechanisms has evolved, allowing consumption of unnecessary activated oxygen species endangering cell survival. *Photooxidative stress is only a particular component of the various aggressions that supercede homeostasis of the oxidizing species normally produced during cell life.* Both enzymatic and nonenzymatic defenses against the damaging effects of oxidative stress are present in the normal cell. With respect to the main factors promoting or sustaining peroxidative reactions in cells, the cell defense system can be separated into three major mechanisms, i.e., (1) protection by the so-called antioxidant molecules, (2) protection by detoxifying enzymes, and (3) protection against activation of peroxides or superoxide radical-anion (O_2^-) by redox metal ions. Unfortunately, these antioxidant defenses may be insufficient to withstand the overload of oxidizing species that damage essential cell constituents. This protective first line of antioxidant defense is then reinforced by a secondary line constituted by repair systems, such as the DNA repair enzymes that have been examined in this chapter (Section 12.2) or the proteolytic enzymes that degrade oxidized proteins.

12.4.2 Biological Antioxidants

Biological antioxidants are compounds that protect biological systems against the potentially harmful effects of processes and reactions that can cause excessive oxidations. Such oxidations may be induced by photodynamic photosensitizers (see Chapter 5).

The different mechanisms for protection against oxidants, enzymatic processes and nonenzymatic processes, will be examined in order.

12.4.2.1 Enzymatic Processes

Several enzymes have evolved to serve a protective function with respect to biological oxidants. Superoxide dismutase catalyzes the reaction between the superoxide free radical (O_2^-) and protons (H_3O^+) to produce molecular oxygen and hydrogen peroxide:

$$O_2^- + 2H_3O^+ \longrightarrow O_2 + H_2O_2 + 2H_2O$$

12.4.2.1.1 Catalase   Catalase detoxifies one of the reaction products of superoxide dismutase, i.e., hydrogen peroxide:

$$2H_2O_2 \longrightarrow 2H_2O + O_2$$

In the absence of superoxide dismutase and catalase and in the presence of transition metals such as iron and copper, the extremely reactive hydroxyl radical ($OH\cdot$) can be generated (see Chapter 5) by the following reactions:

$$O_2^- + Fe^{3+} \longrightarrow O_2 + Fe^{2+}$$

$$H_2O_2 + Fe^{2+} \longrightarrow OH\cdot + OH^- + Fe^{3+}$$

12.4.2.1.2 Selenium Glutathione Peroxidase Selenium glutathione peroxidase (Se-GSH-Px) is currently postulated to act together with phospholipase A_2 in converting potentially harmful phospholipid hydroperoxides (PLOOH) to free fatty acid hydroperoxides (LOOH) and lysophospholipids, and then to harmless fatty acid alcohols:

$$PLOOH \longrightarrow lysophospholipid + LOOH$$

$$LOOH + 2GSH \xrightarrow{\text{Se-GSH-Px}} LOH + GSSG + H_2O$$

where GSH is reduced glutathione and GSSG is oxidized glutathione.

12.4.2.1.3 Phospholipid Hydroperoxide Glutathione Peroxidase A recently described enzyme, phospholipid hydroperoxide glutathione peroxidase (PLOOH-GSH-Px), acts directly on phospholipid hydroperoxides without the necessity of hydrolyzing the free fatty acid:

$$PLOOH + 2GSH \xrightarrow{\text{PLOOH-GSH-Px}} phospholipid-LOH + GSSG + H_2O$$

In this way the membrane-perturbing properties of the PLOOH are removed without the requirement for activating or mobilizing phospholipase A_2, which might have the undesirable consequence of liberating excessive amounts of substrate for prostanoid synthesis. In addition, PLOOH-GSH-Px can also reduce membrane-associated cholesterol hydroperoxides, thus further diminishing the amount of potentially harmful lipid hydroperoxides.

12.4.2.2 Nonenzymatic Processes

Hydroxyl (OH·), peroxyl (LOO·), alkoxyl (LO·), or alkyl (L·) radicals can initiate and/or propagate the radical chain reaction of lipid peroxidation (Chapter 5). *Any compound that can react with these initiating radicals without generating a reactive radical species can be considered an antioxidant.*

12.4.2.2.1 Lipid-Soluble Antioxidants Lipid-soluble antioxidants include tocopherols, carotenoids, quinones, and bilirubin. Each tocopherol molecule can react with two peroxyl radicals: the first reaction, α-tocopherol + LOO·, gives α-tocopherol· (i.e., the α-tocopheroxyl radical, a resonance-stabilized, oxygen-centered radical) + LOOH; the second reaction is α-tocopherol· + LOO· → LOO–α-tocopherol.

The relative effectiveness of α-tocopherol as an antioxidant in liposome and membrane preparations is only 1–2% of that in homogeneous solutions, due to its lower mobility in the membranes and to the greater probability of chain propagation in the more tightly structured environment of the membranes.

12.4.2.2.1.1 Carotenoids In addition to deactivation of photosensitizer triplet states and of singlet oxygen by energy transfer (see Chapter 4), members of this family of conjugated polyenes have antioxidant activity. They are bleached when exposed to radicals such as those arising during lipid peroxidation. Their long, conjugated double-bond systems make them excellent substrates for radical attack and chain-breaking reactions. β-Carotene (Car) is a better antioxidant at 15 torr (2% oxygen) than at 150 torr (20% oxygen).

β-Carotene might react directly with the peroxyl radical LOO·, the propagation step (see Chapter 5):

$$Car + LOO· \longrightarrow LOO–Car·$$

$$LOO–Car· + LOO· \longrightarrow LOO–Car–OOL$$

The free radical quenching ability continues with the formation of multiple resonance-stabilized, carbon-centered radicals on a single carotene molecule, followed by radical–radical quenching with the addition of another peroxyl:

$$LOO–Car–OOL + LOO· \longrightarrow (LOO)_2–Car–OOL$$

The products formed from the interactions of Car and radicals are primarily carbonyl derivatives of Car, along with some epoxides. The oxidation of carotenoids by radical species can be nonenzymatic or enzymatic. Thus, lipooxygenases and peroxidase of thylakoid membranes, alone or in combination, can oxidize carotenoids.

Canthaxanthin (4,4'-diketo-β,β-carotene) behaves both cellularly and in animals similar to β-carotene. Until new evidence is presented, any chemopreventive action of carotenoids will have as its basis one of the known functions that these molecules possess: photoprotection, radical quenching, or antioxidant activity. Some examples of *in situ* biological activities of carotenoids are presented in the following discussion.

Salmonella typhimurium was exposed to the mutagenic action of 8-methoxypsoralen (8-MOP) and UVA (330–400 nm). The addition of β-carotene dissolved in DMSO (10–100 μg/ml) decreased the number of histidine revertants in 8-MOP/UVA-exposed bacteria, thus acting as an antimutagenic compound. β-Carotene injected intraperitoneally into mice as a solution of water-dispersible beadlets delayed the appearance of UVB-induced (290–320 nm) skin tumors and reduced their growth rate. In 1980, Mathews-Roth reported that β-carotene and canthaxanthin administered orally (6.7 g/kg diet) or phytoene injected intraperitoneally could delay the appearance of UVB-induced skin tumors in mice. When the chemical carcinogen dimethylbenzanthracene (DMBA) and low doses of UVB light were used, both β-carotene and canthaxanthin (at a dose of 33 g/kg diet) were protective with respect to the number of tumors produced in mice. Mathews-Roth (1980) and Krinsky (1991) have looked at the effect of dietary carotenoids on UVB-induced skin tumors in mice. They obtained protection at both 7 g/kg and 700 mg/kg of either β-carotene or canthaxanthin, but at levels of 70 mg/kg neither of these carotenoids was effective.

In an attempt to determine whether the effects of carotenoids in this system occurred during the initiation phase or the promotion–progression phase, Mathews-Roth and Krinsky used a single large dose of UVB light to both initiate and promote skin cancer formation in hairless mice. Both β-carotene and canthaxanthin were fed as beadlets at 1.0 g/kg for either a 6-week period before irradiation or for 24 weeks after irradiation. When canthaxanthin was fed before or after irradiation, they reported a significant decrease in the number of animals with tumors. The same effect was only noted with β-carotene when it was fed after the irradiation period. The difference from their prior studies may be due to the fact that a single large dose of UVB light induces papillomas (benign epidermal tumors), as opposed to the squamous cell carcinomas detected following a multiple suberythemic dosing schedule.

Santamaria *et al.* (1981) have reported that both canthaxanthin and β-carotene were protective against mouse tumors induced by a combination of UVA and benzo[a]pyrene. Very similar results were reported for the inhibition by dietary β-carotene or canthaxanthin (500 mg/kg) of skin tumors induced by near-UV irradiation of mice painted with 8-methoxypsoralen.

Immunoenhancement has been reported after oral administration of carotenoids. Dietary β-carotene and canthaxanthin were shown to enhance the responses of isolated T and B lymphocytes to a variety of mitogens. Oral administration of carotenoid pigments results in an enhanced immunoresponse of mice rechallenged with a syngenic tumor. When lymph cells and tumor cells from animals inoculated with a syngenic tumor and on a diet supplemented with β-carotene were injected into recipient mice, the tumor growth was only one-seventh of that of cells obtained from animals that were not on carotenoids. When splenocytes from UVB-irradiated C3H/HENCRLBR mice were transferred to recipients challenged with a syngenic UV-induced tumor, in the group supplemented with either canthaxanthin (10 g/kg) or retinyl palmitate (100,000 IU/kg) there was no significant decrease in the number of animals growing tumors, but in the group supplemented with both canthaxanthin and retinyl palmitate, a significant decrease was observed. Canthaxanthin could have prevented the oxidation of sterols, which makes these compounds strong immunosuppressants.

The *in vitro* and *in vivo* antioxidant effect may well be attributable to the intact carotenoid molecule, although metabolites of these compounds may be exerting this important biological activity. β-Carotene has proved to be nontoxic in humans based on 20 years of administration to patients with erythropoietic protoporphyria. The most recent compendium of carotenoids lists 563 pigments that has been isolated and characterized.

12.4.2.2.1.2 *Quinones* Coenzyme Q, ubiquinone, in its reduced form (UQH_2), inhibits lipid peroxidation, either by reducing the α-tocopheroxyl radical, or by acting directly on either peroxyl or alkoxyl radicals.

12.4.2.2.1.3 *Bilirubin* This end product of heme metabolism is almost as effective an antioxidant as α-tocopherol in liposomes.

12.4.2.2.2 Water-Soluble Antioxidants Water-soluble antioxidants include ascorbic acid, uric acid, metal-binding proteins, and binding proteins for heme and heme-containing proteins. The major extracellular antioxidants are albumin, ascorbic acid, ceruloplasmin, heptaglobulin, hemopexin, lactoferrin, transferrin, and uric acid. The metal-binding proteins reduce the effective concentration of transition metals capable of reacting with hydroperoxides (see Chapter 5).

The iron-binding protein transferrin, which exhibits very high affinity to iron, normally carries only 20–30% of its full capacity for iron. Lactoferrin, made by neutrophils and released into plasma, has very similar properties. Ceruloplasmin appears to have two antioxidant properties: (1) it first binds copper ions and prevents this transition metal from catalyzing hydroperoxide decomposition, and (2) it oxidizes Fe^{2+} to Fe^{3+} and concomitantly converts molecular oxygen to water.

Both heme and heme proteins (hemoglobin and myoglobin) are prooxidants; they react with hydrogen peroxide, forming a ferryl species that can initiate lipid peroxidation. In addition, hydrogen peroxide can cause the release of free iron from heme or heme proteins, which in turn can stimulate peroxidative reactions. Protection from such reactions is obtained with haptoglobin which binds hemoglobin, and hemopexin, which binds free heme. Both complexes are then rapidly cleared from the circulation.

12.5 MELANOGENESIS

Melanin and its distribution in human epidermis is an example of an adaptive response to the detrimental effects of solar UV radiations. White skin shows biological inadequacy in terms of protection against solar environments. Melanin pigmentation is a complex phenomenon involving a highly specialized cell, the melanocyte, which secretes an effective solar light filter. Traditionally, all pigments found in granules (melanosomes) within melanocytes are called melanins.

12.5.1 Biosynthesis of Melanins

In human beings, two kinds of melanins are found: the brown-to-black *eumelanins* and the light-brown, reddish *pheomelanins* (Figure 12.7). These chemically different melanins are the end-products of two diverging metabolic paths after the enzymatic oxidation of tyrosine by tyrosinase: the key copper-containing enzyme of melanogenesis which produces, in a two-step catalytic reaction, dopaquinone, the common precursor of both types of melanin. In fact, tyrosinase activity is increased by the presence of another enzyme called TRPI (or catalase b). The eumelanins thus arise from the copolymerization of tyrosine oxidation and cyclization intermediates including DOPA (5-dihydroxyphenylalanine), dopaquinone, dopachrome, 5,6-dihydroxyindole, 5,6-indole-quinone, and pyrrolecarboxylic acid (Figure 12.8). The pheomelanins are reddish-brown pigments which result from the copolymerization of tyrosine and/or dopaquinone with cysteine or glutathione. The most common monomer units are the 6,7,8,9-tetrahydro-4-hydrothiazolo-(4,5-H)-isoquinoline-7-carboxylic acid and the cysteinyldopas (Figure 12.8). The colored polymers of melanin are bound to a protein forming a melano–protein complex.

12.5.2 Photobiology of Melanins

12.5.2.1 Melanosomes

The melano–protein complex is found in organelles, the *melanosomes*, whose size, shape, and melanin content are genetically determined. In caucasian skin, the melanosome size is about $0.5 \times 0.1 \ \mu\text{m}$; melanosomes are aggregated and pheomelanin pigments predominate. In negroid skin, melanosomes are large ($0.7 \times 0.4 \ \mu\text{m}$), fully melanized, and nonaggregated with eumelanin pigments prevailing.

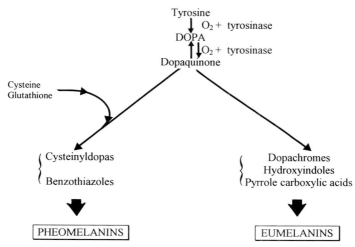

FIGURE 12.7 Scheme of melanogenesis. Note the possible interconversion between dopaquinone and dihydroxyphenylalanine (DOPA).

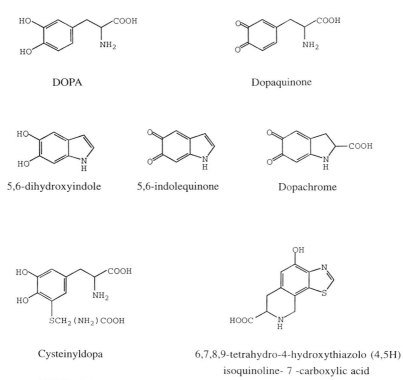

FIGURE 12.8 Chemical structures of important melanin constituents.

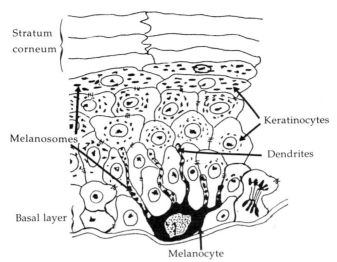

FIGURE 12.9   Scheme showing the dendritic melanocyte surrounded by keratinocytes in the basal and suprabasal layers of epidermis (epidermal melanin unit). Note the position of the melanosomes around the nucleus of the keratinocytes in the suprabasal layer.

There are no more melanocytes in negroid than in caucasian skin. However, in negroids, melanocytes are more actively producing melanin and melanin granules are found in all layers of the epidermis including the stratum corneum. In contrast, in caucasians, melanins are only found in the basal and suprabasal layers of the epidermis. In the epidermis, melanocytes are involved in the "epidermal melanin unit" consisting of one melanocyte surrounded by about 36 keratinocytes (Figure 12.9). Along the basal layer, the melanocyte/keratinocyte ratio is 1/10.

Exposure of human skin to sunlight induces tanning. There are two photobiologically distinct mechanisms involved in the tanning process: (1) immediate pigment darkening (IPD) and (2) delayed tanning (DT) as described in Section 12.3.5. One minimal tanning dose (MTD) of UVA results in a migration of melanosomal complexes within keratinocytes from a perinuclear position to a more distinct location near the cell periphery. This phenomenon is specifically induced by UVA, with 1 to 2 MTDs of UVB resulting in erythema but no redistribution. The redistribution of melanosomes within the keratinocyte is readily observable 18–20 hr after exposure to UVA and persists through the DT process.

12.5.2.2 Photoprotection by Melanins

A number of possible mechanisms by which melanins protect skin from UV photodamage have been suggested. They include filtering and attenuation of incident radiation by scattering, absorption, and heat dissipation. However, in the past two decades, besides this *photoprotective* role, melanins have been shown to be photosensitizing agents. As a result they may not be the inert material that they were believed to be.

(a) Melanins protect from deleterious effects of solar UV radiation which have been transmitted through the stratum corneum by direct UV photon absorption or UV light scattering by melasnosomes.

(b) Melanins absorb visible and infrared photons. The absorbed energy is released as heat; Africans do not enjoy sun-bathing because the absorbed energy raises their cutaneous temperature.

(c) Melanins act as an amorphous semiconductor. Both eumelanins and pheomelanins are stable free radicals exhibiting electron spin resonance signals of semireduced quinoid species. These free radicals are produced during the interconversion between quinoid and hydroxylated derivatives, as shown in Figure 12.7 for dopaquinone and DOPA. Quinoid structures must react with reducing radicals to form hydroxylated derivatives or, conversely, oxidizing radicals reacting with DOPAlike structures (e.g., dihydroxyl derivatives) produce quinones as end-products. In this manner, melanins are able to deactivate harmful free radicals formed by UV light in cells.

Melanins are double-edged swords because they are susceptible to photooxidation by wavelengths as long as 530 nm. This photooxidation generates the activated oxygen species which can in turn induce the lipid peroxidation of polyunsaturated fatty acids. With arachidonic acid as a substrate, this reaction produces hydroperoxides of eicosatetranoic acid (HPETE) which are potent inflammogens, platelet aggregation blockers, and chemotactic agents. Thus, the possible role of melanin photochemistry should not be overlooked in human skin response to solar UV light.

Bibliography

Anderson, R. R. and Parrish, J. A. (1982). *In* "The Science of Photomedicine" (J. D. Regan and J. A. Parrish, eds.), p. 165. Plenum, New York and London.

Aubin, F., and Kripke, M. L. (1992). Effects of ultraviolet A radiation on cutaneous immune cells. *In* "Biological Responses to Ultraviolet A Radiation" (F. Urbach, ed.), pp. 239–247. Valdenmar Publ. Co., Overland Park, KS.

Bissett, D. L., Hannon, D. P., McBride, J. F., and Patrick, L. F. (1992). Photoaging of Skin by UVA. *In* "Biological Responses to Ultraviolet A Radiation" (F. Urbach, ed.), pp. 181–187. Valdenmar Publ. Co., Overland Park, KS.

Cesarini, J. P. (1992). Immediate pigment darkening; an useful epidermal response to monitor UVA aggression. *In* "Biological Responses to Ultraviolet A Radiation" (F. Urbach, ed.), pp. 139–143. Valdenmar Publ. Co., Overland Park, KS.

Chardon, A., Cretois, I. and Hourseau, C. (1992). Color changes induced on various skin categories by repeated exposures to UV rays. *In* "Biological Responses to Ultraviolet A Radiation" (F. Urbach, ed.), pp. 159–175. Valdenmar Publ. Co., Overland Park, KS.

Chedekel, M. R. (1992). Photochemistry and photobiology of epidermal melanins. *Photochem. Photobiol.* **35**, 881–885.

De Fabo, E. C., Reilly, D. C., and Noonan, F. P. (1992). Mechanisms of UVA effects on immune function: preliminary studies. *In* "Biological Responses to Ultraviolet A Radiation" (F. Urbach, ed.), pp. 227–237. Valdenmar Publ. Co., Overland Park, KS.

de Gruiji, F. R., and van der Leun, J. C. (1992). Action spectra for carcinogenesis. *In* "Biological Responses to Ultraviolet A Radiation" (F. Urbach, ed.), pp. 91–97. Valdenmar Publ. Co., Overland Park, KS.

Eckardt-Schlipp, F., Ahne, A., Obermaier, S., and Wendel, S. (1991) UV-inducible repair in yeast. *In* "Photobiology: The Science and Its Applications" (E. Riklis, ed.), pp. 155–162. Plenum, New York and London.

Epstein, J. H. (1977). Effects of beta-carotene on ultraviolet induced cancer in hairless mouse skin. *Photochem. Photobiol.* **25**, 211–213.

Estevenon, A. M., and Sicard, N. (1991). Excision-repair capacity in UV irradiated strains of *S. pneumoniae*. *In* "Photobiology: The Science and Its Applications" (E. Riklis, ed.), pp. 149–154. Plenum, New York and London.

Ferguson, J., and Johnson, B. (1992). Recently developed photosensitizing agents. *In* "Biological Responses to Ultraviolet A Radiation" (F. Urbach, ed.), pp. 107–119. Valdenmar Publ. Co., Overland Park, KS.

Frederick, J. E., and Alberts, A. D. (1992). The natural UV-A radiation environment. *In* "Biological Responses to Ultraviolet A Radiation" (F. Urbach, ed.), pp. 7–18. Valdenmar Publ. Co., Overland Park, KS.

Godar, D. E., and Beer, J. Z. (1992). UVA1-induced nuclear damage in mammalian cells. *In* "Biological Responses to Ultraviolet A Radiation" (F. Urbach, ed.), pp. 65–73. Valdenmar Publ. Co., Overland Park, KS.

Hariharan, P. V., Remsen, J. F., and Cerutti, P. A. (1975). Excision-repair of gamma-ray damaged thymine in bacterial and mammalian systems. *Basic Life Sci.* **5**, 51–59.

Kim, S. -T., Malhotra, K., and Sancar, A. (1992). Functional and structural characterization of *E. Coli* photolyase by biophysical methods and genetic engineering. *In* "Biological Responses to Ultraviolet A Radiation" (F. Urbach, ed.)., pp. 121–129. Valdenmar Publ. Co., Overland Park, KS.

Kligman, L. H. (1992). UVA-induced biochemical changes in hairless mouse skin collagen: a contrast to UVB effects. *In* "Biological Responses to Ultraviolet A Radiation" (F. Urbach, ed.), pp. 209–215. Valdenmar Publ. Co., Overland Park, KS.

Kochavar, I. E. (1992). Mechanisms of phototoxicity for nonsteroidal anti-inflammatory drugs. *In* "Biological Responses to Ultraviolet A Radiation" (F. Urbach, ed.), pp. 101–106. Valdenmar Publ. Co., Overland Park, KS.

Kohen, E., Kohen, C., Reyftmann, J. P., Morliere, P., and Santus, R. (1984). Microspectrofluorometry of fluorescent photoproducts in photosensitized cells. Relationship to cellular quiescence and senescence in culture. *Biochim. Biophys. Acta* **805**, 332–336.

Kollias, N. (1992). UVA melanogenesis–spectral observations. *In* "Biological Responses to Ultraviolet A Radiation" (F. Urbach, ed.), pp. 151–157. Valdenmar Publ. Co., Overland Park, KS.

Kramer, G. F., Baker, J. C., and Ames, B. N. (1988). Near-UV stress in *Salmonella typhimurium*: 4-thiouridine in tRNA IppGpp and ApppGpp as components of an adaptive response. *J. Bacteriol.* **170**, 2344–2351 (cited in Urbach, 1992).

Krinsky, N. L. (1991). Effects of carotenoids in cellular and animal systems. *Am. J. Clin. Nutri.* **53**, 238S–246S.

Krinsky, N. L. (1992). Mechanism of action of biological antioxidants. "Proc. Soc. Exptl Biol. Med. **200**," 238–254.

Ley, R. D. (1993). Photorepair of pyrimidine dimers and suppression of ultraviolet radiation

(UVR)-induced pathologic changes in the skin and eyes. *Photochem. Photobiol.* **57,** Suppl., 45S.

Ley, R. D. (1992) Photo reactivation in tissues. *In* "Biological Responses to Ultraviolet A Radiation" (F. Urbach, ed.), pp. 131–135. Valdenmar Publ. Co., Overland Park, KS.

Mathews-Roth, M.M. (1980). Carotenoid pigments as antitumor agents. *In* "Current Chemotherapy and Infectious Diseases" (J. D. Nelson and C. Grassi, eds.), pp. 1503–1505. Am. Soc. Microbiol., Washington, DC.

McEntee, K. (1982). Molecular aspects of errorprone repair in *Escherichia coli. In* "Trends in Photobiology" (C. Hélène, M. Charlier, T. Montenay-Garestier and P. Laustriat, eds.), p. 193. Plenum, New York and London.

McGregor, W. G., Wang, Y. -C., Chen, R. -H., Maher, V. M., and McCormick, J. J. (1993). Spectra of UV-induced mutations arising in the coding region of the human hypoxanthine (guanine) phosphoribosyltransferase (HPRT) gene: Biological effect of preferential repair of the transcribed strand. *Photochem. Photobiol.* **57,** Suppl., 43S–44S.

McKinlay, A. F. (1992). Artificial sources of UVA radiation: Uses and emission characteristics. *In* "Biological Responses to Ultraviolet A Radiation" (F. Urbach, ed.), pp. 19–38. Valdenmar Publ. Co., Overland Park, KS.

Morliere, P., Moysan, A., Gaboriau, F., Santus, R., Maziere, J. C., and Dubertret, L. (1992). Ultraviolet A et peau: Implication d'espèces activées de l'oxygène. Tendances actuelles et resultats recents. *Pathol. Biol.* **40,** 160–168.

Noonan, F. P., and De Fabo, E. C. (1991). Immune suppression by ultraviolet b irradiation of urocanic acid. *In* "Photobiology: The Science and Its Applications" (E. Riklis, ed.), pp. 763–768. Plenum, New York and London.

Ortel, B., and Gange, R. W. (1992). UVA action spectra for erythema and pigmentation. *In* "Biological Responses to Ultraviolet A Radiation" (F. Urbach, ed.), pp. 79–81. Valdenmar Publ. Co., Overland Park, KS.

Pathak, M. A., and Carbonare, (1992). Reactive oxygen species in photoaging and biochemical studies in the amelioration of photoaging changes. *In* "Biological Responses to Ultraviolet A Radiation" (F. Urbach, ed.), pp. 189–205. Valdenmar Publ. Co., Overland Park, KS.

Peak, M. J., Peak, J. G., and Churchill, M. E. (1992). Cellular and molecular effects of UVA radiation and visible light in mammaliam cells. *In* "Biological Responses to Ultraviolet A Radiation" (F. Urbach, ed.), pp. 39–43. Valdenmar Publ. Co., Overland Park, KS.

Sancar, A., and Rupp, W. D. (1983). *Cell (Cambridge, MA)* **33,** 249–260.

Santamatria, L., Bianchi, A., Arnaboldi, A., and Andreoni, L. (1981). Prevention of the benzo[a]pyrene photocarcinogenic effect by beta-carotene and canthaxanthin. *Med. Biol. Environ.* **9,** 113–120.

Sayre, R. M., and Kligman, L. H. (1992). Action spectra for photoelastosis: a review of experimental techniques and predictions. *In*: "Biological Responses to Ultraviolet A Radiation" (F. Urbach, ed.), pp. 83–89. Valdenmar Publ. Co., Overland Park, KS.

Smith, K. C. (1971). *Photophysiology* **6,** 209–278.

Smith, K. C. (1989). UV radiation effects DNA repair and mutagenesis. *In* "The Science of Photobiology" (K. C. Smith, ed.), 2nd ed., pp. 111–133. Plenum, New York and London.

Sutherland, B. M., Hacham, H., Gange, R. W., and Sutherland, J. C. (1992). Pyrimidine dimer formation by UVA radiation: implication for photoreactivation. *In* "Biological Responses to Ultraviolet A Radiation" (F. Urbach, ed.), pp. 47–57. Valdenmar Publ. Co., Overland Park, KS.

Tyrrell, R. M. (1992). Inducible responses to UVA exposure. *In* "Biological Responses to Ultraviolet A Radiation" (F. Urbach, ed.), pp. 59–64. Valdenmar Publ. Co., Overland Park, KS.

Urbach, F. (1992). Introduction. *In* "Biological Responses to Ultraviolet A Radiation" (F. Urbach, ed.), pp. 1–6. Valdenmar Publ. Co., Overland Park, KS.

Urbach, F. (1992). Discussion. *In* "Biological Responses to Ultraviolet A Radiation" (F. Urbach, ed.), pp. 99–100, 137, 177–179, 281–283. Valdenmar Publ. Co., Overland Park, KS.

Wei, Q., Matanoski, G. M., Farmer, E. R., Hedeyati, M.A., and Grossman, L. (1993). Epidemiology of UV-DNA repair capacity of human skin cancer. *Photochem. Photobiol.* **57**, Suppl., 43S.

Young, A. R., Plastow, S. R., Harrison, J. A., Walker, S. L., and Hawk, J. L. M. (1992). UVA-induced changes in mouse skin collagen. *In* "Biological Responses to Ultraviolet A Radiation" (F. Urbach, ed.), Valdenmar Publ. Co., Overland Park, KS.

CHAPTER 13

Optical Properties of the Skin

13.1 DEFINITION OF OPTICAL AND STRUCTURAL PROPERTIES OF THE SKIN

Skin consists of three main layers: (1) the epidermis, separated from the underlying dermis by a basement membrane, (2) the dermis (typical thickness 1000–4000 μm), with collagen and elastic fibers produced by fibroblasts, blood and lymph vessels, hair follicles, sweat and sebaceous glands, erector pilae smooth muscles, and nerves, and (3) the subepidermal tissue, consisting of a fat layer.

The epidermis of the skin consists of four layers: (1) the basal layer (stratum germinativum, where division occurs (5–10 μm thick), (2) the *stratum spinosum,* consisting of keratinocytes with intercellular bridges (50–150 μm), (3) *the granular layer (stratum granulosum),* where keratohyalin granules are formed (3 μm), and (4) the *stratum corneum,* where anucleated cells form a protective layer (8–15 μm).

As to what is called the Malpighian layer, there seems to be a certain ambiguity in the term depending on which source is used. An histology textbook from the sixties omits mention of the stratum spinosum and refers to the Malpighian layer as the stratum germinativum. In the Webster's Medical Dictionary (1986) the Malpighian layer is defined as the deeper part of the epidermis consisting of cells whose protoplasma has not yet changed into horny material—also called stratum germinativum. In the Textbook of Histology by Junquiera *et al.* (1992) "All mitoses are confined to what is termed the Malpighian layer which consists of both the stratum basale and the stratum spinosum. And in the Stedman's Medical Dictionary (25th ed.), the Malpighian layer is defined as the living layer of the epidermis comprising the stratum basale, stratum spinosum, and stratum granulosum.

Melanocytes are dendritic cells located in the basal layer (see Chapter 12, Section 12.5). Their processes extend up into the malpighian layer. Melanocytes and melanin-loaded malpighian cells (keratinocytes) compose the epidermal melanin unit.

When radiation strikes the skin, part is absorbed, part is remitted, and part is transmitted through successive layers of skin, until the energy of the incident beam has been dissipated (Figure 13.1).

The *transmittance* of a planar sample is defined as that fraction of the radiation incident on one side of the sample that passes through and emerges from the other side of the sample. *Remittance* is defined as that fraction of the radiation incident on one side of a sample that returns from or through the same side. The term *diffuse reflectance* is spatially isotropic remitted radiation (Figure 13.1).

The *regular reflectance* of an incident beam normal to skin is 4–7% over the spectrum from 250 to 3000 nm. At lower angle incidence, higher reflectance is described by the Fresnel equations. The nonplanar surface topology of the stratum corneum is conceptually similar to that of translucent ground glass. The fraction of normally incident radiation entering the skin is 93–96%. Within any layer of the skin this radiation may be *absorbed* or *scattered*.

The *action spectrum* wavelengths of photobiologic responses are those absorbed by specific chromophores that initiate the photochemical reactions. One may correct the biological action spectrum (Figure 13.2) for tissue optics.

13.1.1 Perception of Skin Color

Fair Caucasian skin remits about half of the incident radiation. UV radiation shorter than 320 nm is absorbed by proteins, DNA, and other constituents of epidermal cells (Figure 13.3).

Perception of skin color is determined by remittance in the visible range. The 5 to 10% of incident light that is reflected by the outer surface

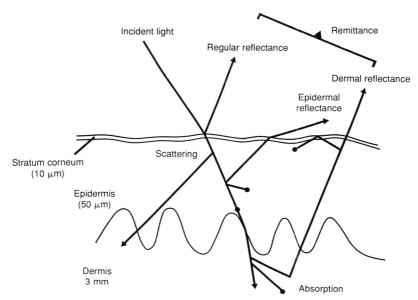

FIGURE 13.1 Schematic representation of optical pathways in human skin. Adapted from Kochevar *et al.* (1993) in "Dermatology in General Medicine" (T. B. Fitzpatrick *et al.*, eds.). With permission of McGraw-Hill, New York.

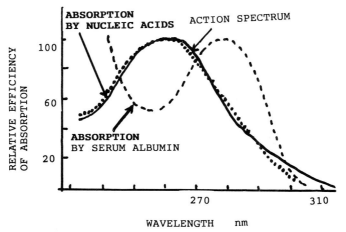

FIGURE 13.2 Comparison of the action spectrum for cell killing with the absorption spectra for nucleic acids and serum albumin. Adapted from Kochevar *et al.* (1993), Section 22, Chapter 131. With permission of McGraw-Hill, New York.

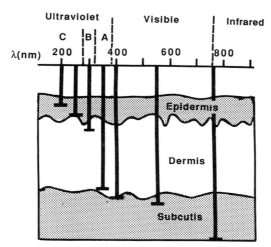

FIGURE 13.3 Depth of penetration of ultraviolet, visible, and infrared wavelengths in the epidermis, dermis, and subcutis.

accounts for the surface appearance of the skin (glossy or rough). In full-thickness fair Caucasian skin, the bulk of the visible remittance scattered from various depths of the dermis returns back through the skin surface. Skin with scales (e.g., psoriasis) scatters more light than normal skin.

Melanin, which absorbs rather uniformly over the visible wavelengths, acts as a neutral density (gray) filter to diminish dermal remittance. Blood within the dermis scatters the longer wavelengths and absorbs the blue–green wavelengths, resulting in a reddish hue.

13.1.2 Relative Roles of Absorption, Scattering, and Transmission

In Caucasian skin, at least 20–30% of the incident radiation in the sunburn spectrum (290–320 nm) reaches the malpighian cells, and probably 10% penetrates to the upper dermis. The stratum corneum of black skin absorbs a greater amount of UVB radiation due to melanin. A larger proportion of UVA and visible wavelengths can penetrate and be absorbed by photosensitizers in the dermis.

13.1.2.1 Measurement of Spectral Transmittance and Remittance through Stratum Corneum or Dermis

Integrating spheres covered with white lusterless paint on the inside are used to measure spectral transmittance and remittance (Figure 13.4). For transmittance, a layer of epidermis or skin is held in a hemispherical

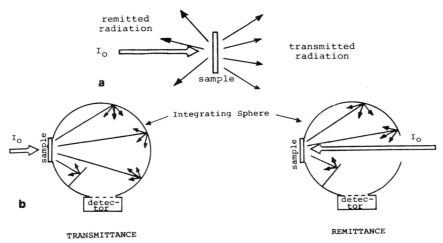

remitted
radiation

I_o

transmitted
radiation

sample

a

I_o

sample

Integrating Sphere

I_o

sample

detec-
tor

detec-
tor

b

TRANSMITTANCE

REMITTANCE

FIGURE 13.4 Determination of transmittance and remittance of a sample, using an integrating sphere. The interior surface of the sphere is coated with $BaSO_4$ or MgO, which have nearly 100% diffuse remittance over the entire optical spectrum. a (upper figure) remitted and transmitted radiation b (lower figure) integrating spheres. From Anderson and Parrish (1982), p. 153. With permission of Plenum Press, New York.

chamber, at the entrance of the integrating sphere. The shape of the chamber and the presence of saline beneath the sample minimize errors due to reflectance at the bottom surface of the sample. The radiation crossing the sample undergoes multiple reflections until it is absorbed by the photodetector.

For remittance measurements, the skin sample diametrically (see Fig. 13.4) faces the entrance window. The measuring beam then strikes the sample through the window. Most of the light transmitted is absorbed by a black carrier. Radiation remitted undergoes multiple reflections until it is absorbed by the photodetector.

13.1.2.2 Transmission of UV Radiation through the Skin

UVB radiation (280–320 nm) produces immediate and delayed erythema, pigmentation, actinic keratosis, and skin cancer. UVA radiation can produce the same effects, but with almost one order of magnitude less efficiency. A decrease in the erythema action spectrum around 280 nm is ascribed to *absorption by the aromatic amino acids of keratin.* Urocanic acid and peptides in the stratum corneum absorb wavelengths shorter than 290 nm. The nucleoproteins in the Malpighian layers absorb around 260 nm. The biologic effects of relatively small exposures to wavelengths shorter than 320 nm are significant.

TABLE 13.1   Approximate Penetration of Incident
Light Energy Density (100% at Skin Surface)

Wavelength (nm)	Remaining energy at indicated depth (μm)		
	37%	10%	1%
250	2	4.5	9
280	1.5	3.5	7
300	6	14	28
350	60	140	260
400	90	210	240
450	150	350	700
500	230	530	1100
600	550	1300	2500
700	750	1700	3400
800	1200	2700	5400[a]

[a] Subcutaneous.

Transmission through fair Caucasian skin is approximately 50% for
UVA radiation and remains high in the visible and infrared regions,
while the photobiologic effectiveness is lower than UVB it is still important, particularly with much higher sun emission in this range.

The calculated values for optical penetration into very fair skin are
listed in Table 13.1.

13.2 THE MODELING OF SKIN OPTICAL PROPERTIES

13.2.1 The Kubelka–Munk Model for Radiation Transfer in a Scattering, Absorbing Medium

Radiation within the sample is broken simply into two opposing
diffuse fluxes, I and J (Figures 13.5 and 13.6). The sample's backscattering
(S) and absorption (K) coefficients for diffuse radiation are defined by
two differential equations:

$$dI = (-KI - SI + SJ)dx \qquad (13.1)$$

$$-dJ = (-KJ - SJ + SI)dx \qquad (13.2)$$

Equations (13.1) and (13.2) simply state that the change dI in flux I
(or J) over some layer of thickness dx is equal to that fraction of I (or J)
removed by absorption and scattering plus some fraction contributed to I
(or J) by backscattering from J (or I). The minus sign in Eq. (13.2) is
necessary because J is defined as positive in the $-x$ direction.

In order to use the Kubelka–Munk model practically, one must also
account for regular reflection occurring at both sample boundaries and

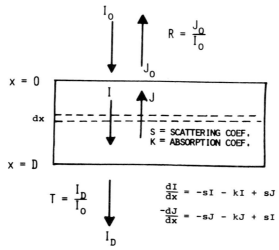

FIGURE 13.5 The Kubelka–Munk model for radiation transfer in a scattering, absorbing medium. I and J are defined as diffuse fluxes, and S and K are backscattering and absorption coefficients for diffuse radiation, respectively. From Anderson and Parrish (1992), p. 158. With permission of Plenum, New York and London.

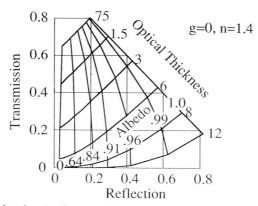

FIGURE 13.6 A plot showing how optical properties of tissue can be derived from reflection and transmission. Albedo and optical thickness are also plotted. The scattering coefficient μ_s is the reciprocal of the average distance traveled by light before being scattered. The absorption coefficient μ_a is the same for absorption. The dimensionless albedo is $a = \mu_s/\mu_a + \mu_a$ and the optical thickness is $\tau = d(\mu_s + \mu_a)$, where d is the geometrical thickness. From Prahl *et al.* (1992), addendum to Vol. 55 Suppl, distributed at the 20th Annual Meeting of the American Society of Photobiology.

adhere to the use of diffuse incident radiation. This requirement has been accomplished for dermal samples *in vitro*, and S and K for *human dermis devoid of blood* has been calculated. *In vivo*, blood exerts a major influence on K, which thus far has been only crudely estimated from *in vivo* remittance spectra of depigmented skin. Once S and K are known, the two fluxes I and J can be constructed. *It is the algebraic sum of* I *and* J *at a certain depth that determines optical dosimetry within the layer.*

For an infinitely thick sample in which T approaches zero, the depth at which the radiation density is attenuated to $1/e$, or about 37%, of its incident value, is derived from $I + J$. Near the front surface of a scattering sample, the sum of I and J can easily exceed I_o, the incident density of optical radiation. In the extreme case, $I + J$ just inside the surface can be twice I_o.

The density of optical radiation in the superficial layers of a scattering sample can easily exceed that of the incident beam. This is particularly the case for fair Caucasian skin, especially over the visible and near-infrared regions.

In the Kubelka–Munk model, variations in the coefficients S and K are essentially independent of one another if the absorbing chromophores scatter weakly compared with surrounding nonabsorbing but strongly scattering structures. The structures of skin that lead to strong scattering, and thereby determine S, are, in general, different from the chromophores that determine K.

In the Kubelka–Munk model, for samples with $S = 0$, the flux at different depths is simply given by the Beer–Lambert law, which states that radiation traveling through a purely absorbing sample is attenuated as a simple exponential function of an absorption coefficient times the pathlength (see Chapter 2).

The resolution of the simultaneous differential equations (13.1) and (13.2) leads to Eq. (13.3):

$$K/S = [(1 + R^2 - T^2)/2R] - 1 \qquad (13.3)$$

which relates the ratio of absorbance over scattering, i.e., K/S to the transmittance T and remittance R. According to the same model, for samples where $K = 0$, i.e., there is no absorption, the equation for the K/S ratio becomes

$$K/S = 0 = [(1 + R^2 - T^2)/2R] - 1$$
$$= (1 + R^2 - T^2 - 2R)/2R$$
$$= 1 + R^2 - T^2 - 2R$$
$$(1 - R)^2 = T^2$$
$$1 - R = T$$

and $R + T = 1$, which indicates that no radiation is lost.

If a sample is infinitely thick, or simply thick enough that T approaches 0, Eq. (13.3) for K/S can be rewritten as:

$$K/S = [(1 + R^2)/2R] - 1$$

$$K/S = (1 + R^2 - 2R)/2R$$

$$K/S = (1 - R)^2/2R$$

which means that the remittance of a thick sample depends solely on the ratio of its absorption and scattering coefficients. For all optical wavelengths less than 600 nm, the dermis can be thought of as infinitely thick so that Eq. (13.3) applies, greatly simplifying the analysis of skin remittance spectra as related to changes in dermal pigments. It is apparent that as the thickness of any particular sample decreases, R always decreases and T always increases.

13.2.2 Charting of Tissue Optical Properties Derived from Reflection and Transmission

Tissue optical properties can be charted by derivation from reflection and transmission coefficients. The plots given in Figure 13.6 are based on the following assumptions:

1. The tissue is a uniform, flat slab (i.e., a plate or slice).
2. The index of refraction is about 1.4.
3. The tissue is not bounded by glass slides.
4. Internal reflection follows the Fresnel formulas.
5. The anisotropy is zero.
6. The irradiance on the sample is perpendicular.
7. Polarization effects are negligible.
8. All the light is collected for reflection and transmission measurements.

The four parameters plotted are transmission, reflection, albedo, and optical thickness. The albedo is the ratio of light remitted to light incident. The scattering coefficient μ_s used in the calculation of albedo is the reciprocal of the average distance traveled by light before being scattered. The absorption coefficient μ_a also used in the calculation of albedo is the same as above for absorption (and is equal to K above). The dimensionless albedo a is equal to

$$\mu_s/\mu_s + \mu_a \tag{13.4}$$

The optical thickness τ is equal to

$$d(\mu_s + \mu_a) \tag{13.5}$$

where d is the physical thickness of the sample.

If reduced optical properties are used, in which g the cosine of the scattering angle is introduced (equal to 0 for diffuse scattering and to 1 for exclusively forward scattering), the above parameters (now designated μ'_s, μ'_a, a', and τ') need to be redefined, i.e.:

$$\mu'_s = \mu_s(1 - g) \qquad a' = \mu'_s/\mu'_s + \mu_a \qquad (13.6)$$

$$\tau' = d(\mu'_s + \mu_a) \qquad (13.7)$$

13.3 OPTICS OF THE STRATUM CORNEUM AND EPIDERMIS

The UV–visible transmission of the Caucasian stratum corneum and epidermis is affected by tryptophan, tyrosine, and other aromatic chromophores that absorb near 280 nm. Nucleic acids (absorption maxima near 260 nm) and urocanic acid (absorption maximum at 277 nm at pH 7.4) also contribute to the 280-nm absorption band. Melanin plays a major role in determining transmission through the epidermis, while the high degree of absorbance of the stratum corneum and epidermis below 250 nm is largely due to peptide bonds.

In transmission measurements, a "solar-blind" detector insensitive to wavelengths over 320 nm is used to eliminate autofluorescence due to tryptophan (excitation, 280 nm; emission, 330–400 nm, see Chapter 2). Total internal reflection from the epidermal sheet or the quartz slide at the entrance port of the integrating sphere is overcome by floating the sample over a hemispherical quartz chamber containing normal saline.

In Caucasians, 1% of incident radiation below 300 nm that reaches the dermis produces a delayed erythema by direct damage of dermal vessels and a transient decrease in circulating lymphocytes.

Melanogenesis and epidermal hyperplasia are UV-inducible photoprotection mechanisms. Even mild hyperplasia protects significantly against UVB and UVC radiation. For UVA protection, induction of melanogenesis is critical. *Melanin absorption steadily increases from 250 to 1200 nm.* Beyond 1100 nm, both transmittance and remittance are unaffected by melanin. *The prevalence of sunburn, abnormal photosensitivity, skin cancer, and cutaneous "aging" decreases with increasing melanin concentration.* UV absorption primarily accounts for melanin's photoprotective effect, although melanin may protect keratinocytes by free radical quenching (see Chapter 12, Section 12.5).

Supranuclear melanin "caps" of pigment granules may produce better photoprotection of basal cells against DNA damage. Differences between Caucasoid and Negroid skin affect the melanosome distribution, the rate of melanin production, and transfer to keratinocytes, but not melanocytes per unit area. Melanin is a remarkably stable protein–polymer complex, with a chromophoric backbone resistant to proteases, acids, and bases. Compared to Negroid and Mongolian melanosomes, Caucasian melanosomes (1) contain less melanin and more granules and (2) suffer greater degradation with keratinocytes. Free *"melanin dust"* (outside melano-somes in the stratum corneum) affords a greater physical cross-section and more protection against UV radiation compared to intact melano-somes. Caucasian skin sites exposed to psoralen plus UVA produce Negroid-type single-granule melanosomes.

The 20- to 30-fold differences observed between Caucasian and Ne-groid sensitivity to UV radiation correlates little to the 2-fold difference noted in corneal transmission (5-fold for whole epidermis).

13.4 OPTICS OF THE DERMIS

The transmission of dermis suggests multiple scattering by structures with dimensions of the same order as, or greater than, the wavelength of light, most probably collagen. Longer wavelengths exhibit both greater and more forward-directed (less diffuse) transmission.

Light scattering in the dermis varies inversely with wavelength, which means that remitted blue light will traverse a short average pathway until exiting the dermis and will have fewer encounters with absorbing mela-nin granules. For this reason, the Ota nevus, an abnormal melanin de-posit in the dermis, will have a blue color. High blue scattering is the only means by which melanin, which absorbs preferentially shorter wave-lengths, can produce a blue skin color.

Vitiligo skin yields estimates of the depth for attenuation to $1/e$ (37%) of incident energy without interference of melanin (Table 13.1). A pig-ment's physical cross-section and stratification at different depths are important: (1) Absorption by hemoglobin–oxyhemoglobin is less than if these chromophores were uniformly distributed throughout the dermal tissue, and (2) blue light in the major absorption band of β-carotene cannot reach the upper limits of the fat layer wherein this pigment is located. In carotenemia, yellowing of the palms and soles is seen because the thick corneum can retain enough carotenoids.

Two optical "windows" exist in skin: between 600 and 1200 nm (high dermal scattering and absorption for λ less than 600 nm) and at 1600–1850 nm, between the two water absorption bands. *At the windows, the volume and depth of tissue affected by phototoxicity will be large [e.g., 630-nm radiation, in hematoporphyrin-derivative (HPD) photoradiation therapy for cancer].* The energy in the 600- to 1300-nm band is enough to photodisso-ciate CO and O_2 liganded to hemoproteins.

13.5 *IN VIVO* REMITTANCE SPECTROSCOPY

The thin epidermis is an optically absorbing element overlying the thick dermis, a diffuse reflector. Absorption bands of major chromo-phores appear as minima in the spectral remittance of the Caucasian skin *in vivo.* Epidermal transmittance above 330 nm can be assumed to depend only on content and distribution of melanin, which decreases remittance at shorter λ. Remittance of the dermal element depends mainly on scattering by collagen and absorption by the dermal pigments.

Over the entire 350- to 1500-nm region, the fraction of light reaching the dermis is T_e; the dermis then returns a fraction R_D of T_e, such that the outward-directed *diffuse flux* at the bottom of the epidermis is $T_e R_D$. The *effective pathlength* passing back through the full thickness of the epidermis is about twice as great for the now diffuse radiation as compared to the collimated radiation in its first passage through the epidermis. If the fraction of light reaching the dermis from the epidermal surface was T_e, the fraction reaching back from the dermis to the top of the dermis through an effectively doubled pathlength will be $T_e \times T_e = T_e^2$ for the epidermal trajectory only, and accounting for the dermal attenuation, the overall skin remittance will be $T_e R_D \times T_e^2 = T_e^3 R_D$. This oversimplified model must be corrected for the effects of regular reflectance occurring at the skin–air interface, and an infinite series of terms representing multiple passes of radiation through the epidermis.

For the model to be practically useful, one must be able to measure the melanin-dependent epidermal transmittance, i.e., the term T_e independently of the effects of dermal chromophores. In the regions 330–400 nm and 650–700 nm, there are two advantages: (1) melanin absorbs, but none of the dermal chromophores absorbs or influences *in vivo* remittance; (2) the dermal penetration depth at these wavelengths is superficial, i.e., even the blood present in superficial capillaries can significantly affect the remittance.

Absorption by oxyhemoglobins (HbO_2) is several orders of magnitude higher in the 540- to 575-nm region than at 640–700 nm. Thus subtraction of green remittance from a red remittance provides a scale for monitoring vasodilatation, independently of melanin pigmentation.

The depression of neonatal skin remittance over the absorption band of bilirubin, centered near 460 nm, apparently correlates with bilirubin levels in jaundiced infants. Epidermal melanin is visualized by UVA photography of remittance and deeper cutaneous vasculature by near-infrared (900-nm) photography.

13.6 MANIPULATING THE OPTICS OF SKIN

13.6.1 Photoprotection

Excitation fluorescence spectroscopy or diffuse reflectance spectroscopy helps to assess noninvasively the attenuation efficiency of sunscreen products. The effective protection factor is calculated by convoluting the absorption features of the product with the action spectrum of the desired end point (e.g., erythema or pigmentation).

Two methodologies are used to quantify *in vivo* the efficacy in UVA protection: (1) phototoxic protection using 8-MOP and (2) immediate pigment darkening (IPD). The activity of the UVA sunscreen is expressed as phototoxic protection factor (PPF) or protection factor from UVA (PF_A):

$$PF_A = \frac{\text{minimal phototoxic dose of protected skin}}{\text{minimal phototoxic dose of untreated skin}}$$

or

$$PF_A = MPD_p/MPD_u. \qquad (13.8)$$

The action spectrum of IPD, a photooxidative process involving melanins and precursors, extends throughout the UVA region, from UVB to the short visible.

13.6.2 Increasing Photobiologic Sensitivity

Because of its high extinction coefficient, the small quantity of urocanic acid present accounts for approximately 75% of the optical absorbance of water-extracted materials in the skin surface.

If there are multiple air–tissue interfaces along the path of the incident radiation as it enters the stratum corneum (cf. psoriasis plaques), the total regular reflectance is considerably greater than for a single air–tissue interface. Mineral oil applied to scaly plaques to fill air spaces between superficial flakes of corneocytes provides a reasonable match of refractive index: a broad-spectrum decrease (5–25%) of remittance occurs within seconds. Thus, the effectiveness of light transmittance for psoriasis phototherapeutic treatment is enhanced by minimizing remittance.

13.7 PHOTOMEDICAL TREATMENTS AND CUTANEOUS OPTICS

In skin, action spectra resemble absorption spectra for a putative chromophore if one corrects for the spectral attenuation of incident radiation by the tissue before reaching the chromophore. The distinct minimum in delayed erythema action spectra is due to absorption of incident radiation by the stratum corneum. The absorption spectrum for phototherapy of psoriasis vulgaris is similar to the delayed erythema spectrum, except for wavelengths under 295 nm (ineffective), through the stratum corneum of psoriatic plaques. At 275 nm, the apparent absorbance ($-\log T$) of psoriatic stratum corneum is triple that of normal, fair-skinned Caucasian patients. The net result would be several orders of magnitude difference in the phototoxic dose at 275 nm in keratinocytes of psoriatic versus normal skin. With phototoxic drugs exhibiting wide action spectra, the depth of photochemotherapy may be controlled by varying the wavelength.

13.8 *IN VIVO* FLUORESCENCE OF SKIN: WOOD'S LAMP

Wood's lamp was developed in 1908 by placing a UVA-transmitting, visible-absorbing filter over a mercury discharge lamp used by dermatologists for the luminescence diagnosis of tinea capitis, erythrasma, and some *Pseudomonas* infections, as well as the appearance of porphyrins in hair or skin. Subtle changes in epidermal melanin pigmentation are more apparent due to greater absorption by melanin in the UV versus visible region. Lesions due to dermal pigmentation, such as the blue nevus of Ota, vanish entirely when viewed under Wood's lamp.

13.9 PHOTON MIGRATION IN TISSUES

13.9.1 Tissue Optical Properties in Relation to Light Propagation Models and *In Vivo* Dosimetry

The choice of irradiation model is strongly dependent on tissue optical properties, particularly on the tissue albedo a,

$$a = \mu_s/(\mu_s + \mu_a) \tag{13.9}$$

where μ_s and μ_a are the scattering and absorption coefficients, respectively, and the scattering anisotropy g equals the average cosine of the scattering angle (see Section 13.2.2).

There are two approaches to the *problem of determining the spatiotemporal distribution of light in tissue.*

13.9.1.1 Optical Fibers Coupled to a Photodetector

With a single cut-end fiber (Figure 13.7), the limited numerical aperture of the detector provides a directional dependence that can be used to determine the *angular radiance* at depth. However, in a region where the radiance is highly anisotropic (e.g., near the surface of the irradiated tissue), multiple insertions of the same fiber to the same point in tissue at different angles are required, so that the true fluence rate can be obtained by integrating over the solid angle. Optical fibers with a small scattering sphere at the tip have recently become available, so that the fluence rate at the point can be determined by a single measurement.

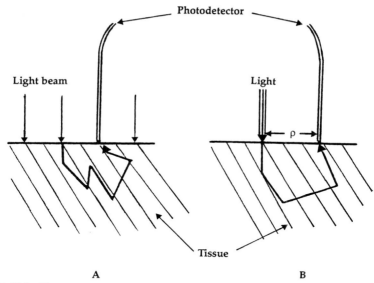

FIGURE 13.7 Illustration of measurements of (A) the total diffuse reflectance, R, from an irradiated tissue and (B) the local diffuse reflectance, $R(\rho)$ using two main geometries. $R(\rho)$ is measured as a function of radial distance ρ from an incident pencil beam. Modified from Wilson *et al.* (1989), p. 34.

For laser surgery or photodynamic therapy, two main geometries are used: (1) measurement of total diffuse reflectance, R, from irradiated tissue and (2) measurement of the local diffuse reflectance, $R(\rho)$, as a function of radial distance from an incident pencil beam.

13.9.1.2 Use of Irradiation Geometry, Tissue Geometry, and Optical Properties of the Tissue to Calculate the Spatial Distribution of Photon Fluence

Three regimes may be identified: (1) absorption dominated, where μ_a is larger than $10\mu_s(1-g)$; (2) scatter dominated, where $\mu_s(1-g)$ is larger than $10\mu_a$; and (3) comparable scatter and absorption.

Absorption is dominant in the UV region at wavelengths below 300 nm and in the far-IR region. The scattering-dominated regime holds only in the "therapeutic window" in the far-visible and near-IR regions (about 600–1200 nm). In the UVA and most visible regions, and also in the IR band from 1300 to 2500 nm, scatter and absorption are comparable.

13.9.1.2.1 Monte Carlo Simulation   The Monte Carlo simulation is a computational model in which individual photon trajectories are traced as each photon undergoes successive scattering interactions until it exits

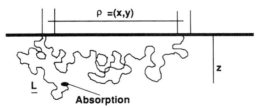

FIGURE 13.8   A schematic of light migration between two points defined by optic fibers on the tissue surface (in the x, y plane) separated by $\rho = (x^2 + y^2)^{1/2}$. The photon direction is randomized over a distance L considerably smaller than ρ (the photon takes steps L at times $n = 1$, 2, and 3. Two photon paths are shown: one reaching the surface at distance ρ and detected as surface emission, the other absorbed at depth z. From Bonner *et al.* (1989). With permission of Plenum, New York.

the tissue or is locally absorbed. The interaction paths and the scattering angles are randomly sampled. The chief advantage of the Monte Carlo method is that there are no *a priori* constraints on the tissue optical properties or irradiation and tissue geometries. Monte Carlo modeling may be used to calculate fluence distribution close to sources and boundaries where diffusion theory fails, and diffusion theory may be used to extrapolate the spatial distributions at depth within tissue.

A model proposed by Bonner *et al.* (1987) is based on a random walk formalism (Figure 13.8) whereby the photon distribution is scored on a three-dimensional lattice, with appropriate weighted probabilities for photon transport from one lattice point to its nearest neighbors. This model is particularly suited to handle calculations in the scatter-dominated regime and is complementary to the Monte Carlo technique.

A photon penetrates the tissue at one point and travels within the tissue, being continuously deflected through repeated encounters with tissue particles and ending up absorbed or reemerging, exiting the irradiated surface. Along the photon's path each lattice step from a given encounter to the next is designated L. The pathlength traveled by the photon in tissue before exiting can be converted by calibration to the time of flight knowing the velocity of light.

With the advent of picosecond lasers and fast time-resolved spectroscopy, additional information can be derived from the photon time of flight. The reflectance R is plotted as a function of pathlength traveled by the photons in tissue before exiting the irradiated tissue surface. It is seen that R depends on both albedo and scattering anisotropy.

13.10 EXPERIMENTAL TIME-RESOLVED METHODS

In principle, the probability distribution of time of flight from one point on a tissue surface to another is determined by *in vivo* spectroscopy, using picosecond laser pulses directed via optical fibers onto a tissue. The attenuation of the signal due to absorption is simply the product of the molar extinction coefficient, the pathlength, and the concentration of the absorbed species. There are two complications as outlined in the preceding section of this chapter:

1. The distribution of the transit times also depends on the tissue scattering properties and the measurement geometry (a complex function even in the absence of absorption).
2. Emergent photons may sample various regions of the tissue that contain different concentrations of the absorbing species.

The probability distribution of optical paths for different measurement geometries depends both on absolute tissue absorption and the relative sampling of different depths. Tissues are such strong diffusers of light that the mean free path between anisotropic scatterings is of the order of 50 μm, and the light is completely randomized within about 1 mm.[1] If the direct path from two points on the surface is much greater than 1 mm, the migration path of the photons may be adequately described by a *random walk of many steps*. Mean transit times between points separated by about 40 mm can be of the order of 1 ns in tissue. With the speed of light, 0.23 mm/ps, mean path lengths are of the order of 230 mm or about $n = 200$ isotropic step pathlengths. Recent experiments involving light scattering and diffusion in tissues show a substantial signal emerging from large tissue masses on the order of 2–10 ns following entry of a 10- to 15-psec laser pulse. The use of pulsed lasers and computer-assisted data acquisition is part of the method called time-correlated single-photon counting (TCSPC), which is applicable to tissue scattering experiments.

13.10.1 Photon Migration in Tissues Studied by Time-Resolved Wave Spectroscopy

In TRS, the delay between the initial picosecond pulse injected and the signal emerging from the surface due to multiple scattering is measured (Fig. 13.9), from which absorbance can be extracted.

In TRS measurements determining the characteristics of photon migration, the distance between input and output fibers (e.g., 2, 4, and 6 cm) has little effect on the decay slope μ (i.e., decrease of absorbance at wavelength λ per centimeter of path) of the curve of photon migration (see Fig. 13.9).

Using 630 nm laser excitation, the decay slope shows concentration change per pathlength or flight time change. The concentration change provides a measure of hemoglobin deoxygenation, and thus tissue anoxia. Fig. 13.10 demonstrates the effect of breathing a hypoxic gas mix-

---

[1] Remember that laser light is polarized.

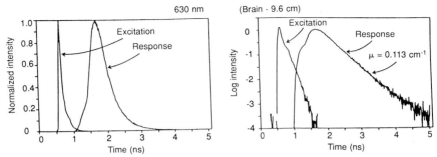

FIGURE 13.9   Photon migration in adult brain (two representations of the same, i.e., normalized intensity and log intensity), measured at 9.6 cm fiber separation at 630 nm. The exponential decay can be observed over four decades of intensity. Slope μ (expressed in cm^{-1}) has a reciprocal relation to the mean free path; the larger the coefficient of absorption, the shorter the mean path crossed within a time t [path (cm) equivalent to time (nsec)]. Hemoglobin absorption was measured at 630 nm. From Chance *et al.* (1989), p. 131. With permission of Plenum, New York..

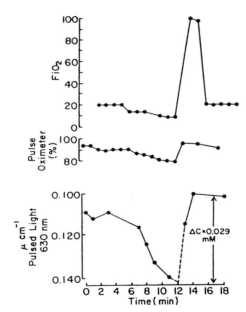

FIGURE 13.10   Decrease in the perfused oxygenation after 6 minutes of breathing a hypoxic gas mixture [i.e., $FiO_2 = 10\%$ in the upper panel] is barely visible with the pulse oximeter. The TRS equipment clearly detects the progressive depletion of O_2 in the brain (μ is the parameter defined in Figure 13.9 as shown in the lower panel). At t = 12 min switching to pure oxygen (100% FiO_2, upper panel) restores the original TRS signal. The ΔC value represents the variation in the brain oxyhemoglobin concentration in the hypoxic and normally oxygenated brain. Modified from Chance (1989). With permission of Plenum, New York.

ture in an adult human volunteer. The hypoxia is barely detectable by conventional finger pulse oximetry, but clearly appears applying TRS measurements to the brain. The change in decay slope $\Delta\mu$ is proportional to the product of the change in concentration ΔC_{TRS} and change in extinction:

$$\Delta\mu_{630} = \Delta C_{TRS} \times \Delta\varepsilon_{630} \qquad (13.10)$$

$$\Delta C_{TRS} = \Delta\mu_{630}/\Delta\varepsilon_{630} \qquad (13.11)$$

As can be seen in Fig. 13.10 the slope μ increases when the oxyhemoglobin concentration decreases.

Although these studies have focused on hemoglobin concentration changes in the brain, many other problems can be studied by the same method, specifically porphyrin distribution in skin and other tissues in connection with photodynamic therapy (B. Chance, personal communication).

Bibliography

Anderson, R., and Parrish, J. A. (1982). Optical properties of the human skin. *In* "The Science of Photomedicine" (J. D. Regan and J. A. Parrish, eds.), pp. 153–169, 174–185. Plenum, New York and London.

Bonner, B. F., Nossal, R., Havlin, S., and Weiss, G. H. (1987). Model for photon migration in turbid biological media. *J. Opt. Soc. Am. A* **4**; 423–432 (in Wilson *et al.*, 1989, p. 35).

Bonner, R. F. (1989). *In* "Photon Migration in Tissues" (B. Chance, ed.), pp. 11–23. Plenum, New York and London.

Chance, B., Smith, D. H., Nioka, S., Miyake, H., Holtom, G., and Maris, M. (1989). Photon migration in muscle and brain. *In* "Photon Migration in Tissues" (B. Chance, ed.), pp. 121–135. Plenum, New York and London.

Chardon, A. (1992). Method of UVA protection assessment based on immediate pigment darkening. *Protocol* No. 3.

Chardon, A. (1992). "Procedure for Using Neutral Physical Filters." L'Oreal, Clichy, France.

Epstein, J. E. (1989). Photomedicine. *In* "The Science of Photobiology" (K. C. Smith, ed.), 2nd ed. pp. 155–192. Plenum, New York.

Greenfeld, R. L. (1989). A tissue model for investigating photon migration in trans-cranial infrared imaging. *In* "Photon Migration in Tissues" (B. Chance, ed.), pp. 147–168. Plenum, New York and London.

Holtom, G. R. (1989). Experimental time resolving methods. *In* "Photon Migration in Tissues" (B. Chance, ed.), pp. 139–146. Plenum, New York and London.

Kochevar, I. E., Pathak, M. A., and Parrish, J. A. (1993). Photophysics, photochemistry and photobiology. *In* "Dermatology in General Medicine" (T. B. Fitzpatrick, A. Z. Eisen, K. Wolf, I. M. Freedberg, and K. F. Austen, eds.), Vol. 1, pp. 1632–1634. McGraw-Hill, New York.

Kollias, N. (1992). Non-invasive *in vivo* evaluation of UVA sunscreens. *Photochem. Photobiol.* **55**, Suppl., 958.

Mascotto, R. F., and Gonzenbach, H. U. (1992). Effectiveness of UV-A sunscreens. *In* "Biological Responses to Ultraviolet A Radiation" (F. Urbach, ed.), pp. 409–418. Valdenmar Publ. Co., Overland Park, KS.

Prahl, S. A., Kollias, N., and Anderson, R. (1992). A couple of cool graphs. *Photochem. Photobiol.* **55,** Suppl.

Wilson, B. C., Patterson, M. S., Flock, S. T., and Wyman, D. R. (1989). Tissue optical properties in relation to light propagation models and *in vivo* dosimetry. *In* "Photon Migration in Tissues" (B. Chance, ed.), pp. 25–42. Planum, New York and London.

14

Photocarcinogenesis

14.1 ENVIRONMENTAL RISK FACTORS FOR SKIN CANCER

The recent significant decrease in stratospheric ozone, the natural barrier against short-wavelength UV radiation, may significantly increase the incidence of skin cancer in the future.

14.1.1 UV Spectrum

UV wavelengths shorter than 320 nm are very detrimental to living cells and tissues. The major effects of UVB radiation (320–280 nm) range from mild redness to second-degree (blistering) burns. The acute effects of single overdoses of UVB radiation heal without scarring. Repeated UVB exposure, prolonged over many years, can result in skin photoaging (chronic degenerative changes in the skin) and premalignant and malignant lesions, i.e., nonmelanoma skin cancer (NMSC) and malignant melanoma.

There are several arguments for a causal role of UV radiation in NMSC production: NMSC is frequently located on exposed body sites; in pigmented races fewer cases of NMSC occur, and NMSC does not affect the sun-exposed areas. There is greater risk among white-skinned people who spend time outdoors. Premature NMSC occurs in genetic diseases that enhance sensitivity to solar UV radiation (xeroderma pigmentosum, albinism).

UVC (200–280 nm) radiation is very efficiently absorbed by nucleic acids. In skin, protection is provided by the overlying keratin layers. Absorbed by ozone and oxygen, solar radiation from natural sources below 200 nm does not reach living organisms.

14.1.1.1 UVB Effects

Most detrimental effects are due to UVB radiation (280–320 nm). Living organisms are protected by feathers, fur, or pigments, behavioral patterns that lead to avoiding sunlight exposure, or molecular DNA repair mechanisms. The action spectrum of UVB shows that wavelengths at 290 nm are 1000–10,000 times as effective at producing thymine dimers, single- and double-stranded breaks in DNA, mutations, cell killing, skin erythema, and skin cancer, compared to wavelengths at 330 nm. The action spectrum for induction of skin tumors in hairless mice is shown in Figure 14.1.

The ozone layer that absorbs UV radiation primarily below 315 nm has a maximum density at an altitude of about 25 km (see also Chapter 7). The balance between the increasing biological effectiveness of decreasing wavelengths in the UV spectrum and their increasing absorption by ozone bears critically on the biological implications of ozone changes (see Chapter 7). A 1% decrease in total ozone column would increase biologically effective (action spectrum-weighted) UVB radiation by 1.6%. This could result in an increase of 2.7% for basal cell carcinoma and 4.6% for squamous cell carcinoma. Should the total ozone column be reduced by 5%, the carcinogenic irradiance would increase by 8%. This could increase basal cell carcinoma by 14% and squamous cell carcinoma by 25%.

The amount of UV radiation over continents is lower as a rule than over clear oceans or over tops of mountains. For every 1000 m in elevation, UVB increases by about 15%. UVB radiation reaching our body will also be altered by reflection from environmental surfaces with significant UVB albedo. The UVB reflectivity of any surface is determined by the combined effects of reflection, refraction, and diffraction. Over UVB-permeable materials such as water, there is a mirrorlike reflectivity that follows Fresnel's law: *the lower the angle of the sun over water, the higher the reflectivity.*

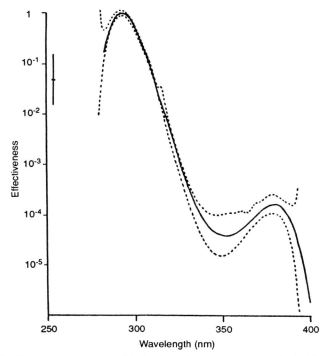

FIGURE 14.1 The action spectrum of induction of cutaneous cancers in hairless mice. The solid-line curve was obtained by averaging the actual experimental curves (dashed lines). From De Gruijl *et al.* (1993), Cancer Research **53**, 53–60, Fig. 1. With permission of the American Association for Cancer Research, Inc. Philadelphia, PA.

14.1.2 Role of Sunlight in Skin Cancer

With regard to the role of sunlight, skin cancers of Caucasians are most prevalent in the geographic areas with the greatest insolation and among people who receive the most exposure. Skin cancers are rare among black-skinned and other deeply pigmented individuals (their skin cancers are unrelated to sun exposure and are most commonly stimulated by trauma, chronic ulcers, or skin irritation); however, in Bantu albinos and in cases of xeroderma pigmentosum, skin cancer follows sun-exposure patterns. Individuals with the lightest complexion (of Scottish and Irish descent) are the most susceptible to increased sun exposure.

14.1.3 Epidemiology of Skin Cancer

Patients with skin cancer have a much greater frequency of certain genetically transmitted traits: light eyes, complexion, and hair; and easier sunburning and poor tannability. The Celts appear to inherit such traits, but are by no means the sole possessors of a sun-sensitive, poorly adapting skin. The prevalence of skin cancer rises with advancing age at about the same rate in Queensland, Australia, and in Galway, Ireland, where the population is mainly Celtic. In El Paso, Texas, where the population consists of few Celts and probably a large Mexican admixture, the slope of the prevalence curve is much lower.

Virtually all human squamous cell carcinoma (SCC) but only two-thirds of basal cell carcinoma (BCC) are related to the total accumulated lifetime dose of biologically effective solar radiation. Much of the variation in the sensitivity to UV radiation among white people is genetic, highly associated with "ease of sunburning" and inability to tan. *Given sufficient exposure for a sufficient time, most white people can develop NMSC.*

14.1.4 Artificial Sources of UV Radiation

Most commercial tanning equipment (solaria) exploits the divergence in the erythema and pigmentation action spectra. Very fair persons can develop mild erythema from the pigmentation dose. Daily sunbathing (e.g., 11AM–3PM) during a 2-week holiday in sunny climates presents a much more significant carcinogenic burden for the skin than does the moderate use of an UVA solarium.

14.1.5 Phototoxicity and Carcinogenicity

Psoralen–UVA combinations are much more carcinogenic than the UV radiation alone; psoralens alone are not carcinogenic. The cross-links formed in DNA by the combination of psoralens + UVA are widely assumed to be the basis for subsequent carcinogenic changes. *o*-Aminobenzoic acid, a phototoxic isomer of the common sunscreen *p*-aminobenzoic acid, may also enhance photocarcinogenesis.

During simultaneous carcinogen and UV radiation exposure, potentiation (due to photodynamic activity) or protection (due to photodecomposition) may occur. Exposure to UV radiation for a long period after an initial exposure to X-rays increases the carcinogenic effects of ionizing radiation.

14.1.6 Effects of Infrared Radiation

Infrared radiation comprises about 40% of the irradiance reaching the earth's surface. SCC can arise from burn scars and chronic heat exposure. An example is *kangri cancer* in India. A kangri is an earthenware pot, covered with wicker work, with a jute thread handle. Burning charcoal or dried leaves are put in this pot, which is tied to the abdomen under the clothes to produce warmth. The temperature of the skin in exposed areas ranges from 65–95°C. In the areas of contact, every stage of chronic

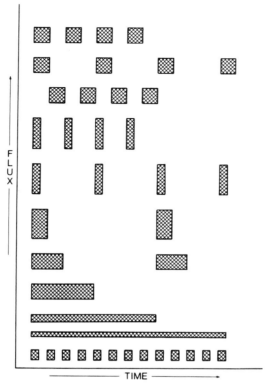

FIGURE 14.2 Geometric model of dose accumulation under several different conditions of flux, time, and fractionation. Each horizontal line of figures totals the same dose. The figures could represent varying "rest" intervals, day–night cycles, and numbers of exposures, as well as differences in flux and total exposure time. From Urbach (1982), p. 268. With permission of Plenum, New York.

dermatitis is seen, from ulcerating or fungating carcinoma with metastases to the inguinal lymph nodes.

Erythema ab igne and skin carcinoma result in cold regions of China and Japan from sleeping on beds of hot bricks or from the use of underfloor braziers. Turf or peat fire cancer has been frequently reported on the lower legs of Irish women who had extensive exposure to open hearths burning peat.

Infrared radiation is capable of causing significant elastic fiber hyperplasia, and greatly enhances that due to UV irradiation of the skin.

14.1.7 The Law of Reciprocity

The second law of photochemistry (law of Bunsen and Roscoe, i.e., the law of reciprocity) states that *photochemical action depends only on the product of light fluence rate and duration of exposure.* This law, however, holds only for primary photochemical action, and *does not apply* to the end points observed in photobiology, such as erythema, pigmentation, and skin cancer, which are all indirect effects.

Blum (1950) found that, within relatively narrow limits (approximate factors of five), fluence rate or interval between doses did not alter the shape or slope of tumor incidence curves, but only their positions on a log time axis. Figure 14.2 shows several ways in which the total dose can be delivered. To test reciprocity the following experiments were made to assess the relative effect of a prolonged continuous dose versus that of fractionation:

1. Mice given the total UV dose in 5 min developed tumors later and in smaller numbers than did those given the total dose in 50 or 500 min.
2. The total UV radiation was delivered either in a single 5-min period or in five 1-min periods, 1 hr apart, each day the mice were exposed. The rationale was to separate the possible effect of flux from that of the delivery time. The longer delivery time yielded a somewhat steeper tumor accumulation rate, but with a "longer latent period."

14.2 MOLECULAR PATHOLOGY OF SUNLIGHT-INDUCED SKIN CANCERS

The cutaneous response to UVB energy in solar radiation can be divided into four stages: (1) immediate tanning response, (2) erythema as a preamble to local inflammation, (3) melanogenesis and more long-lasting tanning, and (4) keratinocyte hyperplasia and peeling.

14.2.1 Cellular and Vascular Responses to Sunlight (Pigmentation, Erythema) and the Minimal Erythemal Dose

In normal skin that has been shielded from recent sun exposure, clustered melanin granules, primarily found around basal keratinocytes, form a "cap" around the external pole of the nucleus. Following acute, sustained exposure longer than 10 min to direct sunlight, the melanin granules become dispersed throughout the apical cytoplasm of the basilar keratinocytes. An erythematous response develops within 60–120 min. The minimal irradiation dose required to develop a still-visible erythematous response within a fixed time (e.g., 8–24 hr) is called the minimal erythemal dose (MED). MED strongly depends on the type of skin subjected to the irradiation. According to the generally used system of classification, four types of skin are described: *type 1,* skin of individual who never suntans and always burns; *type 2*, skin of individual who suntans at times and always burns; *type 3*, skin of individual who never burns, always suntans; *type 4*, skin of Africans. For a type 1 individual, the MED at UVB (300 nm/8 hr) is about 350 J/m^2; and at UVA (365 nm/8 hr) about 600,000 J/m^2.

If the exposure to sunlight is of sufficient duration, melanocytes begin to synthesize new melanin granules at an accelerated rate and may even undergo mitosis. Melanin granules are transferred to basal keratinocytes (suntanned skin). Basal keratinocyte replication results in a greatly thickened epidermis, despite sloughing (peeling) in large sheets as a result of keratinocyte death caused by UV radiation.

14.2.1.1 Responses of Keratinocytes: Carcinogenesis

Actinic degeneration and keratosis (i.e., thickening of the overlying epidermis), which follow repeated episodes of sunburning, may result in (1) basal cell carcinomas that are characterized by slow growth and rarely metastasize or invade locally and (2) squamous cell carcinomas that grow rapidly, invade extensively, and metastasize to distant organs.

Although the incidence of both basal and squamous cell carcinomas rises with increasing age, a third cutaneous malignancy, malignant melanoma, is primarily a disease of younger people. Malignant melanoma may be causally linked to one or only a few episodes of severe acute sunburn, rather than to a lifetime of prolonged exposure. At the present, there is no evidence that sun exposure elicits malignant degeneration among Langerhans cells, Merkel cells, or lymphocytes.

14.2.2 Host Factors Important in Skin Cancer Pathogenesis

Host factors important in the pathogenesis of sunlight-induced skin cancer are (1) indigenous melanin content in the skin and (2) different susceptibilities to developing skin cancer in individuals of the same skin type, which implies that polymorphic gene loci contribute to the overall risk of developing skin cancer.

14.2.3 Genetic Loci Associated with Immune Responsiveness

Studies of renal transplant recipients imply that genetic loci associated with immune responsiveness, such as HLA, may contribute to a particularly high incidence of skin cancers.

14.2.4 UV-Induced Injury to Langerhans Cells: Epidermal Target

Following UVB irradiation, deleterious changes can be observed within the epidermis (keratinocytes, Langerhans cells, and melanocytes) and the dermis (endothelial cells and mast cells). Langerhans cells are particularly susceptible to UV-induced injury: loss of dendrites, reduced surface expression of marker molecules, extensively vacuolated cytoplasm, and progressive loss of epidermal Langerhans cells as the UVB dose is increased.

14.2.5 UV-Induced Injury to Mast Cells: Dermal Target

In the dermis, mast cells undergo degranulation, releasing mediators (histamine) and cytokines (IL-1, TNF-α) that are important in generating the erythema, edema, and inflammation of the sunburn reaction.

14.2.6 UVB Radiation Protocols for Experimental Studies in Mice

There are two types of UVB radiation protocols for experimental studies in mice:

1. *Chronic high-dose model*: Mice are under exposure to high doses of UVB, delivered over intervals ranging from 3 to 4 months. The aggregate dose delivered is several orders of magnitude greater than that ever experienced by humans.
2. *Acute low-dose model*: UVB radiation is delivered on four consecutive days.

Using the high-dose protocol, a very high number of animals develop cutaneous tumors; the tumors are chiefly sarcomatous, they are ultimately lethal, and yet they are highly immunogenic, because tumor tissue implanted into non-UVB-exposed, *syngeneic* recipients is swiftly rejected (tumor grafts are not rejected if implanted in syngeneic animals already bearing their own UVB-induced tumors). Even before evidence of cutaneous tumors, a generalized UV-induced antigen-presenting cell defect robs exposed animals of the capacity to reject UVB-induced tumors from syngeneic donors. The capacity is lost to develop contact hypersensitivity (CH) to haptens, e.g., dinitrofluorobenzene (DNFB), even on skin shielded from UVB radiation. Mice that are not globally immunodeficient can reject allogenic skin grafts and tumors.

Using the low-dose protocol, no normal Langerhans cells are present at the irradiated site; hapten [e.g., trinitrochlorobenzene (TNCB)] engenders only feeble (if any) CH. Mice exposed to hapten, through UVB-treated skin, develop hapten-specific "tolerance" to subsequent exposure, in association with the *induction of hapten-specific suppressor T cells*.

The most important differences between the chronic high-dose and the acute low-dose protocols are (1) that susceptibility to the effects of the acute low-dose regimen on induction of CH is genetically determined and (2) that at least two, and perhaps three, independent genetic loci are involved in producing the UVB-susceptibility trait.

14.2.7 Relationship between UVB Susceptibility and Contact Hypersensitivity

14.2.7.1 Studies Using Human Subjects

On exposure of untanned (buttock) skin of normal, Caucasian subjects to four daily doses of UVB (144 mJ/cm^2), virtually *no CD1$^+$ or HLA-DR$^+$ surface antigen-containing cells remained at the site.* Following this regimen, irradiated skin painted with 2000 μg of dinitrochlorobenzene within 30 min of the terminal exposure was evaluated for (1) phototoxicity (immediately prior to hapten painting), (2) hapten-induced toxicity (48 hr after hapten application), and (3) "primary allergic" reaction, typically within 6–14 days after a hapten was painted on normal human skin.

The forearm skin of these individuals was then challenged with 50 μg of DNCB 30 days after the first exposure to hapten. Of 10 individuals, 6 developed DNCB-specific CH and 4 failed to develop contact hypersensitivity. These results suggested that UVB-susceptibility and UVB-resistance may be a clinically identifiable trait in humans and that the frequency of the susceptibility trait may be rather high, perhaps approximating 40% of normal, adult Caucasians.

A more extensive study was conducted with three panels of Caucasian subjects: young, healthy adults (19–49 years), older, healthy adults (50–80 years), and patients who had been diagnosed as having one or more previous biopsy-proved basal and/or squamous cell carcinomas (no cancerous lesions being present at the time of study). The conclusions were as follows: (1) both UVB-susceptible and UVB-resistant individuals were present in the two panels of healthy subjects (approximately 35–40% UVB-susceptibility trait in both groups); (2) thus, susceptibility and resistance to UVB-impaired induction of contact hypersensitivity are polymorphic traits in humans; and (3) only one of the skin cancer patients proved to be UVB resistant, whereas the remainder proved to be UVB susceptible.

To prove that the prevalence of UVB susceptibility in the cancer group was not secondary to an underlying immune deficiency, unirradiated buttock skin of UVB-susceptible cancer patients was painted with a sensitizing dose (2000 μg) of an unrelated hapten, diphencyprone (DPCP). When the forearm skin was challenged with dilute DPCP 30 days later, all members of the cancer group displayed vigorous CH. It was concluded that the failure of the vast majority of skin cancer patients to develop CH when hapten is painted on UVB-exposed skin is the consequence of a genetic trait that renders these individuals susceptible to this particular effect of UVB radiation.

Skin biopsies from black individuals exposed to UVB radiation indicate that this radiation depletes Langerhans cells from black skin as efficiently as it does from Caucasian skin. The amount and type of melanin pigment in the skin have little or no influence on the deleterious effect of UVB on Langerhans cells.

14.2.7.2 Studies Using Mice

Experiments on the effects of UVB in CH mice revealed that animals that failed to develop CH also displayed hapten-specific tolerance. If genetically determined, it is anticipated that the "UVB-tolerance" trait (revealed in the hapten-specific tolerance) will be independent of, and in addition to, the "UVB-susceptibility" trait that is reflected in impaired induction of contact hypersensitivity.

14.2.7.3 Conclusions of Studies

The conclusions of studies using human subjects are as follows:

1. UVB susceptibility is a polymorphic genetic trait, present in the normal population with a frequency of approximately 35–40%.
2. UVB resistance and susceptibility are unrelated to melanin content of the skin.
3. UVB susceptibility expressed in a minority of normal individuals is a characteristic of virtually all patients with biopsy-proved skin cancers; thus, it may be a risk factor in skin cancer.
4. The incidence of hapten-specific tolerance is considerably higher in UVB-susceptible patients with a history of skin cancer.

14.2.8 UVB Susceptibility and Tumor Necrosis Factor

It is hypothesized that alleles that promote the production of tumor necrosis factor-α are activated in response to UVB and that this cytokine mediates the deleterious effects of UVB on the induction of CH.

UVB susceptibility in mice is dictated in part by the wild-type *Lps* locus, which quantitatively governs the response to bacterial lipopolysaccharide (LPS). When LPS-sensitive murine cells are exposed to low doses of LPS, secretion of tumor necrosis factor-α, interleukin 1, and interleukin 6 is massively up-regulated. The UVB-susceptibility/resistance *polymorphism* in mice is a *polygenically inherited trait that is dictated by the interaction of alleles at two unlinked loci: Tnfα and Lps.* It is hypothesized that the *Tnfα(b)* and *Lps(n)* alleles cooperate to cause excess production of TNF-α in epidermis exposed to low doses of UVB, and that it is TNF-α

released within the epidermis that mediates the effects of UVB on induction and expression of CH.

In mice receiving murine recombinant TNF-α by intradermal (ID) injection in body wall skin, the CH responses to dinitrofluorobenzene are low, compared to mice injected with bovine serum albumin. Injection of 50 ng ID TNF-α seem to be equally effective in UVB-susceptible and -resistant mice. If the injected TNF-α dose is decreased from one-fifth to one-fiftieth, UVB-susceptible mice are more sensitive to the effects of TNF-α on CH than are resistant ones.

Display of vigorous CH in mice that received anti-TNF-α antibodies occurred, leading to the inference that TNF-α alone is responsible for the block in sensitization. That TNF-α, unlike UVB irradiation, does not promote tolerance induction leads to the suspicion that another set of genes is involved in this process. TNF-α induces changes in Langerhans cells (reduction of Ia$^+$ cells, withdrawal of dendrites) as early as 5 min after injection. After UVB exposure, similar changes take 60–120 min to appear: UVB must first activate an epidermal source of TNF-α production.

Within 12–24 hr after TNF-α treatment, the density of Ia$^+$ Langerhans cells returns to normal, suggesting that *the change in numbers represents alterations in expression of surface antigen, rather than loss or death of the cells.*

Langerhans cells treated with TNF-α, and then painted with hapten do not migrate to the draining lymph node (alterations in cytoskeletal elements, e.g., withdrawal of dendrites). Thus the hapten-specific immunogenic signal required for induction of CH never leaves the epidermis.

14.2.9 Possible Molecular Pathogenesis of UVB-Induced Contact Hypersensitivity to Haptens and the Risk of Developing Skin Cancer

UVB radiation presumably activates, within keratinocytes, transcriptionally silent *Tnfα* genes; the UVB-susceptibility trait depends on the amount of TNF-α produced as determined by (1) the ease of *Tnfα* transcription and (2) the capacity of UVB-exposed cells to stabilize TNF-α mRNA from degradation by RNAse. Langerhans cells express surface receptors for TNF-α and thus activation of hapten-specific T cells does not take place. The transient inability of Langerhans cells within UVB-exposed skin of susceptible individuals to carry neoantigens to the draining lymph node may delay the onset of antitumor immunity. This may allow the tumor sufficient time to establish an intraepidermal foothold that becomes progressively more difficult to eradicate by immunologic mechanisms.

14.3 XERODERMA PIGMENTOSUM

Xeroderma pigmentosum (XP) is a paradigm for the idea that carcinogenesis involves DNA damage from radiation and carcinogenic chemicals, and that DNA repair plays a major protective role in human and other species. Xeroderma pigmentosum, ataxia telangiectasia (AT), and Cockayne syndrome (CS) are human diseases that cause increased sensitivity to environmental carcinogens (e.g., UV and ionizing radiations, chemicals) because of genetic defects in the patients' capacity to repair and replicate DNA damage accurately. The clinical picture is shown in Figures 14.3 and 14.4 and the histological appearance is seen in Figure 14.5.

XP is an autosomal recessive disease associated primarily with degenerative changes of sun-exposed regions of the skin and eyes that often leads to neoplasia within the first decade of life. The major defect in XP is a failure to repair UV damage to DNA; in AT, the failure is in repair or replication of double-strand breaks in DNA; in CS, the failure is in recovery of DNA replication after irradiation. Cancer is a major clinical feature of XP and AT, but not of CS. Each disease is complex, with multiple groups defined by complementation in cell–cell hybridization. Overlap is reported between some XP and CS groups.

By as early as 2 years of age, the skin of individuals with XP displays striking evidence of both hyper- and hypopigmentation as a direct consequence of photosensitive reactions. Basal and squamous cell cancers form on sun-exposed skin surfaces before the age of 10. Keratinocytes, fibroblasts, and leukocytes display hypersensitivity to UVB, i.e., markedly increased chromosomal breaks and aberrations. XP displays *heterosis,* i.e., a variety of mutant alleles having the property of interfering, at one or another step, with the complicated process by which cells are able to detect and repair mutations. Little is known about the relationship between UVB-induced mutations and the eventual emergence of the malignant phenotype in a cell. The rapid development of skin cancers in XP patients implies that unrepaired mutations eventually do lead to malignancy.

UV-sensitive hamster cell mutants that complement XP groups are known, as well as a human gene on chromosome 19 that can correct the defect in hamster mutants, but not XP.

The parental consanguinity ratio of XP is about 30%. There may be 5–10 gene loci in XP. There are 9 *complementation* groups: A to H and variant. Because the XP loci of different complementation groups are not allelic, the XP trait being recessive, hybrids of different complementation groups are symptomless. Similarly, in cultures of XP cells, hybrids of cells from different complementation groups will show some correction of the DNA repair defect.

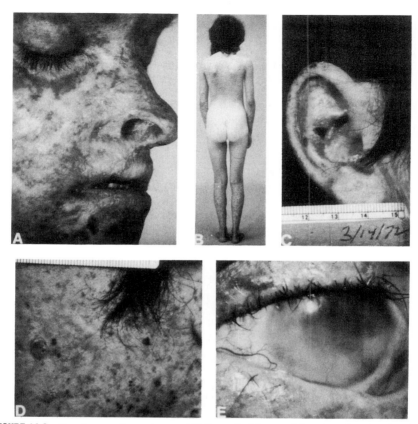

FIGURE 14.3 Xeroderma pigmentosum. (A) Face of a 16-year-old patient showing dry skin with hyperpigmentation, hypopigmentation, atrophy, and cheilitis. (B) Posterior view of the same patient showing absence of pigmentary changes on areas protected from sunlight. (C) Pinna of a 22-year-old patient showing pigmentary abnormalities and a crusted squamous cell carcinoma. (D) Face of a 14-year-old patient showing freckles with different amounts of pigmentation, an actinic keratosis, a basal cell carcinoma, and a scar with telangiectasia at the site of removal of another neoplasm. (E) Eye of the 22-year-old patient showing secondary telangiectasia invading the cloudy cornea, and atrophy and loss of lashes of the lower lid. Courtesy of K. H. Kraemer (1977), p. 40. From the Third Annual Workshop sponsored by the Institute for Medical Research and the National Institute on Aging. Symposia Specialists, Inc. Miami, FL.

The genetic structure is probably more complex than presently identified with a higher heterozygote frequency in the population. Other features of the syndrome with a complex relationship to repair deficiency are (1) neurological disorders in groups A, B, D, and G, (2) overlap of XP with Cockayne syndrome, with evidence for reduced repair of a transcriptionally active sequence in Cockayne syndrome cells, and (3) overlap

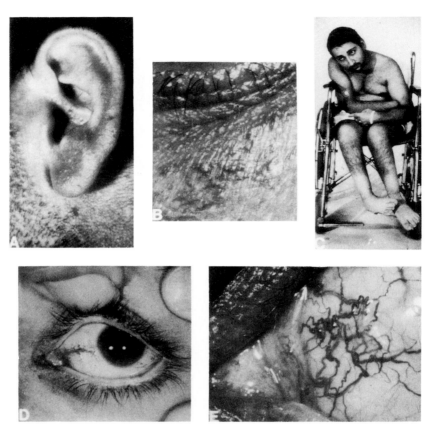

FIGURE 14.4   Ataxia telangiectasia. (A) Pinna of 22-year-old patient showing many fine telangiectatic vessels. (B) Telangiectasia of the malar area and lower lid in another patient. (C) The 22-year-old patient, who is unable to walk unaided. (D) His eye shows interpalpebral telangiectasia. The tortuous vessels do not invade the cornea. (E) Interpalpebral telangiectasia. Note the extreme tortuosity of the vessels and the sausagelike dilatations. From Kraemer (1977), p. 41. Photographs A and C courtesy of Dr. William Lewis; B, D, and E, courtesy of Dr. David Cogan.

of XP group D with trichothiodystrophy, an abnormal UV response of the hair bulb cells.

Phenocopies of XP include patients whose skin symptomatology and neurological disorders may be indistinguishable from XP but who display neither excision repair nor replication defects.

XP, as suggested, might involve multiple gene loci simultaneously, or the many phenotypes of XP may to some extent be explained on the basis

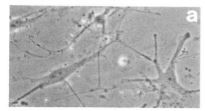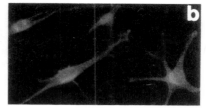

FIGURE 14.5 Culture of melanocytes from nevus. (a) Light microscopy shows typical dendritic morphologic features. (b) Fluorescent staining with TMH-1 monoclonal antibody to tyrosinase demonstrates the presence of tyrosinase dendritic cells (×312). From Kraemer *et al.* (1989), p. 264.

of a series of single gene defects involving DNA repair inherited in a simple Mendelian fashion.

14.3.1 XP and Rodent Complementation Groups

The excision repair deficiency of XP following UV irradiation is recognized as involving complementation groups A to H and variant. The human gene *ERCC1*, which corrects group 1 of the hamster, has sequence similarity to *UVRA* and *UVRC* of *E. coli* (see Chapter 12) and to *RAD10* of yeast at the amino acid levels but is not represented among XP groups. This may indicate that some mutations that remain viable in tissue culture compromise development and viability in the whole organism.

A continuing puzzle are XP groups B and H, which exhibit a complex phenotype ascribed to both XP and Cockayne syndrome. If both groups B and H could be ascribed to a *single* gene defect, and if the corresponding cloned genes fully correct UV sensitivity, the patients could then be regarded as suffering from a complex phenotype due to loss of a *single* vital component of the nucleotide excision repair system. It is conceivable that when a complete set of all UV-sensitive mammalian cells has been defined, XP, the different syndromes, and the *ERCC* series of genes will represent subsets with various degrees of overlap between them.

Compared to rodents, humans might have a globally enhanced repair system that superimposes a level of increased efficiency on top of the minimal rodent system. XP revertants, derivatives of XP group A, have acquired a repair system similar to the rodent system that is specific for [6,4] photoproducts (see Chapter 5).

ERCC1 may therefore correspond to structural or essential genes of the excision process whose absence would impair embryonic development. Each of the *ERCC* complementation groups appears to correspond to a specific gene on a specific chromosome, confirming the identity of complementation groups with deficiencies in a single gene for this series.

14.3.2 The Problem of Partial Correction by Gene and Chromosome Transfer

The most striking observation with transfection studies in XP cells is that, unlike what is seen in hamster cells, *transfer of DNA or even whole chromosomes has rarely produced full correction of the UV-sensitive phenotype.* If XP cells exhibit a low level of survival (e.g., 10^{-3}) after UV irradiation at a fixed dose, but 10^{-1} after transfection, there is a dramatic 100-fold increase in survival. However, compared to normal cells that exhibit full survival, the degree of correction by transfection still has a long way to go.

At a fixed dose of UV fluence (3 J/m^2), survival is 80% for GM637 (normal human), 7% for XP4PA (XP human), and 20% for the XP hamster hybrid XP4PA/3. For a constant level of survival slightly below 50%, the distance along the x axis (i.e., UV dose) between the XP4PA curve and the hybrid curve is one-fifth of the distance between the XP4PA curve and the normal curve, which shows that the hybrid-corrected XP has only gone 20% of the way required for full correction.

Partial complementation by chromosomes or DNA segments may reflect any or all of the following situations:

1. *Incomplete or partial expression of the transfected genes.*

2. *Expression of a gene in an incompatible context.* The human gene and yeast gene, *ERCC1* and *RAD10,* show significant sequence similarity (35% over 108 amino acids). When the yeast gene *RAD10* was expressed in the hamster CHO group 1 cells, the correction of UV sensitivity was only 7%. If the yeast gene does not exactly fit into the niche usually occupied by the *ERCC1* gene product, a fully functional repair system is impossible and the cell remains highly UV sensitive.

3. *Inappropriate quantitative levels of expression.* Not only underproduction but also, under some circumstances, overproduction of individual gene products could lead to suboptimal repair by causing damage instead of mending it.

4. *Codominance of the transfected wild-type gene and the two resident defective genes.* Heterozygote XP cell lines have not consistently demonstrated any UV sensitivity or repair deficiency. However, an XP-expressing cell transfected with wild-type gene is not strictly analogous to a heterozygous XP cell. Most transfected gene-recipient XP cells are transformed and therefore the precise number of defective XP genes may be above diploid; a *single* normal gene has been added. Thus, the transfected cell may resemble multinucleate hybrids.

5. *The gene expressed may not be the homolog of the defective XP gene in recipient cells.* A different gene, such as a cell cycle gene and a gene that regulates nucleotide metabolism, could affect cell survival. Transfer of

genes for glycosylases or apyrimidinic endonucleases could cause small changes in survival by initiating repair of nondimer damage.

6. *More than one gene may be required for full complementation.*

7. *The gene transformed may be a subunit of the repair system with a narrow substrate range.* Nucleotide excision repair in mammalians is likely to involve a complex set of interacting polypeptides causing the fine-tuning of the repair system for different photoproducts.

There may also be fine tuning for different functional states of the genome.

8. *Redundance of repair pathways.* Partial cross-correction of UV sensitivity by human and hamster genes does not necessarily mean that a gene correcting a UV-sensitive mutant codes for exactly the function that is deficient. A transfected gene may bypass instead of correct a deficiency.

14.3.3 XP and the Gene Products of Mammalian Cell DNA Repair

A "forest" of replicative proteins or polypeptides, rather than a single protein acting alone, may be necessary to amplify an initial recognition event on a region of DNA, or to interact in regulatory mechanisms. The more rapid excision of cyclobutane dimers from active genes exemplifies a finely tuned regulatory process in the search along DNA for photoproducts and in the competition with a panoply of chromosomal proteins, to coordinate repair with transcription and replication. The major role for gene products involved with repair may lie in searching, recognizing, unwinding, and preparing damaged regions for eventual excision and resynthesis.

An accessory protein in XP group E resembles, in DNA-binding properties, the yeast enzyme, photolyase. In mammals, photolyase may have lost its photolytic properties but retains the capacity to bind to DNA photoproducts and plays a role as a component of the nucleotide excision-repair system.

14.3.4 XP and Neurological Disorders

XP patients in groups A, B, D, and G have associated neurological problems. The DeSanctis–Cacchione form of XP is associated with mental retardation, decreased deep tendon reflexes, abnormal electroencephalogram, a progressively deteriorating course of cerebellar ataxia, choreoathetosis, sensorineural deafness, and eventual quadriparesis with shortening of the Achilles tendon. The other features of the DeSanctis–Cacchione form of XP include facial freckling, early development of cutaneous neoplasms, microcephaly, gonadal underdevelopment, dwarfism, and congenital deformities. *Increased sensitivity to endogenous oxi-*

dative damage could have two neurological consequences: premature death of neurons and interference with cell elimination processes in neurological development.

14.3.5 XP and Immunodeficiency

In about 5% of XP cases, there is impaired cell-mediated immunity, natural killer cell deficiency, and combined immunodeficiency. The increased UV sensitivity of XP lymphoid cells could result in a phenotypic immunodeficiency under the enhanced influence of sunlight.

14.3.6 Cancer and XP

Many XP patients die of complications from neoplasia. The age-specific skin cancer rates in XP, compared to lymphoid cancers in Bloom syndrome, have been used to suggest that skin carcinogenesis involves a uniquely mutagenic mechanism. XP patients do have significantly elevated rates of internal cancers, but *these must be related to the varied repair competence for different kinds of DNA damage from exposure to exogenous agents or endogenously produced genotoxic agents (i.e. oxidative damage due to such agents).* Increased sensitivity to DNA damage in XP is also associated with induction of damage-inducible or stress-responsive genes at lower doses than in normal tissue.

The observation that skin cancer rates in XP patients can be reduced with oral isotretinoin implies that *a major factor in the onset of clinically significant cancers occurs at the promotion stage of carcinogenesis, where oxidative damage may be involved.*

14.3.7 Case Reports

14.3.7.1 Case Report 1: Xeroderma Pigmentosum Variant with Multisystem Involvement

The XP variant includes patients whose cells show levels of UV sensitivity and repair that are close to normal, but that replicate their DNA after UV irradiation in abnormally short fragments.

A 12-year-old boy was referred to the Ohio State University (Columbus) Division of Dermatology in February, 1987, for a right preauricular basal cell carcinoma. He was the offspring of a consanguinous mating between his retarded mother and her half-brother. At birth he had intrauterine growth retardation, small facies, micrognathia, prominent occiput, abnormal ear placement, widely spaced nipples, short sternum, right simian crease, proximally placed thumbs, and prominent heels.

Physical examination revealed a left nasal telangiectatic papule, a left facial crusted lesion, a right muchal pedunculated papule, asymmetric facies, stellate freckling, microstomia, microcephaly, and dwarfism. Neurologic examination revealed many, but not all, features of the DeSanctis–Cacchione syndrome. Biopsies of nasal, facial, and nuchal lesions revealed basal cell carcinoma. Subsequently, 24 additional basal cell carcinomas and one squamous cell carcinoma on and above the neck developed.

In fibroblast cultures from a punch biopsy of unexposed skin regions followed through 15 passages, the formation and repair of DNA photoproducts (i.e., [6,4] photoproducts) was determined in DNA isolated post-UV irradiation. The results were consistent with the XP variant, i.e., synthesis of low-molecular-weight DNA.

14.3.7.2 Case Report 2: Patients from Egypt

Sixteen XP patients were studied in Egypt, where XP occurs frequently, and biopsies from eight were analyzed for unscheduled DNA synthesis, strand breakage during pyrimidine dimer excision, and complementation groups. The patients were equally distributed between groups A and C. Unscheduled DNA synthesis and strand breaks were significantly higher in group C than in group A cells. Central nervous system disorders were found in all of the group A patients and in none of the group C. No clinical symptoms were observed in the heterozygotes. A 2-month-old sibling of an XP patient was free of symptoms, but unscheduled synthesis and strand breakage in cultures from this sibling were the same as in the related XP homozygote.

In the Egyptian series, group A patients, with one exception, showed very early onset of sun sensitivity and development of skin cancers, microcephaly, and mental retardation. The exceptional group A patient was 35 years old, with normal stature and intelligence, and he had two normal children. He had invasive carcinoma of the nose. The DNA repair level of his fibroblasts (XP13CA) was as low as other group A cells.

Cell cultures from both normal and XP patients reached *in vitro* "senescence" at similar passage levels.

14.4 MALIGNANT MELANOMA

14.4.1 Epidemiology of Melanoma

The past several decades have seen dramatic increases in the incidence of cutaneous melanoma. Melanoma frequently affects young people. The rate in children under 10 is 1 in 1,000,000 per year, and the incidence of melanoma increases 100-fold by the mid-teen years to 10 in 100,000 per year.

14.4.2 Distinctive Clinical Features

Early malignant melanomas are often asymmetric in shape and have irregular borders. Compared with benign pigmented lesions, which are more uniform in color, macular malignant melanomas are usually variegated, ranging from various hues of tan and brown to black, sometimes intermingled with red and white. The "ABCD" characteristics of early malignant melanomas are asymmetry (A), border irregularity (B), color variegation (C), and diameter (D) generally greater than 6 mm.

The diagnosis of malignant melanoma is based on (1) a significant change in a preexisting melanocytic nevus, (2) the development of a new pigmented lesion, particularly in patients over 40; and (3) changes in color, size, shape, elevation, surrounding skin, or consistency. A culture of melanocytes from nevus is shown in Fig. 14.5. A variety of benign and malignant pigmented lesions are shown in Figures 14.6–14.11. Corresponding histological images are shown in Figures 14.12–14.19. Figures 14.15–14.19 show four of the five levels of the Clark classification: level 1, melanoma entirely *in situ;* level 2, lesion has extended to the papillary dermis but does not fill it; level 3, the lesion fills the papillary dermis; level 4, melanoma that extends to the reticular dermis; level 5, melanoma extends beyond the reticular dermis into the subcutaneous tissue.

Dysplastic nevi (precursors to melanoma, or markers of increased risk) may be genetically transmitted within melanoma-prone families as an autosomal dominant gene. Dysplastic nevi can be used in screening to distinguish individuals to be monitored for future development of melanomas in themselves or in their family members. Patients with melanomas thinner than 0.76 mm have 5-year survival rates approaching 100%, compared with only 40% for those with melanomas thicker than 4.00 mm.

14.4.3 Common Benign Pigmented Lesions

14.4.3.1 Simple Lentigo

Simple lentigo is a macular, sharply defined, round, brown to black lesion of 1–5 mm with smooth or jagged edges, and pigmentation in a reticulated pattern. This lesion generally arises in childhood on sun-exposed sites.

14.4.3.2 Junctional Nevus

Junctional nevus is small, less than 6 mm, macular, and well-circumscribed, with smooth surface and relatively uniform pigmentation. It may appear at any site but more often occurs in sun-exposed areas. It develops during childhood, adolescence, or young adulthood.

14.4.3.3 Compound Nevus

Compound nevus is well-circumscribed, less than 6 mm, and is a raised papule with a smooth or rough surface. "Compound" refers to the fact that the nevus cells are both along the dermo-epidermal junction and within the dermis. The nevus may have excess hair.

14.4.3.4 Intradermal Nevus

The nevus cells are wholly in the dermis. The nevus is generally a small, well-circumscribed papule, from skin-colored to tan to various shades of brown, and may have an excess of hair.

14.4.3.5 Solar Lentigo

Known to the lay public as a "liver spot," solar lentigo is a uniform tan to brown macule commonly located on the face, chest, back, and dorsa of the hand.

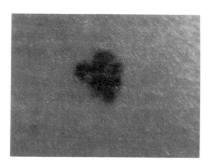
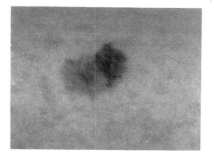
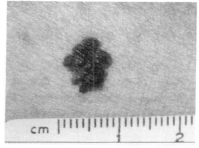

FIGURE 14.6 Border irregularity of early malignant melanoma. Courtesy of R. J. Friedman *et al.* (1991) p. 205. With permission of J. B. Lippincott Co., Philadelphia, Pennsylvania.

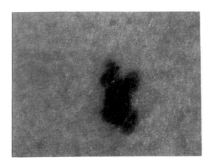
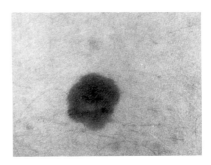
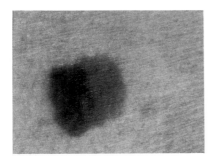
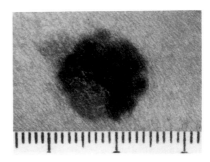
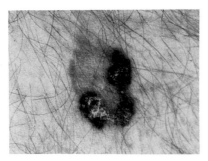
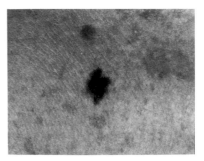

FIGURE 14.7   Color variegation of early malignant melanoma. Irregularly marginated pigmented lesion and prominent lighter components are indicative of spontaneous partial regression. Courtesy of R. J. Friedman *et al.* (1991), p. 206. With permission of J. B. Lippincott, Co., Philadelphia, Pennsylvania.

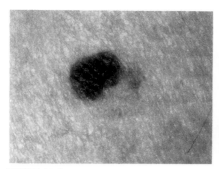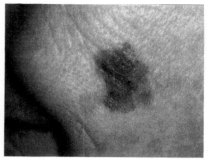

FIGURE 14.8 Early malignant melanoma, approximately 6 mm in diameter. Courtesy of R. J. Friedman *et al.* (1991), p. 207. With permission of J. B. Lippincott Co., Philadelphia, Pennsylvania.

14.4.3.6 Seborrheic Keratosis

Seborrheic keratosis is a generally verrucous, round or ovoid, variably raised, sharply demarcated papule or plaque. The growth is primarily composed of keratinocytes rather than melanocytes, and it is commonly situated on the face, neck, and trunk.

14.4.4 Wavelengths Effective in Induction of Malignant Melanoma: Action Spectrum for Melanoma Induction

Individuals defective in the repair of UV damage to DNA are several thousandfold more prone to the disease compared to the average population. Heavily pigmented backcross hybrids of the genous *Xiphophorus* (platyfish and sword tails) are very sensitive to melanoma induction by single exposures to UV radiation. Groups of five 6-day-old fish were irradiated with narrow wavelength bands at 302, 313, 365, 405, and 435 nm, and scored for melanomas 4 months later. Several exposures at each wavelength were used to obtain estimates of sensitivity for melanoma induction as a function of exposure and wavelength. The action spectrum for melanoma induction shows appreciable sensitivity at 365, 405, and probably 436 nm, suggesting that wavelengths not directly absorbed by DNA are effective in induction. In natural sunlight, 90–95% of melanoma induction may be attributed to wavelengths longer than 320 nm, in the UVA and visible spectral regions (light energy absorbed in melanin). Figure 14.20 shows that the action spectrum for fish melanomas is not superposable on the action spectrum for mammalian mutagenicity.

Knowing the role UVA radiation has in the formation of free radicals, the effect of these active compounds may be considered as possibly operative in fish melanogenesis.

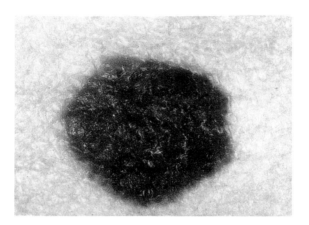

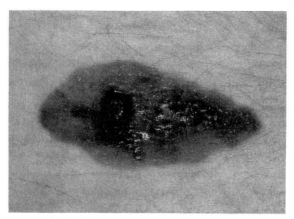

FIGURE 14.9 Progression of malignant melanoma plaque. Progression of malignant melanoma plaque with nodule. Courtesy of R. J. Friedman *et al.* (1991), p. 208. With permission of J. B. Lippincott, Co., Philadelphia, Pennsylvania.

14.4.5 Risk Factors for the Development of Melanoma

From 43 cases investigated, it was determined that six factors influenced the risk of developing malignant melanoma: (1) family history, (2) blond or red hair, (3) marked freckling on the back, (4) three or more blistering sunburns prior to age 20, (5) three or four years of an outdoor summer job as a teenager, and (6) actinic keratosis. Persons who had one melanoma are at increased risk of developing another.

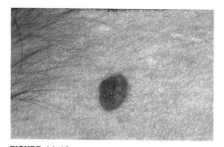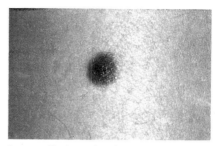

FIGURE 14.10 Compound nevus. Small, well-circumscribed, uniformly pigmented, raised papule. Intradermal nevus. Small, well-circumscribed papule with minimal pigmentation. Courtesy of R. J. Friedman *et al.* (1991), p. 212. With permission of J. B. Lippincott, Co., Philadelphia, Pennsylvania.

14.4.5.1 Pigmented Lesions Associated with a Higher Risk of Malignant Melanoma

Congenital melanocytic nevi and dysplastic nevi are precursors as well as markers for malignant melanoma.

14.4.5.1.1 Congenital Melanocytic Nevi Small (less than 1.5 cm), medium (1.5–19.9 cm), or large (20 cm or more) in size, congenital melanocytic nevi occur in about 1% of newborns. The lifetime risk of malignant melanoma in individuals with this lesion has been estimated to be about 6%, compared to 1% for the general population. Malignant melanoma may begin deep in the dermis, making early diagnosis difficult.

14.4.5.1.2 Dysplastic Nevi Dysplastic nevi (B–K mole, or atypical mole) are acquired pigmented lesions of the skin featuring asymmetry, border irregularity, color variegation, and a diameter of more than 6 mm. They may occur in both familial and nonfamilial settings.

14.4.5.1.3 Familial Atypical Multiple Mole Melanoma Syndrome Familial atypical multiple mole melanoma (FAMMM) is also called dysplastic nevus syndrome. Clinically, patients with the syndrome have a triad of (1) more than 100 moles, (2) at least one 8-mm or larger mole, and (3) at least one mole with "atypical" (malignant melanoma-like) features.

14.4.5.1.4 Common Melanocytic Nevi and Dysplastic Nevi Common melanocytic nevi (CMN) are uniformly tan and brown, round, and have sharp, clear-cut borders; they are flat or elevated, are usually less than 6 mm in diameter, and 10–40 are scattered over the body in the typical adult. Generally they occur on the sun-exposed surfaces of the skin above the waist, with the scalp, breasts, and buttocks rarely involved.

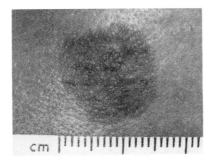
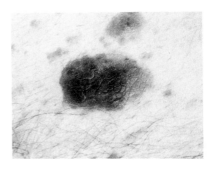

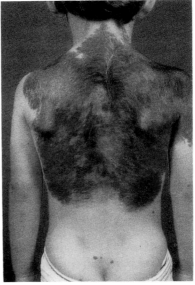

FIGURE 14.11   Solar lentigo, tan patch. Seborrheic keratoses; multiple, tan to darker brown, raised verrucous papillae. Giant congenital melanocytic nevus. Courtesy of R. J. Friedman *et al.* (1991), p. 213. With permission of J. B. Lippincott Co., Philadelphia, Pennsylvania.

Dysplastic nevi (DN) are variable mixtures of tan, brown, black, or red/pink within a single nevus (nevi may look very different from each other); they have irregular borders and pigment may fade off into surrounding skin. They always have a flat portion level with skin, which often occurs at the edge of the nevus. They may be more than 10 mm, often more than 100, and occur on sun-exposed areas. The back is the

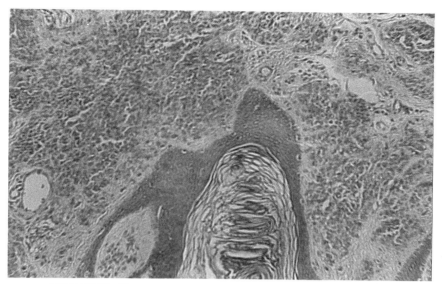

FIGURE 14.12 Photomicrograph of compound nevus by Cahide Kohen.

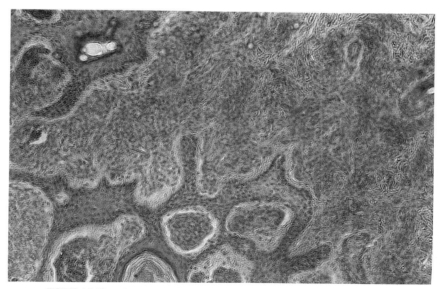

FIGURE 14.13 Photomicrograph of intradermal nevus by Cahide Kohen.

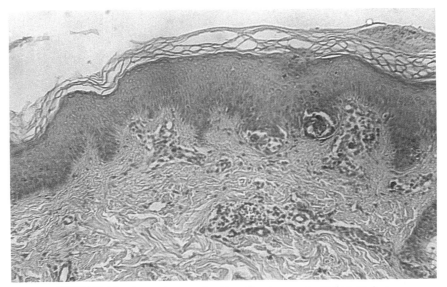

FIGURE 14.14   Photomicrograph of junctional nevus by Cahide Kohen.

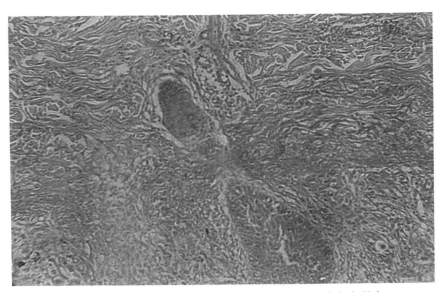

FIGURE 14.15   Photomicrograph of Clark II melanoma by Cahide Kohen.

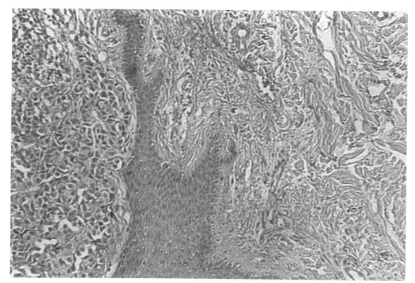

FIGURE 14.16 Photomicrograph of Clark III melanoma by Cahide Kohen.

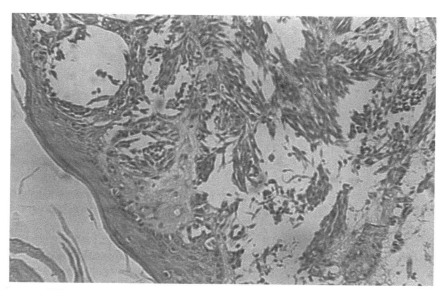

FIGURE 14.17 Photomicrograph of Clark IV melanoma by Cahide Kohen.

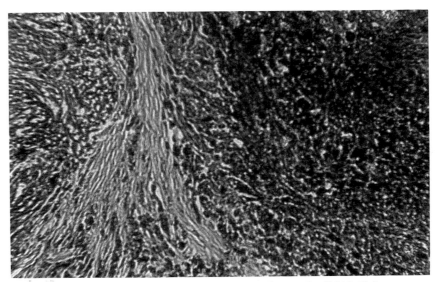

FIGURE 14.18 Photomicrograph of Clark V melanoma by Cahide Kohen.

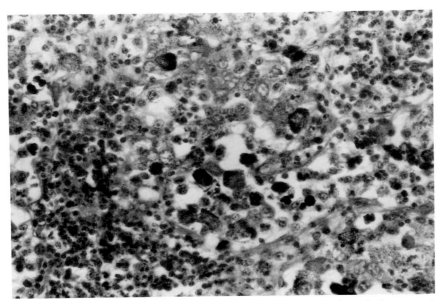

FIGURE 14.19 Photomicrograph of a malignant melanoma of a mandarin duck metastatic to a lymph node. Note dark black pigment, small lymphocytes, and anaplastic melanocytes. H and E stain; ×400. Courtesy of Dr. Alan Herron, Department of Pathology, University of Miami.

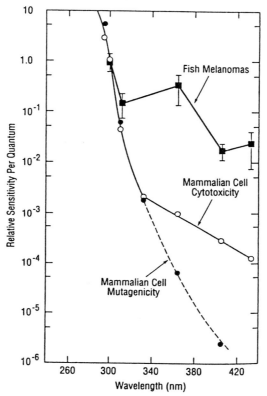

FIGURE 14.20 Action spectrum of melanoma induction. From Setlow, *et al.* (1993), with kind permission of Professor Richard B. Setlow. Copyright permission of Proceedings of the National Academy of Science.

most common site, but they may also be seen on the scalp, breasts, and buttocks.

14.4.6 Computer Applications in the Diagnosis and Management of Malignant Melanoma

A system of computerized-image analysis can in some cases enable differentiation between malignant melanoma and pigmented lesions by evaluation of detail not perceivable by the human eye, in terms of objective criteria related to shape, color, and the border irregularity of melanocytic neoplasms. Digital epiluminescence microscopy obtains images that can be analyzed with computer enhancement. Through the use of three-dimensional computer reconstruction techniques, it has been shown that the tumor volume of malignant melanoma is a better predictor of survival than the tumor thickness.

14.4.7 *In Vivo* Epiluminescence Microscopy of Pigmented Skin Lesions

In vivo epiluminescence microscopy provides for a detailed inspection of the surface of skin lesions. With oil immersion, which renders the epidermis translucent, epiluminescence microscopy accesses a new dimension in skin morphology by including the dermoepidermal junction in the evaluation. In a study of more than 3000 pigmented skin lesions, morphological criteria not readily apparent to the naked eye were defined. These criteria are reliable markers of benign and malignant skin and the configuration, regularity, and other characteristics of both the margin and the surface of pigmented skin lesions.

The established criteria for pattern analysis are (1) the general appearance of the lesion (uniform or heterogenous), its profile (elevated or depressed), and its surface (smooth or rough); (2) the pattern and color of pigmentation (depigmentation, pigment network, brown globules, and black dots); and (3) the margin of the pigmented skin lesions (regular, irregular, pseudopods, and radial streaming).

An irregular network with variable mesh size and configuration may indicate dysplasia or malignancy. The brown globules correspond to nests of melanin-containing melanocytes (they are uniform and regular in benign lesions, of different size and haphazardly spaced in dysplasia and malignancy). The black dots represent focal accumulations of melanin in the cornified layers of the epidermis (only in the center in benign lesions, or in the periphery of the dysplastic and malignant lesions).

In malignant lesions the pigmentation border is always irregular, revealing two types of configurational features: (1) pseudopods are relatively broad and blunt, with ramified extensions of the pigmented margin of the lesion into surrounding skin; (2) radial streaming structures are finely serrate (notched) extensions radiating from the pigmented skin lesions. Both configurations correspond to the radial growth phase of melanoma. The pigment network stops abruptly at the edge of malignant lesions, whereas it thins out in benign lesions.

14.4.7.1 Pattern Analysis of Benign and Malignant Pigmented Skin Lesions

Epiluminescence microscopy is primarily needed in diagnosing clinically equivocal lesions: (1) the differential diagnosis of melanoma from highly pigmented basal cell carcinomas, small angiokeratomas, and smooth and highly pigmented seborrheic keratosis; (2) the diagnosis of very early and small pigmented skin lesions that have not yet sufficiently developed the clinical features used as criteria to define the benign or malignant nature of the lesion; and (3) the assessment of dysplastic nevi.

14.4.7.2 Some Differential Diagnoses of Small Pigmented Skin Lesions by Epiluminescence Microscopy Pattern Analysis

14.4.7.2.1 Dysplastic Nevus versus Superficial Spreading Melanoma Brown granules of different sizes, an irregular pigment network with an abrupt stop in the periphery, irregular borders, and colors varying from black to brown are common to both DN and superficial spreading melanomas. They differ primarily in that the superficial spreading melanoma has a prominent pigment network, black dots at the periphery, radial streaming, and occasional pseudopods at the margin.

14.4.7.2.2 Dysplastic Nevus versus Benign Junction or Compound Nevus In contrast to benign nevi, dysplastic nevi are polymorphous, have an irregular pigment network, brown globules of different sizes haphazardly spaced, irregular outlines, and a focally abrupt termination of the pigment network at the periphery.

14.4.7.2.3 Pigmented Spitz Nevus versus Small Nodular Melanoma The Spitz nevus has a simple oval or round outline and a typical bizarre but regular white or pink region in the central area of the lesion. In contrast, the depigmented areas of melanoma are usually found at the periphery of the tumor, are flat or depressed, and have an irregular outline. Both lesions share features, such as a prominent pigment network, haphazardly placed brown globules of different sizes, and an abrupt termination of the pigment network at the edge. However, in the Spitz nevus, the pigment network is regular, with heavily accentuated rete ridges and meshes consisting of thin lines and dots not encountered in other pigmented skin lesions. Radial streaming and pseudopods are characteristic of melanoma.

14.4.7.2.4 Nodular Melanoma versus Angioma and Angiokeratoma Both lesions are nodular with a variation in color from black to blue, possible erosion, bleeding, and crusts. However, angioma and angiokeratoma have a regular outline and they lack a pigment network, black dots, radial streaming, and pseudopods.

14.4.7.2.5 Superficial Spreading Melanoma versus Pigmented Basal Cell Carcinoma   Malignant melanoma has a pigment network, brown globules, cytoplasmic streaming, and pseudopods at the periphery (all of which are absent from basal cell carcinomas, with telangiectasia a prominent feature).

14.4.8 Immunotherapy of Melanoma

The use of genetically altered tumor-infiltrating lymphocytes (TILs) is under evaluation. TILs from the patient are altered by the insertion of a gene that generates tumor necrosis factor; these are expanded in tissue culture and then infused to the patient. Thus, more than 100 times the amount of TNF that regularly circulates can be generated locally in the metastasis, to which the TIL cells migrate. It is hoped that these cells will act as "guided missiles" on their target, and release enough TNF to destroy the tumor without causing significant untoward effects on the surrounding tissues.

14.5 PHOTOTOXICITY AND PHOTOCARCINOGENESIS IN VETERINARY PHOTOMEDICINE

UVA (320 to 400 nm) radiation is frequently associated with photosensitivity reactions. Photodermatitis can be manifested as phototoxic disorders, photosensitive disorders, photoallergic disorders, and other, miscellaneous disorders.

14.5.1 Phototoxic Reactions

In animals, the effects of UV radiation are attenuated by the hair coat and by reflection and refraction in the stratum corneum. Skin areas with the least hair density are more prone to photodermatitis. Solar dermatitis and subsequent squamous cell carcinoma in white cats and nasal solar dermatitis in dogs are good examples.

In dogs with mottled pigment over their planum nasale, pigmented sites are only mildly affected compared to nonpigmented areas. Chronic solar dermatitis in white cats can be associated with erosions and ulcerations. Such sites are often where more invasive squamous cell carcinomas develop. Carotenoids and surface lipids may act as photoprotective agents as well. Damaged cells release chemical mediators: histamine, serotonine, kinins, prostaglandins, and leukotrienes.

14.5.1.1 Canine Nasal Solar Dermatitis

Affected dogs may be born with poorly pigmented planum nasale or may acquire a noninflammatory depigmentation. Many cases are autoimmune (e.g., lupus or pemphigus erythematosus), although sun exposure appears to be involved in the pathogenesis. Dogs living in snow climates can have aggravated lesions as a result of UV reflection. Severe chronic cases can end with marked nasal ulceration and deformity of the nares and distal planum nasale, rarely progressing to squamous cell carcinoma.

The major differential diagnosis is autoimmune disease.

Diagnosis	Symptom
Lupus erythematosus	More severe generalized skin lesions
	Greater potential for systemic involvement
Pemphigus erythematosus and pemphigus foliaceus	Erosive vesicular or pustular nasal lesions
	Generally also affects the haired portion of the nasal area (with pemphigus foliaceus other body locations can be affected)
Dermatomyositis	Primarily in Shetland sheepdogs and collies
	Can present with nasal lesions, although usually facial and extremity lesions are present as well
Infectious conditions	Bacterial and fungal nasal infections occasionally resemble a nasal solar lesion
Tumors in the nasal region (squamous cell carcinoma, basal cell carcinoma, epitheliotropic lymphoma, fibrosarcoma)	Can resemble a nasal solar lesion
Drug eruption, topical drug hypersensitivity, contact reaction	Can resemble a nasal solar lesion

14.5.1.2 Canine Solar Dermatitis

The clinical disease will be more severe in the flanks and ventrolateral abdomen, also the bridge of the nose and medial hock areas. In advanced cases when squamous cell carcinomas form, other nodular tumors (e.g., lymphosarcoma, mast cell tumor, hemangiosarcomas, and metastatic neoplasia) need to be considered. Primary hemangiosarcomas and hemangiomas developing on solar-damaged skin seem to metastasize at a lower rate than nonsolar hemangiosarcomas.

14.5.1.3 Feline Solar Dermatitis

This is a chronic actinic dermatitis occurring in cats with white-colored areas, liking to sit in the sun on a regular basis. It can progress to squamous cell carcinoma. Affected sites include the ears, eyelids, planum nasale, lips, and face. The margins of the pinnae, sparsely covered by hair, are quite vulnerable to solar radiation.

14.5.2 Photocarcinogenesis

The progression from solar keratosis to carcinoma occurs over several years and in many instances never attains the full status of carcinoma. In carcinoma of the ear pinna and eyelid in Australian sheep, the incidence increases with age, and the prevalence increases with changes in latitude and altitude that favor maximal exposure to sunlight. Hair coat and skin pigmentation are highly protective against tumor development.

14.5.2.1 Squamous Cell Carcinoma

Sunlight is probably the most important carcinogenic stimulus for this locally invasive and occasionally metastatic neoplasm. Sunlight accounts for the localization in the eyelids and conjunctiva of cattle and horses, the ear pinna of cats and sheep, the vulva of cattle, goats, and recently sheared sheep, and the lightly haired, poorly pigmented abdominal and juxtanasal skin of dogs. There is a greater risk of squamous carcinoma at sites of ear notching, branding, and perhaps chronic inflammation.

The microscopic image of the lesion is often dominated by solar keratosis (i.e., hyperkeratosis, parakeratosis, ulceration, acanthosis, and superficial dermal scarring), with only a few foci of unequivocal neoplasia in which squamous cells have invaded across the basal lamina of the hyperplastic epidermis.

The characteristic growth habit of squamous cell carcinoma (Figures 14.21–14.23) distinguishes this tumor from basal cell tumors, including cornifying epithelioma, and reactive pseudoepitheliomatous hyperplasia. The essential diagnosis criterion is the presence of polyhedral cells, resembling those of stratum spinosum, on the dermal side of the basal cell layer and basal lamina. Squamous carcinoma lacks the cushion of basal cells and basal lamina between the tumor cells and the dermis. Keratinization within anastomosing cords of tumor cells results in laminated keratin "pearls" surrounded by tumor cells.

White cats but not dogs show a strong correlation between squamous cell carcinomas and exposure to UV radiation. These tumors may occur multifocally in sparsely haired areas of white-coated dogs who habitually

FIGURE 14.21 Invasive and ulcerated squamous cell carcinoma from the nose and oral cavity in a white cat. Note the unilateral symmetry. Courtesy of Dr. Herron, Department of Pathology, School of Medicine, University of Miami.

sleep in the sun and sustain excessive actinic damage. Dogs with lupus erythematosus, pemphigus erythematosus, and vitiligo may get squamous cell carcinoma in zones of nasal depigmentation (also at the site of injection of a live canine oral papillomavirus vaccine in some dogs).

14.5.3 Case Reports

14.5.3.1 Metastatic Melanoma of Mandarin Duck

A mass on the dorsal surface of the bill in an adult female mandarin duck *(Aix galericulata)* was diagnosed as a malignant melanoma. Two months later the tumor had enlarged considerably, and the duck exhibited open-mouthed breathing. At necropsy metastases were confirmed microscopically, in the mural lymphoid nodes of the jugular lymphatic vessels. The location of the primary tumor on an exposed area is suggestive of UV etiology.

14.5.3.2 Hemangiomas and Hemangiosarcomas of Dogs

Hemangiomas have no predilection for dermis (51% dermal, 47% subcutaneous), but hemangiosarcomas have a marked predilection for dermis (73% dermal, 7% subcutaneous). The increased incidence of dermal hemangiomas and hemangiosarcomas in ventral glabrous skin suggests an association with UV radiation, considering postural habits of

FIGURE 14.22   Photomicrograph of a squamous cell carcinoma of the pinna of the ear of a white cat. Note variably sized nests and chords of tumor cells that are infiltrating and causing focal ulceration. H and E stain; ×40. Courtesy of Dr. Alan J. Herron, Department of Pathology, Division of Comparative Pathology, University of Miami.

dogs. Approximately 18% of dogs with dermal but not subcutaneous hemangiomas and hemangiosarcomas had solar elastosis in skin adjacent to the tumor. There is also evidence that hemangiosarcomas arose within preexisting hemangiomas in sun-exposed skin. In two dogs dead with the histologic diagnosis of cutaneous hemangioma the actual tumor may have been a metastasizing hemangiosarcoma.

14.5.3.3 Squamous Cell Carcinoma of Bengal Tiger and Christmas Cat

The first case involved a 9-year-old female, Kampur breed. There were malignant squamous cell carcinoma tumors in the oral cavity and the conjunctiva. Both of the specimens contained a poorly delineated mass of atypical epithelial cells with abundant eosinophilic cytoplasm, oval marginated nuclei, and prominent intercytoplasmic bridges. The presence of two separate tumors in the same animal is known to occur in white domestic cats from the influence of solar radiation.

The second case, seen at the Sunset Animal Clinic, Miami, Florida, involved a 6-year-old female Christmas cat. The tumor was a squamous cell carcinoma from the ear pinnae. There was a poorly delineated invasive neoplasm composed of nests and chords of basaloid cells with central cores of keratin. This tumor is slow to metastasize, but is invasive.

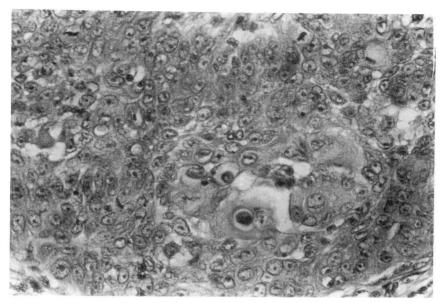

FIGURE 14.23 Photomicrograph of squamous cell carcinoma on the pinna of the ear of a white cat. Note nests of highly malignant cells. Courtesy of Dr. Alan J. Herron, Department of Pathology, Division of Comparative Pathology, University of Miami.

Bibliography

Ames, B. (1989). Mutagenesis and carcinogenesis: Endogenous and exogenous factors. *Environ. Mol. Mutagen.* **13**, 1–12.

Blum, H. F. (1950). On the mechanism of cancer induction by ultraviolet radiation. *J. Natl. Cancer Inst. (U.S.)* **11**, 463–495.

Cascinelli, N., Morabito, A., and Bufalino, R. (1980). Prognosis of stage I melanoma of the skin: WHO collaborating centres for evaluation of methods of diagnosis and treatment of melanoma. *Int. J. Cancer* **26**, 733–739.

Clark, W. H., Jr., Reimer, R. R., and Greene, M. (1978). Origin of familial malignant melanomas from heritable melanocytic lesions: "The B–K mole syndrome." *Arch. Dermatol.* **117**, 732–738.

Cleaver, J. E. (1986). DNA repair and replication in xeroderma pigmentosum. *In* "Antimutagenesis and Anticarcinogenesis Mechanisms" (D. M. Shankel, P. E. Hartman, T. Kada, and A. Hollander, eds.), pp. 425–438. Plenum. New York.

Cleaver, J. E. (1990). *Commentary.* Do we know the cause of xeroderma pigmentosum? *Carcinogenesis (London)* **11**, 875–882.

Cleaver, J. E., Zelle, B., Hashem, N., El-Hefnawi, M. H., and German, J. (1981). Xeroderma pigmentosum patients from Egypt: Preliminary correlations of epidemiology, clinical symptoms and molecular biology. *J. Invest. Dermatol.* **77**, 96–101.

DerGruijl, F. R., Sterenborg, J. C. M., Forbes, P. D., Davies, R. E., Cole, C., Kelkens, G., Van Weelden, G., Slaper, H., and Van der Leun, J. C. (1993). Wavelength dependence of skin cancer induction by ultraviolet irradiation of albino hairless mice. *Cancer Res.* **53**, 53–60.

Friedman, R. J., Rigel, D. S., Silverman, M. K., Kopf, A. W., and Vossaert, K. A. (1991). Malignant melanoma in the 1990s: The continued importance of early detection and the role of physician examination or self-examination of the skin. *Ca—Cancer J. Clin.* **41**, 201–226.

Gianelli, F., Pawsey, S. A., and Avery, J. A. (1982). Differences in patterns of complementation of the more common groups of xeroderma pigmentosum: Possible implications. *Cell (Cambridge, Mass.)* **29**, 451–458 (cited in Cleaver, 1990).

Hargis, A. M., Ihrke, P. J., Spangler, W. L., and Stannard, A. A. (1992). A retrospective clinicopathologic study of 212 dogs with cutaneous hemangiomas and hemangiosarcomas. *Vet. Pathol.* **29**, 316–328.

Hashem, N., Bootsma, D., Keijzer, W., Greene, A., Corielli, L., Thomas, G., and Cleaver, J. E. (1980). Clinical characteristics, DNA repair and complementation groups in Xeroderma pigmentosum patients from Egypt. *Cancer Res.* **40**, 13–18.

Hessel, A., Siegle, R. J., Mitchell, D. L., and Cleaver, J. E. (1992). Xeroderma pigmentosum variant with multisystem involvement. *Arch. Dermatol.* **128**, 1233–1237.

Jubb, K. V. F., Kennedy, P. C., and Palmer, N. (1985). "Pathology of Domestic Animals," Vol. 1. Academic Press, Orlando, FL.

Kenet, R. O., Tearney, G. J., and Kang, S. (1990). *In vivo* characterization of pigment patterns in dysplastic nevi using digital epiluminescence microscopy. Poster exhibit No. 75. *Annu. Meet. Am. Acad. Dermatol.* Abstracts, p. 124.

Kraemer, K. H. (1977). *In* "Cellular Senescence and Somatic Cell Genetics: DNA Repair Processes" (W. W. Nichols and D. G. Murphy, eds.), p. 40. Symposia Specialists, Inc., Miami, FL.

Kraemer, K. H., Herlyn, M., Yuspa, S., Clark Jr., W. H., Townsend, G. K., Neisis, G. R. and Hearing, V. J. (1989). Reduced DNA repair in cultured melanocytes and liver cells from a patient with Xeroderma Pigmentosum. *Arch. Dermatol.* **125**, 264.

Lambert, W. C., and Lambert, M. W. (1985). Co-recessive inheritance: A model for DNA repair, genetic disease and carcinogenesis. *Mutat. Res.* **145**, 227–234 (in Cleaver, 1990).

Muller, G. H., Kirk, W. R., and Scott, D. W. (1989). "Neoplastic Diseases in Small Animal Dermatology," 4th ed., pp. 854–958. Saunders, Philadelphia.

Pehamberger, H., Steiner, A., and Wolff, K. (1987). *In vivo* epiluminescence microscopy of pigmented skin lesions. I. Pattern analysis of pigmented skin lesions. *J. Am. Acad. Dermatol.* **17**, 571–583.

Rosenkrantz, W. S. (1993). Solar dermatitis. *In* "Current Veterinary Dermatology" (C. E. Griffin, K. W. Kwochka, and J. M. MacDonald, eds.), pp. 309–315. Mosby, St. Louis, MO.

Setlow, R. B., Grist, E., Thompson, K., and Woodhead, A. D. (1993). Wavelengths effective in induction of malignant melanoma. *Proc. Natl. Acad. Sci. U.S.A.* **90**, 6666–6670.

Sober, A. J. (1991). Cutaneous melanoma: Opportunity for cure. *Ca—Cancer J. Clin.* **41**, 197–199.

Steiner, A., Pehamberger, H., and Wolff, K. (1987). *In vivo* epiluminescence microscopy of pigmented skin lesions. II. Diagnosis of small pigmented skin lesions and early detection of malignant melanoma. *J. Am. Acad. Dermatol.* **17**, 584–591.

Streilein, J. W. (1990). "Molecular Pathology of Sunlight-induced Skin Cancers," Senior Faculty Forum Seminar. Department of Microbiology and Immunology, University of Miami, Coral Gables, FL.

Urbach, F. (1982). Photocarcinogenesis. *In* "The Science of Photomedicine" (J. D. Regan, and J. A. Parrish, eds.), pp. 261–292. Plenum, New York and London.

Urbach, F. (1993). Environmental risk factors for skin cancer. *Recent Results Cancer Res.* **128**, 243–262.

Wood, R. D. (1989). Repair of pyrimidine dimer ultraviolet light photoproducts by human cell extracts. *Biochemistry* **28**, 8287–8292 (cited in Clever, 1990).

Wood, R. D., Robins, P., and Lindhal, T. (1988). Complementation of the Xeroderma pigmentosum DNA repair defect in cell-free extracts. *Cell (Cambridge, Mass.)* **53**, 97–106.

15

Photoimmunology

Photoimmunology deals primarily with the effect of UV light on the skin's immune system. Sections 15.1 to 15.3 introduce the skin's immune system. The remaining sections discuss the effects of light.

15.1 THE SKIN'S IMMUNE SYSTEM: FROM GOLD IMPREGNATION TO MONOCLONAL ANTIBODIES

The skin and its draining lymph nodes cooperate to provide an effective *immune surveillance system*. The epidermis is the source of sensitizing as well as tolerizing signals. In 1858, Paul Langerhans, a medical student, disclosed a population of dendritic cells in the suprabasal regions of the epidermis by impregnating human skin with gold chloride. Langerhans concluded that he was dealing with a system of intraepidermal nerve cells. Langerhans cells (LCs) present a striking morphological similarity to histiocytosis X cells. Based on immunophenotyping with monoclonal antibodies, they appear to be of bone marrow origin and a unique member of the family of *Ia antigen-bearing cells* necessary for the initiation of T cell-dependent immune responses.

LCs are weak stimulators of resting T cells, but on short-term culture *they develop into potent immunostimulatory dendritic cells*. This is called "LC maturation," mediated by keratinocyte-derived cytokines [granulocyte–macrophage colony-stimulating factor (GM-CSF) and interleukin 1 (IL-1)].

Antigens taken by LCs are processed in specialized compartments of these cells. The resulting antigenic fragment is finally complexed with the major histocompatibility complex (MHC)-encoded antigens. Within a few hours, LCs undergo morphological and functional changes, they leave the epidermis, enter the lymphatics, and migrate to the paracortical areas of the draining lymph nodes. There, they present the antigen/MHC complex on their surface to the T cell antigen receptor (TCR). The T cell blasts, thus generated, find their way back to the skin.

The expression of tissue-specific *homing receptors* is accompanied by expression of corresponding *adressins* on vasculo-endothelial cells. Adhesion molecules of "endothelial cell–leukocyte" and "vascular cell" origin mediate the adhesion of memory T cells to cytokine-activated dermal microvasculo-endothelial cells. This promotes the migration of memory T cells into the dermis. *After renewed stimulation by antigen-presenting cells (APCs),* T cells can undergo cloned expansion. The result is the generation of effector cells and molecules that eliminate the pathogen or aggressor agent.

The hypothesis of immune expression in the skin was originated by Streilein as the skin-associated lymphoid tissue (SALT) concept.

15.2 DENDRITIC CELLS

Dendritic cells (DCs) are a class of leukocyte with specialized immunostimulatory capacity. They induce the mixed-leukocyte reaction, allograft rejection, antibody formation to thymus-dependent antigens, cytolytic T cell development, T cell proliferation in response to antigens and mitogens, and contact sensitivity. When DCs present antigen to small or resting T cells, the latter are transformed into large, functioning helper T lymphocytes. These secrete lymphokines and help the development of other effector cells, such as antibody-secreting B cells and cytotoxic T lymphocytes.

DCs are distinguished by surface markers detected via monoclonal antibodies and specific ligands. DCs express high levels of class I and II histocompatibility antigens (Ia antigens). Members of DC "lineage" include LCs in the epidermis, and some Ia$^+$ interstitial cells in most tissues and "interdigitating" cells in the T-dependent area of lymphoid organs.

The differences between DCs and phagocytic cells are that the former lack *active* endocytic capacity (DCs internalize antigens via the MHC complex) and several macrophage surface antigens; they have abundant motile processes and they also lack cytochemical markers such as abundant acid phosphatase, nonspecific esterase, and myeloperoxidase.

15.2.1 Antigenic Moieties in Dendritic Cells

In humans the antigenic moieties found to be present in LCs, and which are easily detectable, include (1) panhemopoietic marker CD45, (2) class I antigens encoded for by the HLA-A, -B, and -C loci of the major histocompatibility complex, (3) class II alloantigens encoded for by the HLA-D region of the MHC [also termed the immune response-associated (Ia) antigens], (4) CDI antigens, which are also found in cortical thymocytes, (5) the S-100 protein, and (6) the cytoskeletal protein vimentin present on intermediate-sized filaments of LCs.

LCs form an Ia^+ dendritic system in the epidermis. Just after isolation, LCs express several markers in common with macrophages, i.e., ATPase and nonspecific esterase, but these traits are lost *in vitro*.

15.2.2 Mechanisms whereby DC Cells Initiate Immune Responses

DCs actively stimulate T helper cells to release lymphokines and to become responsive to T cell growth factor or interleukin 2 (IL-2). These IL-2-responsive T lymphoblasts are generated from discrete aggregates of DCs and T cells in culture. The aggregates exhibit broad and close contact between the DC lamellipods and T lymphocytes, and many T cells interact with any one DC. In scanning electron micrographs, several lymphocytes are seen in contact with the numerous lamellipods of a single LC.

Enriched populations of cultured epidermal LCs exhibit strong immunofluorescent staining with anti-Ia monoclonal antibodies. Mouse epidermis, in contrast to human epidermis, is very thin, so that the LC system can be photographed in a single plane of focus.

15.2.3 Ultrastructure of the Langerhans Cell

Ultrastructurally, the LCs are recognized by a distinctive organelle, the tennis-racket-shaped Birbeck granule. In activated, dendritic, antigen-presenting cells, the Birbeck granules are thought to be formed from internalization of the cell membrane. These structures may be involved in the intracellular processing and transport of antigens. Recent studies with gold-labeled anti-Ia antibody-treated cells have demonstrated that such cells internalize Ia antigen into the Birbeck granules, which are contiguous with both the cell membrane and endosomes.

In vivo, in lymph node dendritic cells, most of the hapten is contained within discrete cytoplasmic organelles (i.e., lysosomes). Thus some sort of intracellular antigen processing takes place during *in vivo* sensitization.

15.2.4 Cultures of Langerhans Cells

Epidermal cells (ECs) are suspended in keratinocyte growth medium supplemented with recombinant human granulocyte–macrophage colony-stimulating factor and are then plated. ECs are then utilized as freshly isolated ECs (fECs) or are cultured for 72 hr at 37°C in a humidified 5% CO_2 atmosphere (cECs). fECs or cECs are then maintained in the presence of a recall antigen for 18 hr (antigen pulse). During this "pulsing period," fLCs present in the EC population develop a dendritic morphology. The recall antigens used are live influenza virus (FLU), tetanus toxoid (TT), and *Candida albicans* extract (CAND). Following the antigen pulse period, nonadherent ECs are washed, resuspended, and centrifuged over a density gradient. An enrichment of LCs is obtained. When fLCs and cLCs are cultured with T cells for varying periods, significant T cell proliferation is detected by the sixth day of coculture.

LC phenotype changes include enhancement of the class II major histocompatibility complex antigens, with concomitant reduction or loss of Birbeck granules and acidic organelles (endosomes) as well as certain monocyte/macrophage markers, and nonspecific esterase. It appears that the retention or loss of antigen processing is MHC dependent.

It is possible that fLCs and cLCs utilize different pathways of antigen processing, e.g., a different repertoire of endosomal proteases, yielding different or the same epitopes.

When LCs are removed from their usual epidermal milieu, they express increased amounts of Ia and become more potent stimulators of T cell responses. The ability of LCs to process and present intact protein antigens (i.e., hen egg lysozyme, cytochrome *c*, and ovalbumin) to T cell clones and hybridomas was tested. Microfluorometry with fluorescein isothiocyanate (FITC)-labeled MHC antibody was used to quantify MHC expression. Tritiated thymidine deoxyribose incorporation was used as a measure of T cell proliferation in response to antigen presentation. cLCs were, on a per cell basis, somewhat less efficient than were fresh LCs. Class II MHC turnover in cLCs is far slower than in fLCs and this may account, at least in part, for the finding that cLCs are not superpotent APCs for intact protein antigens, as they are for haptens.

The induction, specificity, and expression of T cell-mediated immune responses are critically dependent on the interaction of Ia antigen-positive subclasses (i.e., LCs and L3T4+ cells) of T helper cells (Th). Specific antibodies to class II MHC determinants and L3T4+ cells have been shown *to inhibit significantly the induction, development, expression, and transfer of delayed-type hypersensitivity (DTH).*

In vivo treatment with anti-class II antibody inhibits the induction of contact sensitivity significantly. This inhibition is short-lived and probably not attributable to the induction of antigen-specific suppressor cells. *In vivo* administration of anti-class II antibody, even in extremely high doses, does not affect LC function *in vitro*.

15.3 INTERACTION BETWEEN APCS AND T CELLS

Substantial antigen-specific responses occur only if the immune T cells from one strain are challenged with antigen-pulsed macrophages/monocytes (MM) from the same homologous strain. Mouse T cells sensitized to hapten or virus-modified cells are primarily cytotoxic for similarly modified target cells that are MHC I compatible. The failure of antigen associated with allogeneic macrophages/monocytes to activate immune T cells *in vitro* can be attributable to the fact that these T cells have been primed *in vivo* only to antigen associated with syngeneic MM. *Thus, the genetic restriction of the T cell response is regulated entirely by the histocompatibility type of the APC used for initial sensitization.* The T cell antigen receptor simultaneously recognizes both the nominal antigen and the MHC product.

The responses of T cells to antigens are controlled by specific I region products (i.e., Ir antigens). Two alternative models can account for lack of T cell response to presented antigen:

1. If the antigen and the MHC class molecule fail to interact, an effective complex that can be recognized by the T cell antigen receptor does not form (determinant selection model).
2. Antigen and MHC class molecules interact normally, but T cells capable of corecognizing antigen and a given MHC class molecule are not present among the T cells that are primed. The lack of such T cells might be secondary to absence from the genotype repertoire, elimination during the induction of tolerance, failure to be selected by the thymus, or suppression during the induction of immune responses ("hole in the repertoire" model).

Antigen presentation is a time-dependent process and agents that increase the pH of lysosomes, e.g., ammonia and chloroquine, if added early, interfere with the ability of the cell to present antigen. This suggests that the antigen seen by T cells is not intact, but rather fragmented or denatured intracellularly before expression on the APC surface. The determinant recognized by the TCR on MHC class-specific T helper cells consists of a short oligopeptide bound to the MHC class molecule. In MHC class II-restricted responses, the ligand of the TCR is a peptide bound to class II MHC molecules.

A two-pathway model of antigen processing and presentation to T cells has been developed.

15.3.1 Exogenous Pathway

According to the *MHC class II* presentation pathway for exogenous protein antigens, (1) intact proteins enter the cell by binding to molecules engaged in receptor-mediated endocytosis (RME). Then (2) antigen is internalized into sequential acidic endosomal compartments, where the processing reactions (proteolysis, denaturation, and/or fragmentation) occur. Next, (3) antigen is transported to the endosomes of class II MHC, either newly synthesized or recycled from the cell surface ("rescue" of antigenic peptides from further degradation by binding to class II MHC within endosomes). Finally, (4) from the endosome the antigen goes either to the lysosome, where it is degraded and exocytosed, or back to the cell surface via the trans-Golgi network; it is within this neutral latter compartment that the interaction of the processed antigen with the class II molecule is believed to occur. Once formed, antigen/MHC class II complexes are then recycled to the cell surface. Only the class II, but not the class I, presentation pathway is sensitive to chloroquine.

15.3.2 Endogenous Pathway

MHC class I presentation for endogenous proteins is less well defined but, again, the selection of peptides for surface expression rather than for lysosomal degradation probably depends on their structural features (α-helix?) determining their affinity to the MHC molecule.

Influenza hemagglutinin synthesized endogenously serves as a good target for class I-restricted, but not for class II-restricted, hemagglutinin-specific T cell clones. In contrast, exogenous administration of hemagglutinin protein makes excellent targets for class II-restricted, but not for class-I restricted, cytotoxic T lymphocytes (CTLs).

15.4 EFFECT OF UV RADIATION

15.4.1 Alteration of Contact Hypersensitivity Response

Mice exposed to very low doses of UV radiation fail to develop contact hypersensitivity (CHS) responses to haptens applied onto the UV-irradiated skin. This effect correlates with a reduction in the number of LCs in the epidermis as detected by ATPase staining. Hapten-specific suppressor T lymphocytes are present in the spleens of mice developing hapten-specific immunologic tolerance.

Fluorescein isothiocyanate has proved to be a useful reagent with which to follow antigen-labeled APCs to draining lymph nodes, where they bind to lymphocytes. Dendritic lymph node (DLN) cells are collected 18 hr after sensitization from C3H mice first exposed to low doses of UV, and then sensitized with FITC through the irradiated skin. These DLN cells, most of which migrate from the skin, are injected into the foot pads of normal syngeneic recipient mice to test their ability to induce CHS (as a measure of their antigen-presenting activity). The DLN cells of mice sensitized through UV-irradiated skin fail to induce CHS and instead induce hapten-specific suppressor lymphocytes. Based on this study, the altered activity could result from (1) a change in the antigen-presenting capability of the DLN cells, (2) inactivation of a subset of antigen-presenting cells, (3) the presence of a different type of FITC$^+$ cells, or (4) a combination of these factors.

15.4.2 Alteration of Ability to Activate T Lymphocytes *in Vitro*

UV irradiation of hapten-conjugated epidermal cell suspensions (keratinocytes + LCs) impairs their ability to activate T lymphocytes *in vitro*. Injection of such cells *in vivo* fails to induce CHS and instead results in the induction of immunological tolerance.

The precise nature of the change induced in Langerhans cells by UV remains unclear. The molecular target of the UV irradiation does not appear to be the Ia surface marker, and the expression of class II molecules on antigen-presenting cells, although reduced by UV, does not correlate with loss of function. A more likely target is the expression of adhesion molecules that are down-regulated by UV and are necessary for T lymphocyte activation by antigen-presenting cells; UV-irradiated keratinocytes are known to release a variety of cytokines, including interleukin 1, which then secondarily affect Langerhans cells (see Figure 15.1).

The delivery of the primary signal transduced by action of the MHC/antigen complex of APCs with the T cell receptor complex requires a viable second signal normally delivered by interaction of a costimulatory factor derived from the APC with its surrounding ligand in the T cells.

UVB radiation distorts the antigen-presenting function of LCs. Utilizing four sources of APCs, i.e., epidermal cells (ECs), flow microfluorometry-purified Ia$^+$ ECs (LCs), flow microfluorometry-purified Ia$^-$ ECs, and splenic adherent cells (SACs), it can be shown that (1) irradiated ECs and LCs lose their ability to stimulate T helper cells 1 (Th1), but retain fully their capacity to simulate T helper cells 2 (Th2); (2) irradiated SACs lose the ability to induce proliferation of both Th1 and Th2; and (3) suppres-

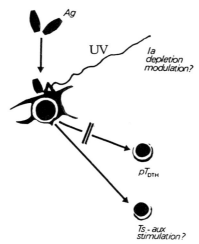

FIGURE 15.1 UV–LC interaction. When an antigen (Ag) is presented to the skin it couples to the Ia cells surface marker-bearing ($\triangle$) LC; this complex stimulates precursor T cells of delayed-type hypersensitivity (pT_{DTH}), but auxiliary T suppressor cells (Ts-aux) are also stimulated. When the skin is irradiated with UV light, the Ia on the LC is modulated and the stimulatory signal of the antigen–Ia complex is abrogated. Stimulators of T suppressors by the LC (see above) or the antigen itself can supervene and result in a tolerogenic effect toward the aggressor, whether it is a pathogen or neoplastic tissue in the process of formation. Modified from K. Wolff (1982). With permission of Plenum, New York.

sion of CHS mediated by UVB-irradiated LCs may result from an alteration of the ratio of and/or activity of Th1 and Th2 cells.

UVB can convert LCs or SACs from immunogenic to tolerogenic APCs. Addition of allogeneic SACs during preincubation prevents the induction of unresponsiveness by UVB-LCs or UVB-SACs.

Local suppression of CHS *in vivo* by low-dose UVB is associated with alterations in the morphology and antigen-presenting function of LCs and the appearance of hapten-specific T suppressor (Ts) cells in lymphoid organs.

Genetic factors play a major role in low-dose UVB-induced immunosuppression. The UVB-susceptible trait appears to be transmitted as an autosomal dominant trait that is regulated in part by the major histocompatibility complex genes.

15.5 UV CARCINOGENESIS AND IMMUNOLOGICAL MECHANISMS

15.5.1 Some Analogies with the Actions of Chemical Carcinogens

In addition to UV radiation, some chemical carcinogens and tumor promoters decrease the number and alter the morphology of antigen-presenting cells in the epidermis. It is not clear whether the carcinogen effects are direct on epidermal LCs or mediated by means of other cells or molecules that are the actual target of the carcinogens. It is also not clear whether the suppressor-inducing, antigen-presenting cells are ordinary LCs that have been altered by the UV radiation, a UV-resistant subpopulation of antigen-presenting cells in the epidermis, or a population of inflammatory macrophages that have infiltrated the skin. It is not known whether chemical mediators, such as cis-urocanic acid formed in the UV-irradiated part of the skin, act by affecting the antigen-presenting activity of the LCs or by contributing to depressed immune reactivity.

UVB irradiation acting at various stages of tumor progression may tilt the balance of local defenses in favor of a developing tumor by generating suppressor cells. The following list is a synopsis of UVB-induced carcinogenesis and immunosuppression mechanisms.

1. UV-induced skin cancers are difficult to transplant, even to isogeneic mice. UV-induced sarcomas have a high level of transplantation resistance, and they are as antigenic as sarcomas induced by potent chemical carcinogens.

2. Most UVB-induced (280–320 nm) squamous cell carcinomas are highly antigenic, at least in some systems. Such tumors regress when transplanted to normal syngeneic mice, but they grow in immunosuppressed (by UV radiation) mice.

3. After removal of UV-induced tumors, primary hosts are susceptible to challenge with their autochthonous tumors, but most of the tumors are rejected by untreated control mice. The same primary hosts are also susceptible to grafts of antigenically dissimilar UV-induced tumors from other syngeneic primary hosts.

4. UV-irradiated non-tumor-bearing mice are susceptible to challenge with syngeneic UV-induced tumors, even when subcarcinogenic doses are given.

5. UV irradiation does not interfere with rejection of UV-induced syngeneic tumors in animals that have been immunized prior to UV irradiation.

6. Normal mice can be rendered susceptible to syngeneic UV-induced tumors by the adaptive transfer of lymphoid cells from either tumor-bearing UV-irradiated mice or from non-tumor-bearing UV-irradiated donors.

7. Lethally X-irradiated mice, their immune system reconstituted with lymphoid cells from normal animals resistant to UV-induced tumors, or an equal number of cells from UV-irradiated and nonirradiated donors, rejected the isogeneic tumor transplants (Figure 15.2). Those receiving cells from UV-irradiated mice did not reject isogeneic UV-induced tumor transplants.

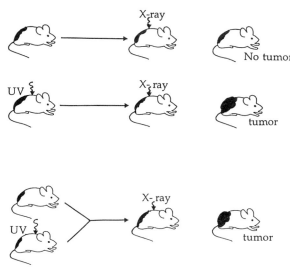

FIGURE 15.2 Experimental protocol used to demonstrate the immunologic basis of the UV effect in radiation-induced systemic alterations. Lymphoid cells from normal mice and UV-irradiated mice were used to reconstitute the immune system of lethally X-irradiated animals. Susceptibility to UV-induced skin cancer was transferred with lymphoid cells taken from UV-induced syngeneic tumors. Mixing studies (lymphoid cells from normal mice and from UV-induced tumor cases) demonstrated that the effectiveness was due to T lymphocytes in the lymphoid cell preparation. Adapted from Granstein R. D. *Dermatology in General Medicine*, Vol. I (J.D. Fitzpatrick et al. eds) Fig. 132.1. McGraw Hill, New York. (1993); reproduced from Kripke (1986).

8. T cells involved in inducing susceptibility to tumors are suppressor cells. Thus, it would appear that a specific UV tumor-associated antigen is recognized by these cells. T suppressor cells able to support growth of such tumors have been cloned. Ts appear to arise in UV-irradiated mice prior to the time the animals develop tumors.

There is additional evidence of Ts existence:

1. Groups of lethally X-irradiated mice had their lymphoid system reconstituted with spleen and lymph node cells from syngeneic normal mice, chronic UV-irradiated mice, and a mixture of both.

2. After 4 weeks, grafts from irradiated areas of skin from mice having received chronic UV exposure developed significantly more tumors in animals reconstituted from irradiated mice or from mixtures of irradiated/normal mice, than in animals reconstituted with lymphoid tissue from normal mice.

3. A second group of mice was intravenously injected with T cells from normal or chronically UV-irradiated syngeneic mice. Recipients of T

cells from UV-irradiated animals developed more tumors than did the recipients of T cells from nonirradiated animals, and the tumors appeared earlier.

T suppressor (Ts) cells induced by UV not only inhibit rejection of tumor transplants, but also play an important role in carcinogenesis.

15.5.2 Pathway of UV Skin Carcinogenesis Formation

UV irradiation has two effects (Figure 15.3). The first effect is transformation of keratinocytes to a malignant phenotype with expression of tumor-associated antigens (TAAs), which may induce Ts cells that prevent immunological destruction of an incipient tumor. A second effect involves changes preventing the effective presentation of antigen for

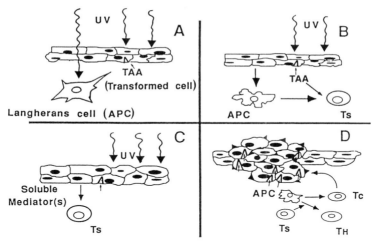

FIGURE 15.3 Model of UV-induced carcinogenesis. Experimental evidence exists to suggest several mechanisms for UV-induced immunologic downregulation with regard to the tumor-associated antigen (TAA). (A) Direct functional derangement of antigen-presenting cells (APC) from the epidermis (presumably Langerhans cells), which are capable of presenting an antigen from transformed cells for CD4-dependent responses. (B) Selection and preservation of the UV-resistant APCs (Ia$^+$ or Thy$^+$) that may present TAA for activation of T suppressor (Ts) cells. (C) UV-induced release of an epidermal factor(s), resulting in preferential activation of Ts cells and immunologic down-regulation. (D) Clonal proliferation of transforming clone cells expressing TAA. Immune rejection is inhibited by lack of functional APCs to induce T helper activity against a TAA and/or the presence of APCs primed for suppression, and of epidermal immunosuppressor release. Tc, T cell; TH, T helper cell. Adapted from Granstein (1993) in "Dermatology in General Medicine" (T. B. Fitzpatrick *et al.*, eds.), McGraw-Hill, New York; reproduced from Gallo *et al.* (1989).

effector immune mechanisms, together with preservation or enhancement of down-regulating suppressor mechanisms.

15.5.3 Further Human Studies on UV Photocarcinogenesis

Exposure to summer light for 1 hr results in a significant increase of
suppressor/cytotoxic phenotype CD8 lymphocytes, and a decrease of
helper phenotype CD4 lymphocytes in peripheral blood. A 30-min exposure to solarium light reduces the cutaneous response to dinitrochlorobenzene (DNCB), also producing a decrease in natural killer cells and a
decrease in the CD4/CD8 ratio. Humans may vary in their susceptibility
to UV-induced inhibition of contact hypersensitivity. Skin cancer patients
have a higher percentage of nonsusceptibility to DNCB and about half
become tolerant to DNCB. No normal subjects become tolerant to DNCB.
Human epidermal cells obtained after exposure to UVB radiation are
capable of activating suppressor–inducer lymphocytes.

Bibliography

Aberer, W., and Katz, S. I. (1989). Effects of *in vivo* administration of anti-*Ia* antibodies on
contact sensitivity. *Arch. Dermatol.* **125**, 280–284

Aiba, S., and Katz, S. I. (1990). Phenotypic and functional characteristics of *in-vivo*-activated
Langerhans cells. *J. Immunol.* **145**, 2791–2796.

Aiba, S., and Katz, S. I. (1991). The ability of cultured Langerhans cells to process and
present protein antigens is MHC-dependent. *J. Immunol.* **146**, 2479–2487.

Cohen, P. J., and Katz, S. I. (1992). Cultured human Langerhans cells process and present
intact antigen. *J. Invest. Dermatol.* **99**, 331–336.

Cruz, P. D., and Bergstresser, P. R. (1990). Antigen processing and presentation by epidermal Langerhans cells. *Dermatol. Clin.* **8**, 633–647.

Cruz, P. D., and Bergstresser, P. R. (1991). The influence of ultraviolet radiation and other
physical and chemical agents on epidermal Langerhans cells. *In* "Epidermal Langerhans
Cells" (G. Schuler, ed.), pp. 253–271. CRC Press, Boca Raton, FL.

Enk, A. H., and Katz, S. I. (1992). Identification and induction of keratinocyte-derived IL-10.
J. Immunol. **149**, 92–95.

Ferreira-Marques, J. (1951). Systema sensitivum intraepidermicum. Die Langerhanschen
Zellen als Receptoren des hellen Schmerzes: Dolorireceptores. *Arch. Dermatol. Syph.* **193**,
191 (cited in Wolff, 1991).

Gallo, R. L., Staszewski, R. and Granstein, R. D. (1989). Physiology and pathology of skin
photoimmunology. *In* "Skin Immune System" (J. D. Bos, ed.), p. 381–402. CRC Press,
Boca Raton, FL.

Granstein, R. D. (1993). Photoimmunology. *In* "Dermatology in General Medicine" (T. B.
Fitzpatrick, A. Z. Eisen, K. Wolff, I. M. Freedberg, and F. Austen, eds.), pp. 1638–1651.
McGraw-Hill, New York.

Inaba, K., Schuler, G., Witmer, M. D., Valinsky, J., Atassi, B., and Steinman, R. M. (1986).
Immunologic properties of purified epidermal Langerhans cells. Direct requirement for
stimulation of unprimed and sensitized T-lymphocytes. *J. Exp. Med.* **164**, 605–613.

Kripke, M. L. (1986). Immunology and photocarcinogenesis. New light on an old problem. *J. Am. Acad. Dermatol.* **14**, 149.

Morrison, W. L., and Parrish, J. A. (1982). Photoimmunology. *In* "The Science of Photomedicine" (J. D. Regan and J. A. Parrish, eds.), pp. 293–320. Plenum, New York and London.

Muller, H. K., Bucana, C., and Kripke, M. L. (1992). Antigen presentation in the skin: Modulation by UV radiation and carcinogens. *Semin. Immunol.* **4**, 205–215.

Pure, E., Inaba, K., Crowley, M. T., Tardelli, L., Witmer-Pack, M. D., Ruberti, G., Fathman, G., and Steinman, R. M. (1990). Antigen processing by epidermal Langerhans cells correlates with the level of biosynthesis of major histocompatibility complex class II molecules and expression of invariant chain. *J. Exp. Med.* **172**, 1459–1469.

Romain, N., Stingl, G., Tschacler, E., Witmer, M. D., Steinman, R. M., Shevach, E. M., and Schuler, G. (1985). The Thy-1-bearing cell of murine epidermis. *J. Exp. Med.* **161**, 1368–1383.

Romani, N., Koide, S., Crowley, M., Witmer-Pack, M. D., Livingstone, A. M., Fathman, C. G., Inara, K., and Steinman, R. M. (1989). Presentation of exogenous protein antigens by dendritic cells to T cell clones. Intact protein is best presented by immature, epidermal Langerhans cells. *J. Exp. Med.* **169**, 1169–1178.

Rosenthal, A. S., and Shevach, E. M. (1973). Function of macrophages in antigen recognition by guinea pig T lymphocytes. I. Requirements for histocompatible macrophages and lymphocytes. *J. Exp. Med.* **138**, 1194 (cited in Stingl and Shevach, 1991).

Schuler, G., ed. (1993). "Epidermal Langerhans Cells." CRC Press, Boca Raton, FL.

Schuler, G., Romani, N., and Steinman, R. M. (1985). A comparison of murine epidermal Langerhans cells with spleen dendritic cells. *J. Invest. Dermatol.* **85**(1), Suppl. 99s–106s.

Shimada, S., Caughman, S. W., Sharrow, S. O., Stephany, D. and Katz, S. I. (1987). Enhanced antigen-presenting capacity of cultured Langerhans cells is associated with markedly increased expression of Ia antigen. *J. Immunol.* **139**, 2551–2557.

Simon, J. C., Cruz, P. D., Jr., Bergerstresser, P. R., and Tigelaar, R. E. (1990). Low dose ultraviolet B-irradiated Langerhans cells preferably activate CD4$^+$ cells of the T helper 2 subset. *J. Immunol.* **145**, 2097–2091.

Simon, J. C., Tigelaar, R. E., Bergerstrasser, P. R., Edelbaum, D., and Cruz, P. D., Jr. (1991). Ultraviolet B radiation converts Langerhans cells from immunogenic to tolerogenic antigen presenting cells. *J. Immunol.* **146**, 485–491.

Simon, J. C., Krutmann, J., Elmets, C. A., Bergstresser, P. R., and Cruz, P. D., Jr. (1992). Ultraviolet B-irradiated antigen-presenting cells display altered accessory signaling for T-cell activation: Relevance to immune responses initiated in skin. *J. Invest. Dermatol.* **98**, 66S–69S.

Steinman, R. M. (1988). Cytokines amplify the function of accessory cells. *Immunol. Lett.* **17**, 197–202.

Steinman, R. M., and Schuler, G. (1985). Murine epidermal Langerhans cells mature into potent immunostimulatory dendritic cells *in vitro*. *J. Exp. Med.* **161**, 526–546.

Stingl, G., and Shevach, E. M. (1991). Langerhans cells as antigen-presenting cells. *In* "Epidermal Langerhans Cells" (G. Schuler, ed.), pp. 159–190. CRC Press, Boca Raton, FL.

Stingl, G., Hauser, C., and Wolff, K. (1993). The epidermis as an immunologic microenvironment. *In* "Dermatology in General Medicine" (T. B. Fitzpatrick, A. Z. Eisen, K. Wolff, I. M. Freedberg, and F. Austen, eds.), pp. 172–192. McGraw-Hill, New York.

Streilein, J. W. (1983). Skin associated lymphoid tissues (SALT). Origin and Function. *J. Invest. Dermatol.* **80**, 12s.

Streilein, J. W., and Tigelaar, R. E. (1983). SALT (Skin-associated lymphoid tissue). "Photoimmunology" (J. A. Parrish, ed.), p. 95, Plenum, New York.

Witmer-Pack, M. D., Valinsky, J., Olivier, W., and Steinman, R. M. (1988). Quantitation of surface antigens on cultured murine epidermal Langerhans cells: Rapid and selective increase in the level of surface MHC products. *J. Invest. Dermatol.* **90**, 387–394.

Wolff, K. (1982). Skin: Structure, natural and therapeutic targets of ultraviolet irradiation. *In* "Trends in Photobiology" (C. Hélène, M. Charlier, T. Montenay-Garestier, and G. Laustriat, eds.), pp. 253–266. Plenum, New York and London.

Wolff, K. (1991). The fascinating story that began in 1868. *In* "Epidermal Langerhans Cells" (G. Schuler, ed.), pp. 1–9. CRC Press, Boca Raton, FL.

Zucker-Franklin, D., Creaves, M. F., Grossi, C. E., and Marmont, A. M. (1988). "Atlas of Blood Cells: Function and Pathology," 2nd ed., pp. 361–377. E. Eedi-ermes, Milano, and Lea & Febiger, Philadelphia.

CHAPTER 16

Photosensitive, Photoallergic, and Light-Aggravated (Photo-Koebner) Photodermatoses

Photodermatoses are light-dependent (light-influenced) skin diseases that are caused, initiated, or aggravated by light (Figures 16.1 and 16.2). Consideration will be first given to phototoxic effects (i.e., *photo-auto reactions*) which are directly due to photochemical reactions of photosensitizing substances leading to immediate acute cellular effects. Subsequently the photoallergic effects (i.e., *photo-hetero reactions*) will be presented, which are responsible for *delayed reactions,* not directly linked to light injury.

16.1 PHOTO-AUTO REACTIONS

16.1.1 Photo-Auto Reactions Resembling Chronic Light Damage

16.1.1.1 Tar and Pitch Skin

The phototoxic effects of this syndrome are due to insoluble cutting oils and chlorinated hydrocarbons. Tar and pitch skin is characterized by

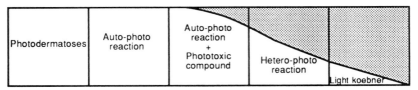

FIGURE 16.1 Significance of light in the different types of photodermatoses. Adapted from Ippen (1982), p. 351. With permission of Plenum, New York.

acne, comedones (blackheads), papules, nodules, pustules, true cysts, and hyperpigmentation on the face.

16.1.1.1.1 Tar Keratosis Tar keratosis is characterized by small, grayish, oval, flat, premalignant papules that can be removed by fingernails without bleeding; they are localized on the face, forearm, the medial aspect of the ankles, and/or the dorsal aspect of the hands.

16.1.1.1.2 Tar Melanosis and Melanodermatitis Toxica Hoffman Haberman Tar melanosis follows a photosensitivity reaction with reticulate hyperpigmentation. There is follicular hyperkeratosis with pigment-laden macrophages and/or perivascular lymphocytic infiltration.

Melanodermatitis toxica Hoffman Haberman occurs in middle-aged women or workers exposed to hydrocarbon-containing cosmetics or and fragrances.

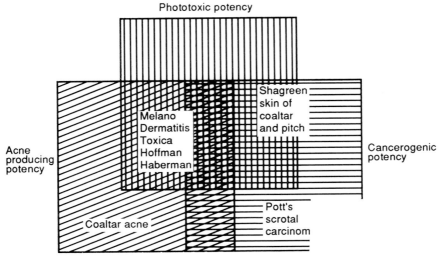

FIGURE 16.2 Cutaneous reactions from coal tar and pitch. Adapted from Ippen (1982), p. 359. With permission of Plenum, New York.

16.1.1.1.3 Keratotic Papilloma (Tar Wart) These lesions take 1 month to 6–20 years to develop; they localize on the cheeks, sides of the neck, forearms, and/or hands. They appear on poikilodermatous skin (i.e., atrophy, telangiectasia, hypo- or hyperpigmentation). Verrucous dermatitis can change into squamous cell carcinoma.

16.1.1.1.4 Favre–Racouchot Solar Comedone These lesions predominate on the temples of older persons. They are characterized by nodular elastosis, cysts and huge open comedones, and atrophic sebaceous glands.

16.1.1.1.5 Pseudomilium Colloidale (Wagner–Pelizarri) A familial disturbance in protein metabolism may be involved in these face and hand lesions. They show chronic light damage, and occur subsequent to the use of mineral oils and hydroquinone. Subepidermal, sharply circumscribed deposits probably form from a structural glycoprotein secreted by fibroblasts.

16.1.1.1.6 Granuloma Actinicum This lesion is a foreign-body granuloma that arises as a reparative process accompanying chronic light damage to the skin.

16.1.2 Photo-Auto Reactions Resembling Acute Light Damage: Sunburn-Type Photodermatoses

16.1.2.1 Dermatitis Phototoxica

Internal administration of a photosensitizer (porphyrin, tetracyclin, or herb liquor) results in acute erythema with or without edema of the skin (Figures 16.3 and 16.4). The reaction occurs after 1–3 min of sun exposure, instead of the typical erythema threshold time of 10–60 min. Shade and window glass afford no protection against UVA radiation.

Photoirritant contact dermatitis (PICD) reactions can be immediate or delayed. Natural plant product ingredients in therapeutic compounds and in fragrances, i.e., furocoumarins (psoralen), cause a much more delayed PICD response compared to tar photosensitization. In nature, psoralens act as antiinfective agents to protect plants from various infectious diseases. In PICD there is inter- and intracellular edema in the epidermis with necrosis of keratinocytes.

Furocoumarins react with UV radiation to induce monoadducts and cross-linking of DNA. Photodynamic photoirritants such as porphyrins and dyes (probably also tars) induce cellular membrane damage through production of $^1O_2$ (see Chapters 4 and 5).

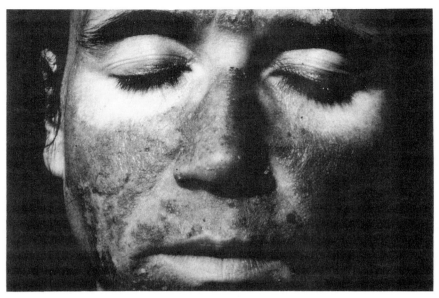

FIGURE 16.3 Phototoxic effect of dechlomycin on the face of a male patient. Courtesy of Professor L. C. Harber, Rhodebeck Professor of Dermatology, College of Physicians and Surgeons, Columbia University, New York.

Other agents inducing photocontact dermatitis include drugs (phenothiazines, sulfonamides), dyes (eosin, methylene blue), and plants (celery, parsley, parsnips, and figs).

Celery infected with fungus contains enough 8-methoxypsoralen (8-MOP) to induce PICD.

From a reader, to the *Miami Herald:*

> Yesterday I squeezed some limes . . . to take on a picnic. Today my hands look as if they have been burned, I have several watery blisters.

The answer by the *Herald's Home & Design* columnists Steve Ritter and Mary Misitis:

> You are describing the symptoms of lime . . . disease . . . caused by sunlight reacting with lime sap on your skin. The sunlight excites the molecules in the sap . . . causing burning, blistering or skin discoloration.

16.1.2.2 Berloque Dermatitis

In this syndrome, very sharply demarcated macules of light brown to dark brown hyperpigmentation frequently occur in the neck (Figure 16.4) in the form of downward-running drops, like necklace pendants. The photosensitizer, dissolved in alcoholic solutions or in plant oils, reaches the deeper keratinocytes and basal cells of the epidermis. Pigment incon-

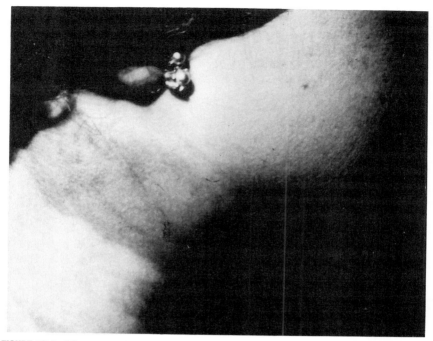

FIGURE 16.4   Phototoxic reaction (redneck) to the psoralen in Shalimar perfume. Courtesy of Professor L. C. Harber.

tinence (melanin phagocytosis in the dermis) may last for months or years. Absence of the early erythematous stage and the marked persistence of the pigmentation distinguish cosmetic-induced chloasma.

16.1.2.3 Phytophotodermatitis

Photodermatitis (bullosa) striata, or meadow dermatitis, is phytophotodermatitis involving penetration of photodynamic sensitizers no farther than the superficial keratinocytes, producing an increased sunburn reaction with erythema and blisters. It is caused by contact with a plant, *Heracleum sphondylium* (meadow parsnip, cow parsnip, or hogweed), and is recognized by the striped or leaf-shaped erythematous reaction, followed by blisters, and ultimately hyper/hypopigmentation. Soldiers on maneuvers may be affected.

16.1.3 Porphyrias

The porphyrias are inherited and acquired disorders. These diseases are classified as either hepatic or erythroid in origin, depending on the principal site of expression of the gene defect.

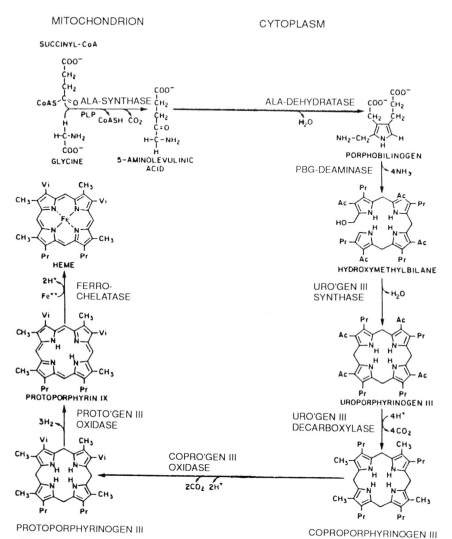

FIGURE 16.5 The mitochondrial and cytoplasmic pathways of porphyrin biosynthesis.

In the biochemical pathway of heme (Figure 16.5), a mitochondrial enzyme, Δ-aminolevulinic acid synthase (ALA synthase: ALAS), condenses glycine and succinyl CoA to form aminolevulinic acid (ALA). The level of ALAS in the liver is regulated by heme, which can directly inhibit the activity of ALAS, bind to the cytoplasmic ALAS precursor (preventing its unfolding, and transfer across the mitochondrial membrane), or repress ALA mRNA synthesis at the transcriptional or posttranscriptional level.

A drug-responsive element in the cytochrome P450 gene promoter increases transcription of this gene, resulting in lower heme levels. Drugs may also increase ALAS gene transcription (up to 20-fold) through a drug-responsive element in the ALAS promoter. Elevated iron levels inactivate the repressor protein bound to the iron-responsive element (IRE) region of ALAS mRNA. Excess heme would negatively control its own synthesis by regulating iron availability and hence ALAS mRNA translation.

ALA moves to the cytosol, where porphobilinogen (PBG) is formed by the action of ALA dehydratase. By condensation of four molecules of PBG in the presence of PBG deaminase, only uroporphyrinogen I is formed, whereas uroporphyrinogen III is also formed when the cosynthase for the latter is present. A cytosolic enzyme, uroporphyrinogen decarboxylase, yields coproporphyrinogen. Coproporphyrinogen moves to the mitochondria, where coproporphyrinogen oxidase yields protoporphyrinogen IX. This is then converted by protoporphyrinogen oxidase to protoporphyrin IX. The final step is the insertion of iron into protoporphyrin IX by ferrochelatase (protoheme-ferrolyase).

The enzyme deficiencies in porphyrias are summarized below:

Enzyme Deficiency	Type of Porphyria
ALA synthase	
ALA dehydratase (PBG synthase)	Acute porphyria with ALA dehydratase deficiency
PBG deaminase	Acute idiopathic porphyria (AIP)
Uroporphyrinogen III cosynthase	Congenital erythropoietic porphyria (CEP)
Uroporphyrinogen decarboxylase	Porphyria cutanea tarda (PCT)
Coproporphyrinogen oxidase	Hereditary coproporphyria (HCP)
Protoporphyrinogen oxidase	Variegate porphyria (VP)
Ferrochelatase	Erythropoietic protoporphyria (EPP)

The phototoxic mechanisms due to photosensitization by porphyrins are shown in Figure 16.6.

16.1.3.1 Congenital Erythropoietic Porphyria

CEP, or Gunther's disease, is an extremely rare autosomal recessive deficiency of uroporphyrinogen III cosynthase, with severe cutaneous photosensitivity and scarring. Uroporphyrins and coproporphyrins accumulate in the bone marrow, red cells, urine, and feces. Characteristic features include (1) cutaneous lesions of sun-exposed areas, termed hydroa estivale, with increased severity during the summer; (2) bullae and vesicles with red fluorescent fluid; (3) porphyrin deposits in the teeth [reddish fluorescence (erythrodontia)]; and (4) fluorescence of late nor-

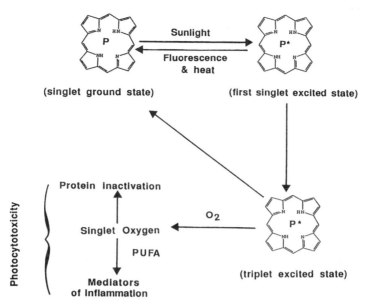

FIGURE 16.6 Scheme showing the ground state and the excited first singlet and triplet states of porphyrin, the formation of singlet oxygen, and its deleterious effects on protein and polyunsaturated fatty acid (PUFA), which are mediators of inflammation.

moblasts (porphyroblasts), early reticulocytes, and circulating erythrocytes.

The real pathogenesis in CEP may be overproduction of porphyrin Is rather than deficiency of uroporphyrinogen III and heme formation. A case history of CEP revealed the following pathogenesis: The affected girl was born in the United Kingdom to Pakistani parents who are first cousins. Blistering scars on the skin and pink urine were noted during the patient's infancy. At the age of 8 she had retarded growth, multiple scars on her face and hands, and pink teeth. Urinary uroporphyrin excretion was 1000-fold normal and coproporphyrin excretion was 50-fold normal. After a bone marrow donation from an HLA-identical brother, erythrocyte uroporphyrinogen III synthase activity was just above normal. Urinary porphyrin excretion fell from the onset of cytoreductive chemotherapy. The transplanted marrow showed all male cells that matched those of the donor. The patient died of severe progressive pneumonitis and nonspecific encephalopathy 9 months after transplantation.

16.1.3.2 Erythropoietic Protoporphyria

Erythropoietic protoporphyria (EPP) is an autosomal dominant disorder, characterized by excess protoporphyrin in red cells and feces. It is probably due to a genetic defect in *ferrochelatase*. Photosensitivity is generally much less severe than in CEP (Figure 16.7). Cross-linking of spectrin reduces erythrocyte deformability. Hydrophobic protoporphyrin (PP) taken up by mitochondria can impair mitochondrial respiration to a greater degree than can hydrophilic uroporphyrin.

The basement membrane of the epidermis is multilayered and fragmented. PP fluorescence is detectable mainly in reticulocytes. Liver function is usually normal, but PP is visible as a dark brown pigment birefringent under polarizing light microscopy. Occasionally, advanced liver disease can be seen with marked deposition of PP in hepatocytes, Kupffer cells, portal histiocytes, and bile canaliculi. Fluorescent gallstones have been reported. Oral administration of β-carotene improves the skin symptoms and enhances tolerance to sunlight.

16.1.3.3 Acute Intermittent Porphyria

Acute intermittent porphyria (AIP; acute porphyria, Swedish porphyria) is an autosomal dominant disorder caused by a deficiency of PBG deaminase. The majority of heterozygotes remain clinically latent.

Increased excretion of ALA and PBG is almost always found, even between attacks. AIP is exacerbated by steroids, drugs, and nutrition, which alter ALA synthase and porphyrin–heme synthesis rates in the

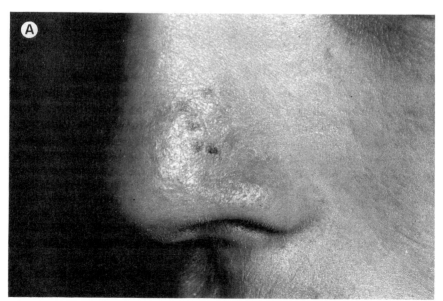

FIGURE 16.7 (A) Photograph of papulo-vesicular eruptions on the nose in erythropoietic protoporphyria. Courtesy of Dr. M. Jeanmougin, Policlinique de Dermatologie, Service du Professeur L. Dubertret, Hôpital Saint Louis, Assistance Publique, Hôpitaux de Paris. (B) Face of child with erythropoietic protoporphyria. Courtesy of Dr. Penneys, Department of Pathology, School of Medicine, University of Miami.

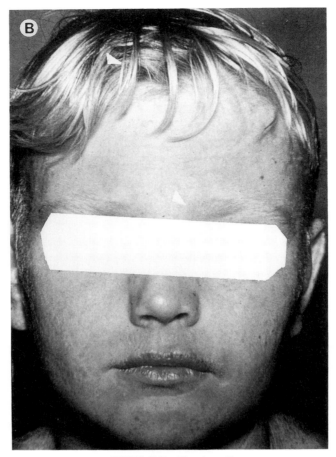

FIGURE 16.7   (Continued)

liver. Intermittent attacks after puberty are more common in women than in men, and may relate to hormonal fluctuations with the menstrual cycle.

Visceral neuropathy that can progress to total paralysis is prominent. Being primarily a motor neuropathy, AIP begins with weakness in the arms (focal symmetrical or asymmetrical paresis). Cranial nerve involvement can progress to death by bulbar paralysis.

During acute attacks, with disorientation, hallucinations, and paranoia, patients may be violent. Abnormal encephalograms even in the absence of seizures suggest an organic brain syndrome. Neuron damage and axonal degeneration probably precede demyelination.

AIP affects hepatic heme synthesis; there are marked increases in liver and urine ALA and PBG, with no impairment of overall hepatic function.

Erythrocyte PBG deaminase activity is most useful in detecting gene carriers in families of known AIP patients. Other biochemical findings of AIP include (1) reddish urine due to increased porphyrins of the isomer III series and porphobilin, a spontaneous degradation product of PBG, and (2) stool porphyrin levels normal or mildly increased.

16.1.3.4 Hereditary Coproporphyria

Hereditary coproporphyria (HCP), much less common than AIP, is inherited as an *autosomal dominant trait*. There is a partial deficiency of coproporphyrinogen oxidase. HCP generally presents with neurovisceral symptoms. Photosensitivity due to coproporphyrin occurs in a minority of patients. Drugs, usually phenobarbital, are responsible for HCP attacks. Attacks in female subjects are associated with menstrual cycle, pregnancy, or treatment with oral contraceptive steroids. Increased production of 17-oxysteroids and reduced metabolism of androgens via the 5-α-reductase pathway in liver occur in clinical exacerbations.

Fecal porphyrins in HCP generally show a great predominance of coproporphyrin, with little increase in protoporphyrin, whereas in variegate porphyria (VG) the amounts of these porphyrins are approximately equal. Increased sensitivity to ALA synthase-inducing factors (drugs, hormones, and nutritional factors) results in ALA and PGB excretion during acute attacks. Coproporphyrinogen and coproporphyrin appear to be efficiently transported into bile or plasma from hepatocytes.

16.1.3.5 Variegate Porphyria

Variegate porphyria (VP) is an autosomal dominant disease that can present *neurovisceral symptoms*, cutaneous photosensitivity, or both. There is either a deficiency of protoporphyrinogen oxidase, or a type of defect in ferrochelatase that is different from that found in EPP.

Cases of VP among South African whites can be traced back to a man, or his wife, who married in 1688 shortly after emigration from Holland.

The main clinical features are cutaneous photosensitivity with increased pigmentation and hypertrichosis, abdominal pain, tachycardia, vomiting, constipation, hypertension, neuropathy, confusion, bulbar paralysis, back pain, renal dysfunction, and exacerbation by oral contraceptives.

In a patient with signs of acute porphyria, normal activity of PBG deaminase and the stool porphyrin pattern (protoporphyrin, coproporphyrin isomer III, fecal X-porphyrin fraction, i.e., heterogeneous porphyrin–peptide conjugates) differentiate from AIP.

16.1.3.6 Porphyria Cutanea Tarda

Porphyria cutanea tarda (PCT) is divided into three types: (1) sporadic cases, (2) familial cases in which there is evidence of autosomal dominant inheritance, and (3) cases due to exposure to halogenated aromatic hy-

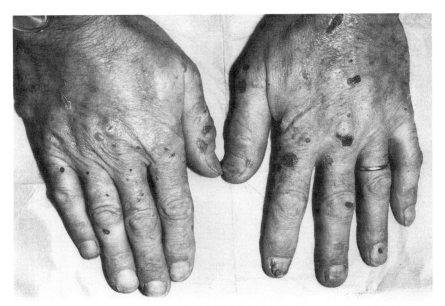

FIGURE 16.8 Postbullous erosions of the back of the hands in a case of porphyria cutanea tarda. Courtesy of Dr. M. Jeanmougin, Policlinique de Dermatologie, Service du Professeur L. Dubertret, Hôpital Saint Louis, Paris.

drocarbons. *Porphyrins accumulate in the liver, pass into the blood, and are excreted in stool and urine.* The pattern of porphyrin excretion is explainable by *a deficiency of uroporphyrinogen decarboxylase* in the liver. People with hepatoerythropoietic porphyria (HEP) are homozygous for the decarboxylase defect.

The characteristic features of PCT are (1) lesions on sun-exposed areas (Figure 16.8); (2) small white plaques, termed milia, which may precede or follow the formation of fluid-filled vesicles and bullae, particularly after minor trauma; *increased friability of the skin over sun-exposed areas is very common*, and progressive thickening, scarring, and calcification of the skin may lead to a scleroderma-like appearance (pseudoscleroderma); (3) red fluorescence can be demonstrated on unfixed sections of liver; electron microscopy shows needlelike lucent areas, probably in lysosomes, and paracrystalline inclusions in mitochondria.

Liver cell damage, iron overload, and estrogens appear to be the most important contributing factors to PCT. Hepatic siderosis is very common, but in biopsies it is not as severe as untreated hemochromatosis. Men who are treated with estrogens for prostatic cancer and women given estrogens alone or as oral contraceptive combinations may develop PCT.

A massive outbreak of a PCT-like syndrome occurred in 1956 in Eastern Turkey when seed wheat to which hexachlorobenzene had been added as a fungicide was used for food by the impoverished population. Children with increased hair growth and hyperpigmentation were described as having "monkey disease." The average interval between exposure and symptoms was 6 months. Breast-fed infants who got hexachlorobenzene from milk died of "pink sore," with convulsions and annular erythema. Older children and adults developed neurological involvement. Even 20 years later, patients with "porphyria turcica" showed scarring, pigment, hypertrichosis, pinched facies, perioral scarring, short stature, and contracted hands and fingers, but porphyrin excretion was near normal without treatment.

The incidence of hepatocellular carcinoma in PCT is from 4 to 47%. The tumors rarely contain porphyrins in large amounts. Porphyrins may bind to DNA and play a role in carcinogenesis. Urine contains mostly uroporphyrin and 7-carboxylate porphyrin. The major porphyrins in bile and stool are often isocoproporphyrins.

16.1.3.7 Porphyria with ALA Dehydratase Deficiency

In this disease there is a marked increase in urinary ALA excretion and markedly reduced erythrocyte ALA dehydratase. The urine contains excess porphyrin, at least 95% of which is coproporphyrin III. Intermediate reductions of ALA dehydratase in some first-degree relatives are consistent with autosomal recessive inheritance.

16.1.3.8 Suggested Reading: Porphyria in History

Porphyria may be called a royal malady. In the first clinical study devoted to George III of England, Ray wrote "Five times" was the king "struck down by mental disease . . . ." Based only on Gunther's triad of abdominal symptoms, polyneuritis, and mental disturbance, the condition is at once recognizable as acute intermittent porphyria. The symptomatology of George III is a textbook case: colic and constipation; painful paresis of arms and legs; vocal paresis, visual disturbances, and other signs of bulbar involvement; radicular pain; and encephalopathy ranging from insomnia to excitement, raging delirium, stupor, and fits.

The physician attending George III recorded that "His Majesty has passed . . . bloody water . . . during the last 16 hours," of which "no tinge remained the following day." These observations were made during paroxysms when the excretion of porphyrin and phorphobilin-like chromogens is known to be the greatest.

Acute Intermittent Porphyria is usually transmitted as a Mendelian dominant. To review the medical history of the royal family is a major task because of the number of probands (the father of George III was one of eight children; George III was one of nine children and the father of 15). The youngest sister of George III, Caroline Mathilda (1751–1775), queen of Denmark and Norway, succumbed to a mysterious illness within 1 week in her 24th year. It started with malaise, followed rapidly by paralysis of legs, arms, and bulbar centers, so that in her last hours she able to communicate only by moving her eyes. The picture of acute ascending paralysis is not uncommon in fulminating porphyria. Caroline Mathilda manifested the disease at the same age as her brother.

The disease is traced to Mary Queen of Scots. It is also possible to trace the disorder from Mary down to two living family members (in the 1960s), through 13 generations spanning more than 400 years. The actual sequence of transmission may have been as follows: From Mary Queen of Scots, to her son James VI(I); from James I through his daughter, Elizabeth, Queen of Bohemia, to her daughter Sophia, wife of Ernst August, Elector of Hanover. From her, foundress of the English House of Hanover, through her son George I to George III and his descendants; and through her daughter, Sophia Charlotte, wife of Frederick I, King of Prussia, in the Brandenburg–Prussian line, to Frederick the Great, cousin-german to George III. In 1938, every ruling family of Europe, with the exception of the king of Albania, traced descent from Elizabeth of Bohemia [daughter of Queen Mary's son James I of England (James VI of Scotland) and great-great-great-grandmother of George III].

Of the two cases documented in the 1960s (living royal family members), the first was a female descendant of George I and II. In her sixth decade she developed pneumonia. Her physician described her urine as dark red and on a number of occasions as containing large amounts of coproporphyrin, uroporphyrin, and porphobilinogen. She was diagnosed as a porphyric in acute exacerbation.

The second case, a woman (descendant of George II) in her fifth decade, had suffered since her early twenties from attacks of colic with pain "unspeakably severe." There was also a history of sun sensitivity and easy blistering of the hands from trauma. She passed dark, reddish urine during exacerbations. The symptoms receded in her seventh decade and eventually ceased. Total fecal porphyrins were over 100 $\mu g/g$ dry weight (if total porphyrins are greater than 75 $\mu g/g$ dry weight, porphyria should be considered).

The original posthumous pathographic diagnosis for George III was acute intermittent porphyria, with periodic exacerbations. A different type of porphyria, termed variegate porphyria, runs a similar course, but in addition there is easy traumatization of the skin and/or intolerance of sunlight. Both cutaneous and abdominal nervous symptoms may occur. During attacks patients may have moderately increased urinary porphobilinogen excretion. A more constant abnormality during the quiescent state is the presence in the feces of large amounts of porphyrin–peptide conjugates (designated X-porphyrin).

Early in the 1778 attack, George III had "great weals [welts], strip-shaped lumps on his arms" and during the attack his face often had a violaceous hue, giving an expression of fury. In the summer of 1790 he was not well and was observed to "doze off" in the sun, so that special carriages of cane were provided to protect him.

A nineteenth century family member had such easily chafed skin that special uniforms were made for him, ingeniously padded to avoid rubbing on neck, shoulders, elbows, and knees.

16.1.3.8.1 James VI of Scotland (James I of England) (1566–1625)

The physician attending James, Mayerne, recorded the following observations:

> The liver . . . is liable to obstructions and generates much bile . . . Sometimes he is melancholy . . . he becomes very irascible . . . often his eyes become yellow but it soon passes off, he glows with heat . . . sleeps badly . . . readily vomits and at times so violently that his face is covered with red spots for two or three days . . . He often suffers bruises . . . Very often he laboured under painful colic . . . preceded by melancholy and nocturnal rigors . . . pain in the chest, palpitation, sometimes hiccough . . .

He often passed urine "red like Alicante wine" (James' own words) but without attendant pain. Mayerne, whose hobby was experimenting with the chemistry of pigments—a field in which he made scientific contributions, watched the color changes in urine with extreme assiduity. From his notes it appears that the urine was discolored most probably by the chromogens of porphyria.

Death came in 1625. James suddenly took a turn for the worse, went into convulsions, and died. At postmortem the left kidney, long suspected by Mayerne to be full of michievous stones, was healthy.

16.1.3.8.2 Mary Queen of Scots (1542–1587)

From her late teens, Mary suffered attacks of excruciating abdominal pain and vomiting, painful lameness, fits, and mental disturbance—a combination suggestive of porphyria. The most severe attack, in which she nearly died, occurred in 1566, when she was 24. Its rapid onset and alarming symptoms followed by quick recovery gave rise to the suspicion that she was poisoned. She was suddenly taken ill and vomited continuously—it is said 60 times. She became delirious; 2 days later she lost her sight and speech, had a series of fits, and remained unconscious. Within 10 days she was up and well again.

Attacks followed in later years. Her mental "instability" together with her recurrent invalidism, her "grievous pain on her side," and her equally inexplicable recoveries impressed those around her as histrionic. Mary Queen of Scots shares with many sufferers from porphyria, living and dead, the fate of being judged hysterical.

16.1.3.8.3 Princess Charlotte Augusta of Wales (1796–1817) Princess Charlotte expected a child in October, 1817. Labor was delayed until November 3. At 10 p.m. on November 5 the delivery of a stillborn male child was announced, but "Her Royal Highness is doing extremely well." Within 6 hours Charlotte was dead. About 3 hours after delivery the nurse brought her gruel, but she could not swallow. She had acute pain in the chest and abdomen; extreme restlessness developed, respiration became labored, she went into convulsions and expired 2 hours later.

It was known "that the whole Royal family are liable to spasms of violent descriptions." The postmortem examination revealed "loss of uterine tone" and "general atony of viscera," as shown by a distended colon and dilated stomach, which contained 3 pints of fluid.

From the age of 16 she had frequent attacks of abdominal pain. She wrote, "I am in a terrible bad state of nerve, spirits . . . & such a nervous headache that I can scarcely open my eyes." She also wrote about her "bilious complaints." During the pregnancy she kept on a "low" diet without meat to subdue "her morbid excess of animal spirits."

Cases have been observed wherein a fulminating attack of porphyria sets in a few hours after confinement, with restlessness, cerebral irritation, difficulty in swallowing, other signs of bulbar involvement, and finally respiratory paralysis. Charlotte could have inherited porphyria from her father, but because her parents were first cousins and her mother possibly was also affected, her risk may have been even greater.

16.1.4 Veterinary Work on Phototoxicity in Animals

16.1.4.1 Photosensitization Problems in Livestock

Photosensitization differs from sunburn in that it requires the presence of a phototoxic agent, the onset of reaction is rapid, and the inciting wavelengths of light often are above 320 nm. Unlike photosensitization (PS), sunburn rarely is reported in domestic animals. Photosensitivity diseases of livestock are grouped into four categories.

16.1.4.1.1 Primary PS The phototoxic compound or its precursor reaches the site of action in the skin either directly by absorption through the epidermis following topical exposure or via the systemic circulation following oral or parenteral exposure. Hypericin, a photodynamic quinone from *Hypericum performatum* (St. John's Wort), is a red pigment ingested by livestock; it has an action spectrum in the 500- to 600-nm region. Fagopyrin, a derivative of hypericin (Chapter 5), is found in buckwheat and produces primary PS in sheep, goats, pigs, horses, and cattle. In Utah, PS of the udder of ewes has resulted in significant loss of lambs because affected ewes will not allow their young to nurse.

Type I cases of PS are due to celery and parsnips, which contain phototoxic quantities of furocoumarins that are produced as a defense mechanism by celery infected with the fungus *Sclerotinia sclerotiorum*. Consumption of moldy leaves of *Cooperia pedunculata* (Giant rain lily) produces major outbreaks in cattle and deer in southeast Texas. The greatest damage occurs on the teats, udder, and muzzle. Keratitis results in corneal scars and blindness. Similar lesions occur in calves and swine treated with phenothiazine, due to the sulfoxide that forms *in vivo*. Their livers have a diminished capacity to metabolize the sulfoxide to the nonphototoxic leukophenothiazone.

16.1.4.1.2 Photosensitization Caused by Aberrant Pigment Synthesis Phototoxic porphyrins photodynamically sensitize the degradation of biomolecules not directly affected by light. Two diseases are placed in this group. The first, pink tooth, occurs in homologous recessive animals. Excessive amounts of uroporphyrin I and coproporphyrin I are formed because of a reduction of uroporphyrin III cosynthase activity. Uroporphyrin I is responsible for a reddish-brown discoloration of teeth and bones. This disease occurs in swine.

The second disease of this group, congenital erythropoietic porphyria (CEP), occurs only in Limousin cattle, with a marked decrease in the activity of ferrochelatase. Photosensitizer-induced dermatitis is suggested by the onset of nervous signs of an epileptic type, together with alopecic lesions and scabs (head and ears) when the cattle are at pasture; symptoms disappear when the cattle are housed. Photosensitization is secondary to a defect of the hepatobiliary excretion of phylloerythrin, or is occasionally acquired as a result of ingestion of photoreactive substances. There are marked increases in the concentration of protoporphyrin IX in heme production sites, i.e., erythrocytes (100 times greater than in controls). PP bound to its transport protein reaches the skin, where it acts as a photosensitizer. Liver fibrosis is not observed, despite accumulation of PP pigment, probably due to the very short time span of porphyria development in cattle.

In cattle, erythrocytic protoporphyria appears to be transmitted in a recessive autosomal manner, whereas in humans the transmission is dominant with variable expression. In a reported study the mothers of affected cattle had been mated with their own father, which may have led to homozygotism, essential to the development of the disease.

A possible explanation of the nervous disorders could be the accumulation of Δ-aminolevulinic acid, which competitively inhibits the binding of the neurotransmitter γ-aminobutyric acid to synaptic membranes.

16.1.4.1.3 Hepatogenous Photosensitization The agent of this third group of diseases is invariably phylloerythrin (PE), a metabolite of dietary chlorophyll. Toxic or physical damage to the biliary apparatus can result in PS through accumulation of PE (Figures 16.9–16.12) caused by: (1) ligation of the common bile duct and exposure to sunlight, (2) an inherited defect in the hepatic uptake of unconjugated bilirubin, and (3) hepatocytes with a diminished ability to excrete conjugated bilirubin and other conjugated metabolites.

Hepatogenous photosensitivity diseases are grouped according to their origin: those resulting from damage primarily to the liver parenchyma and those resulting from damage primarily to the biliary system. In the first group are (1) diffuse degenerative and fatty changes due to *Lantana* poisoning and the hepatoxin of *Microcystis aeruginosa*, and (2) zonally distributed necrosis (*Myoporum* and presumably *Tetradymia* poisoning). In the second group are (1) facial eczema in New Zealand, Australia, and South Africa (caused by the fungus *Pithomyces chartarum*) and (2) a disease called *geeldikkop/dikoor*, in South Africa (caused by *Tribulus terrestris* and certain grasses of the genus *Panicum*).

FIGURE 16.9   Facial eczema of a lamb—acute stage, showing congestion and edema of the ears, muzzle, and eyelids. The disease is the result of a hepatogenic photosensitization. The hepatic lesions are due to ingestion of sporidesmins (mycotoxins of a microscopic fungus *Pithomyces chartarum*, which grows in the fall in haylike pasture fields). The primary result of toxin injection is intrahepatic cholestasis leading to accumulation in the blood and tissues of phylloerythrin (a product of the catabolism of chlorophyll by intestinal bacteria). Courtesy of Professor P. Bezille, Pathologie Médical du Bétail et des Animaux de Basse-Cour, Ecole Nationale Vétérinaire de Lyon, Lyon, France.

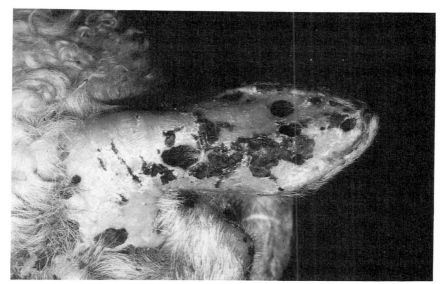

FIGURE 16.10   Stage of pruritus at about 2–8 days, showing loss of hair and crust formation. Courtesy of Professor P. Bezille.

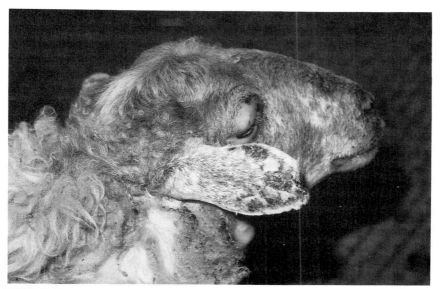

FIGURE 16.11   Postinflammatory lichenoid lesions (facial eczema) at 3–21 days. Courtesy of Professor P. Bezille.

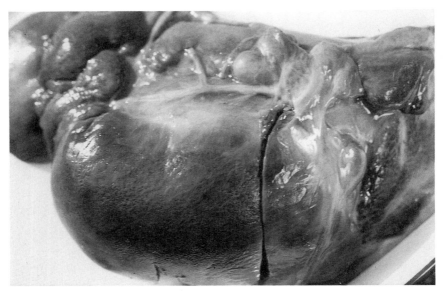

FIGURE 16.12   Hepatic fibrosis with regenerative nodules (terminal stage at 3 to 6 months). Courtesy of Professor P. Bezille.

Lippia icterus is a photodermatosis occurring in sheep. This disease, also called *geeldikkop* ("yellow thick head" in Afrikaaner), consists of icterus and acute phototoxic manifestations, particularly swellings, affecting the less hairy skin, predominantly of the head. Intestinal resorption of phylloerythrin, produced from chlorophyll in the food, causes the acute phototoxic reactions in the skin.

Diffuse changes in the liver parenchyma bring about retention of PE by interference with canalicular drainage of bile as a result of hepatocellular swelling. PS indicates extensive zonal damage in the liver and prognosis is poor.

Facial eczema is caused by ingestion of the mycotoxin sporidesmin, which damages the mucosae of the major intra- and extrahepatic bile ducts and ductus cysticus, with marked proliferation of periductular granulation tissue and fibrosis.

Pathologically, geeldikkop and dikoor are identical. The most conspicuous histopathologic feature of geeldikkop/dikoor is the accumulation of microliths (birefringent crystalloids) in the portal ducts, hepatocytes, and Kupffer cells. Some of the crystalloids consist of a steroidal sapogenin.

The mycogenic photosensitizing diseases of livestock are due to the mycotoxins sporidesmin (facial eczema) and phomopsins A/B (mycotic lupinosis). Steroid drugs that cause temporary hepatic dysfunction should be considered potentially responsible when PS follows their use.

16.1.4.1.4 Photosensitization of Unknown Origin

Trefoil dermatitis (clover disease, trifoliosis) refers to a short-lived PS syndrome of cattle, sheep, horse, and swine feeding on luxuriantly growing legumes (*Trifolium* spp., *Medicago* spp., or *Vicia* spp.). Icterus and other evidence of liver damage are lacking.

16.1.4.2 Clinical Signs of PS in Animals

PS skin lesions are confined to white or lightly colored skin exposed to light. Behavioral abnormalities seen immediately on exposure to sunlight include restlessness, photophobia, seeking shade, scratching or rubbing the pruritic areas and self-inflicted trauma, lacrimation, and keratitis. Animals with liver damage may become anorexic, depressed, and separate from the herd. Animals with all-black coats may develop no lesions other than keratitis.

In cattle, the udders and teats of cows and the scrotum of bulls are particularly vulnerable. White-faced animals suffer greatly, with damage to the whole face or to the muzzle and eyelids; there is scarring, keratitis, and blindness.

In sheep and goats, lips and eyelids are affected first, followed by ears and the remaining areas of the head, all of which become edematous (thus the popular names "bighead" and "swellhead"); the udder and external genitalia are affected last. Swelling of the muzzle and crust formation constrict the external nares, leading to breathing difficulty. Blindness following recovery may occur because of retinal damage or the formation of scar tissue over the eyes.

Oat PS of swine occurs in the southeastern United States, producing swollen ears that eventually wrinkle and in severe cases slough off.

In horses, lesions are generally confined to white or light areas of the face, around the mouth, legs, neck, underside of the abdomen, and udder. Conjunctivitis, photophobia, keratitis, and partial blindness may occur.

The plant *Ammi majus* causes a furocoumarin-induced PS in geese and ducks. There are blisters on the upper beak, acute conjunctivitis, and a brown crust on foot webs. Chronic signs include deformed upper beaks; scarred, thickened, and perforated foot webs; deformed digits, blepharoconjunctivitis; keratitis, synblepharon, and ankyloblepharon; and, in ducks, pigmentary retinopathy.

16.2 PHOTO-HETERO REACTIONS AFFECTING MAN

In Section 16.1 photo-auto reactions were considered (i.e., reactions induced by a direct photochemical action producing a biological response). The present section deals with secondary biological reactions involving substances produced or induced by a primary photochemical reaction having no immediate phototoxic effect.

16.2.1 Criteria for Phototoxicity versus Photoallergy

Allergenic photoproducts (photoantigens) are produced from abnormal skin components or from substances not normally reacting photochemically. Drug-induced phototoxicity (photo-autoreactions) is differentiated from photoallergic reactions (photo-heteroreactions) by the following criteria:

1. The incidence of phototoxicity is theoretically as high as 100%; that of photoallergy is usually very low.
2. Phototoxicity induces an immediate reaction, whereas in photoallergy a photoallergen must first form and trigger immunological sensitization processes; reaction is observed on a subsequent second challenge by the photoallergen.
3. "Flares" at distant previously involved sites are possible only in photoallergy.
4. Cross-reactions to structurally related agents are not seen in phototoxicity, but they are frequent in photoallergy.
5. The concentration of drug necessary for reaction is significantly higher in phototoxicity as compared to photoallergy, which can be induced by minimal amounts of drug.
6. Chemical alterations of photosensitizers sometimes occur in phototoxicity, but invariably take place in photoallergy.
7. There is no covalent binding with carrier protein in phototoxicity, but binding is the first step in formation of the photoallergen.
8. There is passive transfer in photoallergy, but not in phototoxicity.
9. The "lymphocyte stimulation test" and the "macrophage migration inhibition test" are positive in photoallergy.

With topically applied chemicals the responses are either due to direct damage of skin cells by *photoirritant contact dermatitis (PICD)* or to *photoallergic contact dermatitis (PACD)*, with formation of a "photoantigen." The chemicals involved are those that absorb in the solar spectrum, usually in the UVA (320–400 nm) or visible (400–800 nm) ranges. Because of the differences in mechanisms, there are clinical differences between PACD and PICD, as shown in Table 16.1.

TABLE 16.1   Differences between PACD and PICD Contact Reactors

Criterion	PACD	PICD
Occurrence on first exposure	No	Yes
Onset after UV exposure	24–48 hr	Minutes to days
Dose dependence		
Chemical	Not crucial	Important
Radiation	Not crucial	Important
Clinical morphology	Eczematous (erythroderma)	Erythema and edema, bullous, hyperpigmentation, psoriasiform
Histology	Eczema	Necrotic keratinocytes
Action spectrum	UVA	UVA
Diagnosis	Photopatch	Clinical picture

PACD is by definition an *immune-triggered response*. The chromophore gains access to the body via the skin, absorbs radiation, and is converted into a photoantigen. The mechanism of such conversion may likely involve processing by macrophages and complexing or covalent photobinding to proteins, possibly the HLA-DR molecule.

PACD is T cell mediated and will occur in a genetically predisposed part of the population. The response is *eczematous* and does not vary appreciably with agents. Typical cases of PACD are seen with topically applied sulfanilamides, antibacterial soap bars (tetrachloro- or bromosalicylanilide), fragrance ingredients, musk ambrette, sunscreens containing PABA esters, chlorpromazine, and promethazine. Up to 20% of photosensitivity patients suffer from PACD.

16.2.2 Photo-Hetero Reactions of Extraneous Substances in the Skin

16.2.2.1 Argyrosis and Chrysiasis

In *argyrosis* there is deposition of silver in the sun-exposed areas of the skin, with formation of lesions of blue to slate-gray color, as the result of a photoallergic process. The pigmentation is more proportional to the amount of silver than to duration of sun exposure. Refractile granules smaller than 1 μm are seen scattered through the dermis, around sweat glands, and in the papillary dermis.

Chrysoderma, the presence of gold in the tissues as the result of a photoallergic process, may be associated with the parenteral administration of gold. The lesions are prominent around the eyes as oval, irregular-sized patches in sun-exposed areas. Histologically, there are granules within dermal histiocytes localized around blood vessels.

16.2.3 Solar Urticaria

Solar urticaria is probably a nonfamilial rare form of photosensitization in young adults, characterized by an immediate reaction to irradiation. The symptoms are pruritus, erythema, urticaria limited to exposed skin, and "whealing." The reaction lasts for only 1 to a few hours. The action spectrum involves various combinations of UVB, UVA, and visible radiation.

Solar urticaria may be primary or may result from the topical application of chemicals, for example, tar pitch dyes. It may be endogenously mediated, as in erythropoietic protoporphyria, or drug-mediated, as with benoxaprofen. Exposure to radiation at 320–400 nm with preirradiation at 450–500 nm enhances the appearance of urticaria, but postirradiation fails, which shows that energy is absorbed by a precursor of photosensitization.

Hypersensitivity may be to a specific photoallergen present only in the skin of the sensitive patient, or to a nonspecific photoallergen present in all persons. The pathogenesis of solar urticaria is associated with mast cell degranulation and recruiting of neutrophils and eosinophils by release of a chemotactic factor. Tissue distribution of the eosiniphilic major basic protein amplifies the "whealing effect."

16.2.4 Photoallergic Eczema

The morphological and histological changes of eczema spongiosis (epidermal intercellular edema) are manifest in sun-exposed skin. Photosensitivity is confined to UVB radiation, without intervention of externally added chemicals. *A characteristic is the negative photopatch test* (see Section 16.2.6.2). There is no detectable photosensitizing allergen of external source (see Section 16.2.6.1). The disturbance is seen disproportionately in males as a dermatitis in the sixth to seventh decade. Plaques on the cheeks simulate lupus erythematosus.

16.2.5 Photopathology of Oral Contraceptives

Orally administered contraceptives may act to provoke latent dermatoses; they are not thought to be primary causative agents.

16.2.6 Chronic Actinic Dermatitis

16.2.6.1 Persistent Light Reactors

In rare cases patients with PACD continue to exhibit photosensitivity months or years after the last exposure to antigen. Such individuals have been called persistent light reactors (Figure 16.13). Persistent reactors will

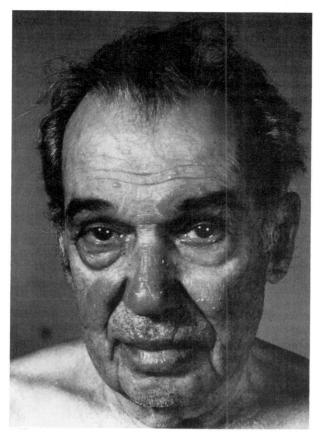

FIGURE 16.13 Case of chronic actinic dermatitis (actinic reticuloid): persistent light reactor, due to musk ambrette. Courtesy of Professor L. C. Harber, Department of Dermatology, College of Physicians and Surgeons of Columbia University, New York.

be found to have abnormal or lowered minimum erythema doses (MEDs) in the UVB range and possibly in the UVA range. These responses will usually persist and will be eczematous.

PACD has been shown to be the final diagnosis in 7–20% of patients referred for evaluation of photosensitivity.

16.2.6.2 Photopatch Testing in PACD

Contact allergy is diagnosed by equally positive reactions at irradiated and nonirradiated sites. Photocontact allergy is diagnosed when the irradiated site is positive together with a negative nonirradiated site. A combination of contact and photocontact allergy is diagnosed when both sites are

TABLE 16.2   Characteristics of Chronic Actinic Dermatitis

Symptom	Spectrum of sensitivity			Photopatch test	Abnormal histology
	UVB	UVA	Vis		
Persistent light reaction	+	+/−	−	+	−
Actinic reticuloid	+	+	+/−	+/−	+
Photosensitivity eczema	+	−	−	−	−
Photosensitivity dermatitis and actinic reticuloid syndrome	+	+/−	+/−	+/−	+/−
Eczematous polymorphic light eruption	+	+/−	−	−	−

positive but with the irradiated site having a greater reaction compared to the nonirradiated site.

16.2.6.3 Chronic Actinic Dermatitis

Chronic actinic dermatitis (CAD), previously known as actinic reticuloid, is an eruption with eczematous morphology of the kind usually associated with exogenous chemicals, either as a photosensitivity to a systemic agent or as a photoallergic contact dermatitis. In CAD, no responsible exogenous chemical can be identified. The pathogenetic mechanism of CAD is poorly understood. Some cases of persistent light reaction have reportedly resulted from persistence of antigen in the skin. In most cases, however, CAD photosensitivity involves the entire skin surface, even areas never exposed to antigen.

The action spectrum of photosensitivity is in the UVA range, whereas the action spectrum of CAD is in the UVB range. In some cases the action spectrum of CAD may extend into the UVA and even visible ranges, but without antigen. CAD characteristics are given in Table 16.2.

Individuals now grouped under the diagnosis of CAD do not clear when photoantigen exposure is discontinued. Some were previously diagnosed as having actinic reticuloid because their clinical photosensitivity was associated with histologic evidence suggestive of lymphoma in skin biopsy. These individuals usually have broad-spectrum sensitivity to UVB, UVA, and visible radiation and in some of them photopatch tests are positive.

It has been suggested than an exogenous photoantigen persists in the skin, but most investigators now favor the theory that development of allergy to an endogenous photoantigen is involved, i.e., some normal component of the skin is made antigenic on exposure to UVB radiation (e.g., a carrier protein that first was involved in production of a complete photoallergen by conjugation with an exogenous hapten).

16.2.7 Chronic Polymorphic Light Eruption

Polymorphic light eruption (PMLE) has been used to *describe any delayed eruption related to sun exposure for which no other diagnosis or cause could be defined.* PMLE is an intermittent, delayed, transient abnormal cutaneous reaction to UV. The onset of PMLE is during the first three decades of life. Women patients outnumber men by two or three times.

PMLE, the most commonly seen photosensitive reaction in the United States, is characterized by small or large papules that appear within hours or days of exposure to sunlight and persist for days, with MEDs within normal limits (Figure 16.14). Lesions are also eczematous, insect-bite type, and erythema multifocal type, and they tend to occur symmetrically on the same skin areas in any one patient. The eruption usually occurs while the patient is on holiday in sunny climates, persists for days, and heals without scarring. Typical sites are arms, thighs, and chest. Facial lesions are more common in the plaque-type disease. Most affected individuals suffer a single episode. If multiple episodes occur, however, during a single season, they will tend to lessen in severity, possibly due to tachyphylaxis (exhaustion of photoallergic response). The patient will eventually no longer respond to radiation during that season, but symptoms may return on exposure during the next year.

PMLE pathogenesis is associated with abnormal immunological response to sunlight-induced cutaneous neoantigen. The action spectrum of PMLE is 56% UVA, 17% UVB, and 27% UVB + UVA. This is explained by cutaneous levels of different UV-evoked inducing antigen.

Broadband UVA radiation may be less effective in provoking PMLE reaction due to the presence of shorter wavelength UVA, which is inhibitory, by interference with the generation of photoallergen. The chromophore involved in PMLE pathogenesis has not been identified. Within 5 hr following irradiation there is T cell-dominated perivascular infiltration (CD4 + helper T dominated), which peaks at 72 hr (CD8 + cytotoxic suppressor T dominated). Intercellular adhesion molecule 1 (ICAM-1)) is expressed in keratinocytes overlying the perivascular region.

Other aspects of PMLE are suggestive of a phototoxic, nonimmune mechanism. Reproduction of lesions requires usage of multiples of the MED in both the UVA and UVB ranges. Irradiation of large areas of previously involved skin is necessary to reproduce the lesions (multiple-insult high-dose testing).

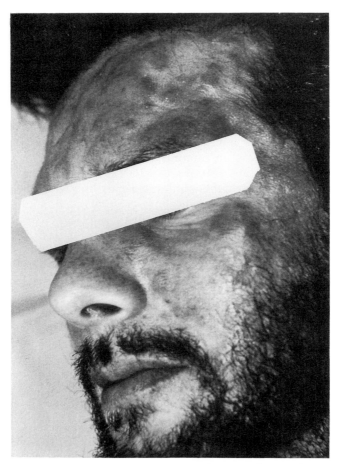

FIGURE 16.14   Chronic polymorphic light eruption. Courtesy of Dr. Neal Penneys, Department of Pathology, School of Medicine, University of Miami, Miami, Florida.

16.2.8 Idiopathic Photosensitivity

16.2.8.1 Actinic Prurigo

Actinic prurigo is a slowly evolving, persistent pruritic, papular or nodular eruption of sun-exposed regions. It is present in childhood and puberty, tending to regress in adulthood. Wavelengths longer than 305 nm are relevant to the pathogenesis of actinic prurigo.

The pathogenesis implicates a disturbance of the lipoxygenase metabolism of arachidonic acid, or an immunologically mediated photo-hetero reaction rather than a photo-auto reaction.

Actinic prurigo is more common in females, with a family history in 15–50% of cases and tolerance at the peak of the summer season that indicates *tachyphylaxis* (diminished symptoms presumably due to exhaustion of antibodies). Lesions last weeks to months. Lower lip cheilitis is common. A thick stratum corneum covers an epidermis exhibiting a rudimentary or pronounced stratum granulosum, acanthosis, and hypergranulosis.

16.2.8.2 Hydroa Vacciniforme

This is a very rare idiopathic disorder, observed principally in childhood and characterized by recurrent crops of vesicles on sun-exposed areas of the skin. Hydroa vacciniforme usually occurs during summer months and generally resolves in adolescence, leaving vacciniform scarring.

16.2.9 Dermatitis Vernalis Aurium Burckhardt

This disease occurs predominantly in children. Lesions reminiscent of chilblains (an inflammatory swelling caused by exposure to cold) follow the first relatively marked exposure to the sun in the spring and are prevalent on the upper edge of the external ears.

16.2.10 Other Photosensitive Skin Diseases

16.2.10.1 Bloom Syndrome

This is an autosomal recessive disease characterized by sun sensitivity, prominent facial telangiectasia (Figure 16.15), immunodeficiency, chromosomal abnormalities, and frequency of malignant tumors, i.e., lymphoma, nonlymphatic leukemia, and lymphosarcoma carcinomas of the skin, breast, oral cavity, and digestive, tract. The frequency of neoplasm is attributable to chromosomal abnormalities and immunodeficiency. The syndrome is more common among Ashkenazi Jews, with a carrier frequency of 1:120.

16.2.10.2 Cockayne Syndrome

This is a very rare autosomal recessive disease associated with cutaneous, ocular, neurologic, and somatic abnormalities and cachectic dwarfism. Birth weight and initial development appear normal, with onset of the disease in the second year of life.

Xeroderma pigmentosum (XP) and Cockayne syndrome present similarities and may even appear together. In cases of the simultaneous presence of both disorders, there is freckling of sun-exposed skin and a

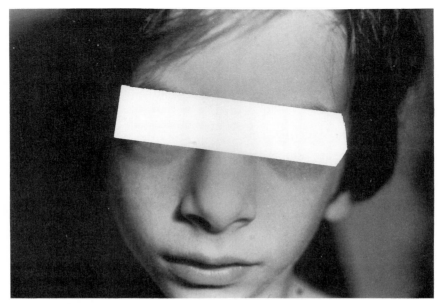

FIGURE 16.15   Bloom's syndrome in a child. Courtesy of Dr. Penneys.

prolonged decrease in the rate of DNA synthesis following exposure to UVB radiation.

In Cockayne syndrome pigmentation is not as pronounced as in XP, and neurological abnormalities that induce *neuronal* degeneration lead to progressive loss of nervous function (in XP, there is axonal degeneration). UV-hypersensitized cell kill is seen in both XP and Cockayne syndrome, but very few neoplasms are seen in the latter. Cockayne syndrome cells are unable to repair photoproducts of the cyclobutane type in UV-treated plasmids, but they are able to repair nondimer photoproducts.

16.2.10.3 Disseminated Superficial Actinic Porokeratosis

In this autosomnal dominant disease, a mutant clone of epidermal cells expands peripherally, leading to cornoid lamella formation at the boundary between the clonal population and normal epithelial cells. Additional triggering factors (e.g., sunlight) lead to clinical manifestations of the disease.

16.2.10.4 Rothmund and Thomson Syndromes

Poikiloderma congenita Rothmund and Thomson develops early in childhood and shows a certain preference for sun-exposed sites. This is an autosomal recessive disease with onset between the third and sixth months. There is initially a diffuse erythematous skin, which usually spares the trunk.

16.2.10.5 Seckel Syndrome

This very rare autosomal recessive deformity is found with proportionate nanosomia and an abnormality of tryptophan metabolism. The disorder is characterized by a bird-headed profile, skeletal defects, hypodontia, pancytopenia, hypersplenism, premature hair graying, and occasionally vitiligo-like hypopigmented macules; trident hands with simian crease are typical.

16.2.10.6 Dermatosis Atrophicans Maculans

Altered skin reminiscent of a combination of ichthyosiform (scaly; fish-scale-like) erythroderma, mild xeroderma pigmentosum, and chronic radiodermatitis is found in all sun-exposed areas. The photodermatosis is probably sex linked. The cases described have been in males who had exhibited altered skin since childhood.

16.3 PHOTO-KOEBNER PHENOMENA

Photo-Koebner phenomena involve the appearance of disease-specific skin changes in response to nonspecific stimuli. Koebner described the phenomenon when his psoriasis became aggravated after he was bitten by a horse. Light energy is *only one* of the many possible provocative factors.

16.3.1 Collagen Diseases

These are any of various diseases or abnormal states characterized by changes in connective tissue, presumably involving destruction of collagen.

16.3.1.1 Lupus Erythematosus

Lupus is a disease in which there are striking changes in the immune system (Figures 16.16–16.18). Because systemic lupus erythematosus (SLE) patients make large quantities of antibodies that react against their own normal tissue (self-reacting antibodies), the disease is called an "autoimmune" disease.

Photosensitivity reactions occur in approximately 40% of patients. UV radiation is believed to disrupt the body's tolerance to DNA and to enhance the immune response toward it. It has been proposed that lupus patients are immunized to UV-altered DNA and that antibodies produced are not specific enough and will react with the DNA of the

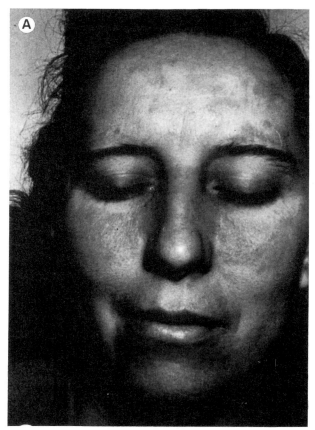

FIGURE 16.16   (A) Systemic lupus erythematosus: butterfly image on face. Courtesy of Dr. Penneys, Department of Pathology, School of Medicine, University of Miami, Miami, FL.

patient's own cells. The adverse reaction is due to UVA (320–400 nm) and UVB (290–320 nm) spectra. Because of its simulation of other dermatological disorders and systemic autoimmune diseases, lupus erythematosus has been called "the great imitator."

16.3.1.1.1 Discoid Lupus Erythematosus   Discoid lupus erythematosus (DLE) involves the skin only. Patchy, crusty red blotches form the classic "open-winged butterfly" (Figure 16.16) over the bridge of the nose and the cheeks (Figures 16.16 and 16.17). The cutaneous lesions, which consist of either poorly defined malar erythema or sharply demarcated erythematous scaling plaques, may occur or worsen with sun exposure.

DLE converts to SLE in approximately 5% of individuals. It can be painful, and in some cases, scarring can result. Whenever lesions are

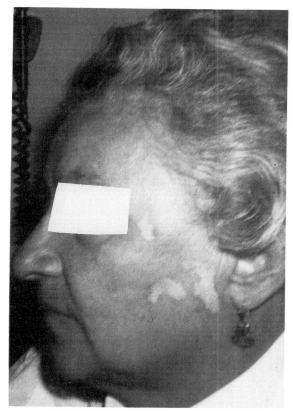

FIGURE 16.17 Discoid lupus erythematosus, lesions in front of the ear and on side of face. Courtesy of Professor Naomi F. Rothfield, Division of Rheumatic Diseases, Department of Medicine, The School of Medicine of the University of Connecticut Health Center, Farmington, Connecticut.

distributed in a generalized fashion, the term *disseminated discoid lupus* is used.

Dilated and tortuous blood vessels are visualized through the thinned epithelium (telangiectasia). With a hand lens, small, keratotic plaques are visualized in follicular ostia. Persistent lesions may result in permanent scarring and disfiguring.

A characteristic granular band of immunoglobulin and complement along the dermo-epidermal and dermal-follicular junctions (so-called lupus band test) is revealed by direct immunofluorescence (Figure 16.18).

16.3.1.1.2 Systemic Lupus Erythematosus Although SLE is often relatively mild, if it is not controlled it can result in damage to vital

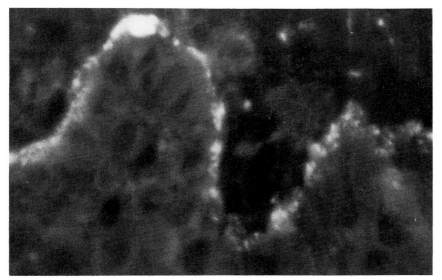

FIGURE 16.18   Fluorescence micrograph of immunofluorescence reaction demonstrating the localization of antinuclear antibodies at the epidermal–dermal junction, in a case of lupus erythematosus. Courtesy of Dr. F. N. Rothfield.

organs, such as the kidneys, brain, heart, and lungs. It is a remitting febrile illness. The periods of improvement or remission may last for weeks, months, or perhaps years before the next flare-up.

SLE has long been considered a prototypic immune disease characterized by multiple autoantibodies, primarily to cellular constituents. It affects people of all races and genes; however, the incidence in women is seven times that of men. In women the initial presentation of the disease is more often during the childbearing years (ages 17 to 39).

The existence of a limitless number of antibodies against self-constituents indicates that the fundamental defect of SLE is a failure of the regulatory mechanisms that sustain self-tolerance.

16.3.1.1.2.1 Clinical Picture of SLE   One SLE patient with a relatively mild disease concluded that he no longer needed to avoid sunlight. He basked in the sun and got a severe sunburn. Within 4 weeks he had arthritis and within 2 months, severe kidney disease. He did respond to intensive therapy with massive doses of drugs, but could have saved himself considerable life-threatening risk by avoiding *photoaggravation of the disease*.

16.3.1.1.2.2 Serology and Immunology of SLE The LE cell seen in the bone marrow aspirate from patients with SLE is a distended polymorphonuclear leukocyte with its nucleus flattened against the cell membrane and a large cytoplasmic inclusion that appears to be a pale, swollen lymphocyte nucleus. The serum component responsible for the LE cell formation is gammaglobulin, which reacts with intact nuclei and nucleoprotein.

The predominant immunological feature of SLE is the presence of heterogeneous antibodies specific for DNA, erythrocytes, leukocytes, platelets, cell nuclei, nucleoli, ribosomes, deoxyribonucleoprotein, histones, soluble nuclear glycoproteins, ribonucleoproteins, and cytoplasmic antigens.

Antibodies to T lymphocytes may be responsible for T cell defects, usually involving a decrease in T suppressor (Ts) activity, which correlates with anti-double-stranded DNA (anti-dsDNA) antibody activity, and exacerbations of the disease. Anti-dsDNA antibodies are present in 50–100% of SLE patients with active disease and in 0–25% of patients with inactive disease.

16.3.1.1.2.3 Genetics of Lupus As many as 80% of SLE patients (and as many as 30% of normal individuals) have genetic factors determining predisposition to SLE on the surface of their antibody-forming cells. The risk of SLE is increased by the inheritance of HLA-DR3 or -DR4. Complement deficiencies, especially of C2 (HLA linked) and C4, are both well documented. In identical twins, SLE is found in both twins only in about one-half of the cases. The causes that can trigger lupus to flare are a severe cold, stressful situations, excessive fatigue, exposure to sunlight, injury, infections, environmental chemicals and certain medications.

16.3.1.1.2.4 Diagnostic Criteria of Lupus Physicians follow 11 criteria established in 1971 by the American Rheumatism Association. An individual with a positive response to four or more criteria is likely to have lupus:

1. Facial rash.
2. An extensive skin problem, including rash, blotches, and thick, raised patches shaped like disks.
3. A significant photosensitivity reaction, even with a minimal exposure to sunlight.
4. Ulcers (sores) in the mouth, nose, or throat.
5. Arthritis, unaccompanied by any noticeable or marked deformity of the affected joints.
6. Pericarditis or pleurisy.

7. Proteinuria or blood cell casts in sputum or urine (a sign of slight internal bleeding).

8. Neurological disorder, seizures, or psychotic problems.

9. Hemolytic anemia, leukopenia, lymphopenia, and thrombocytopenia.

10. Disorder in the immune system; LE cells; false-positive results to the test for syphilis; anti-DNA antibodies; antibodies to nuclear proteins (anti-Sm antibodies),

11. Antinuclear antibodies (ANAs).

16.3.1.1.3 Drug-Induced SLE   IgG reactivity with the H2A–H2B DNA complex, a subunit of the nucleosome, has been detected in many patients with lupus induced by procainamide and quinidine. The native H2A–H2B dimer is highly antigenic for patients with procainamide-induced lupus. The finding that autoantibodies induced by different drugs display the same specificity suggests the existence of a common pathway independent of the known pharmacology or chemical nature of the eliciting drug.

16.3.1.1.4 Lupus Erythematosus in Mice   The development of several inbred strains of mice (NZB/NZW, MRL, BSXB) that spontaneously develop a disease strongly resembling LE in humans provides important opportunities for basic research. Injury to the kidney causes DNA antibodies to be taken up by the kidney tissue during the injury. The NZB × NZW F_1 hybrid develops a syndrome very similar to human SLE, with fatal glomerular nephritis, LE cell factor, antinuclear antibodies, and antibodies to both ssDNA and dsDNA.

Virtually 100% of mice had anti-ssDNA antibodies by 3 months of age. As the disease progressed, these animals appeared to develop decreased suppressor T cell activity.

The NZB/W lupus is mainly a disease of females. Female with ovaries removed before puberty and given injections of male hormone show a marked decrease of disease. In the NZB/W mice, both genetic factors and a virus appear to play an important role in the disease.

A soluble immune response suppressor (SIRS) is released by specially stimulated spleen cells of normal mice, but not by the spleen cells of mice with SLE. Antibody-forming cells of the sick mice can be turned off by SIRS obtained from the spleen cells of young NZB/W mice, not yet affected by the disease. The consequent increase in the survival of NZB/W females supports the possibility of an eventual therapy of SLE based on supplying such missing factors.

16.3.1.1.5 Lupus Erythematosus in Canines Canine systemic LE (predominantly in females) is characterized by autoimmune hemolytic anemia, thrombocytopenia, glomerulonephritis, and polyarthritis. Ovariohysterectomy enhances remission. *Skin lesions may be exacerbated by exposure to sunlight.*

The genetic data implicate genes that predispose to autoimmunity and genes that predispose to the actual expression of autoimmune disease.

16.3.1.2 Dermatomyositis

This disease is associated with a *persistent erythema of sun-exposed areas in the neck and shoulder regions.* The heart, lung, and blood vessels may be affected. The disease has a first peak in childhood and a second around the age of 45 years. Dermatomyositis probably has an autoimmune pathogenesis, like the other collagen diseases. There is, in this disorder, an overrepresentation of the HLA-DR3 phenotypes.

A virutic etiology (influenza, echo, and Coxsackie viruses) is suspected. Echo virus can induce myositis in children with agammaglobulinemia. Cell-mediated immunity is identified by the predominance of lymphocytic infiltrates (of B or T CD4 type).

The symptoms are fiery erythema of the upper chest and neck, multiple telangiectasias, heliotropic periorbital edema, plaques or papules over bony prominences (Gottron papules), and lichen planuslike skin lesions.

There is (Figure 16.19) epidermal atrophy, basal membrane degeneration, vascular infiltrates, subepithelial fibrinoid deposits, mucinous changes in dermis, and poikiloderma.

16.3.2 Viral Infections

16.3.2.1 Herpes Solaris

Large doses of sun exposure are required for photoprovocation leading to cutaneous herpes solaris. The histopathological findings are shown in Figure 16.20.

16.3.3 Psoriasis and Seborrheic Skin Conditions

Administration of an excessive dose during UVB phototherapy may cause psoriasis to worsen and to progress to erythroderma. Erythrodermic psoriasis is confined to exposed sites. The face is rarely involved; relief or exacerbation is reported by patients following sun exposure. In

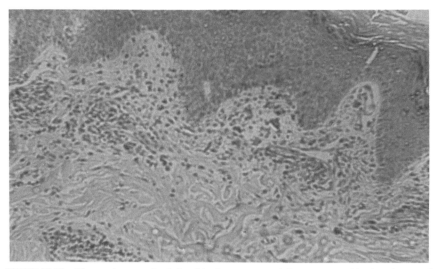

FIGURE 16.19   Photomicrograph of the skin from a case of dermatomyositis showing moderate epidermal atrophy and inflammatory infiltrate in the upper dermis (by Cahide Kohen); ×100. Courtesy of Dr. George Elgart (who kindly provided the microscope slide), Department of Dermatology, School of Medicine, University of Miami.

many patients susceptible to photoprovocation, an exudative-seborrheic psoriasis is present.

16.3.4 Other Dermatoses Provoked by Light

16.3.4.1 Lichen Planus Actinicus

This disease manifests as mild pruritic lesions on exposed sites in the spring and summer, with spontaneous remission in the winter. The affliction is triggered as a Koebner reaction to sunburn injury or UV irradiation in suberythemogenic doses. The clinical pattern is characterized by annular hyperpigmented papules (face and neck) and skin-colored aggregated pinhead papules (face and dorsa of hands).

16.3.4.2 Dyskeratosis Follicularis Darier

This disease, also called Darier–White's disease, is a dominant autosomal disease characterized by a genetic disorder of keratinization (Figure 16.21). *Symptoms get worse in the sunlight.*

Electron microscopy reveals acantholysis, with loss of desmosomes and detachment of tonofilaments. An earlier proteolytic insult may be

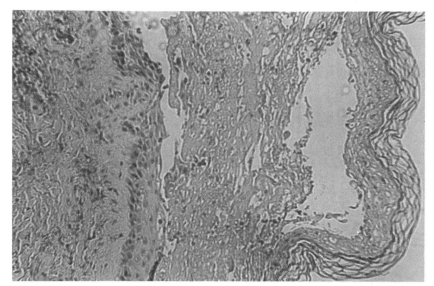

FIGURE 16.20 Photomicrograph of the skin from a case of herpes solaris (by Cahide Kohen). Note the bullous lesion, characterized by a separation of layers in the stratum corneum and the inflammatory infiltrate; ×100. Courtesy of Dr. Elgart, who provided the microscope slide.

involved, for example, due to a hypothetical "cell-dissociating factor." Increased localization of plasminogen in suprabasal cells and of fibrinogen in the keratotic plug, and antigenicity of keratin, have been reported. Together with a combination of genetic and possibly immunologic factors, there seems to be a Koebner-type photoaggravation.

The histological changes are a fissure or lacuna above the basal layer, as a result of faulty epidermal cell adhesion, and dyskeratotic cells of two types, i.e., (1) *corps ronds* (round bodies) and (2) grains (small cells with elongated nuclei and scant cytoplasm).

16.3.4.3 Benign Familial Pemphigus of Hailey–Hailey

This is an inherited autosomal dominant trait characterized by locally recurrent eruptions of small vesicles. Two-thirds of patients report a family history. By peripheral extension the lesions may assume a serpiginous configuration, on the neck and in intertriginous areas (axillae, groin). A defect of the glycocalyx material is suggested by lack of binding of pemphigus vulgaris antibody, and of the fluorescein isothiocyan-

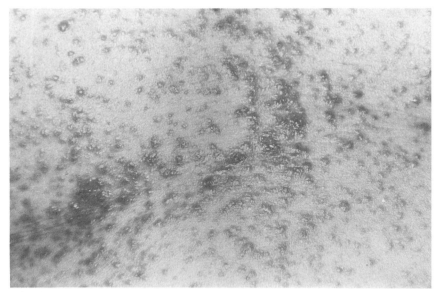

FIGURE 16.21   Dyskeratotic papular lesions in Darier's disease. Courtesy of Dr. Penneys.

ate conjugate of concanavalin A. This results in keratinocytes with defective cell cohesion.

16.3.4.4 Pemphigus Seborrheicus Erythematosus Senear–Usher

The hallmark of this disease is the presence of an IgG antibody against the keratinocyte cell surface. Early blisters in pemphigus foliaceus reveal acantholysis (loss of cell-to-cell contact) just below the stratum granulosum layer. Deeper epidermis below the granular layer remains intact.

16.3.4.5 Dermatoses of Uncertain Classification: Photoprovocation in Isolated Cases

16.3.4.5.1 Purpura Solaris   *Purpura solaris, also called Bateman's senile purpura, occurs frequently following trauma to sun-damaged skin of the dorsal forearm and hands of elderly persons.* Torsional stress rapidly leads to ecchymoses; it may be months before the blood pigment is reabsorbed. The defect is in abnormal platelet function or platelet levels (below 50,000–70,000/ml) or in the formation of fibrin clots, which stabilize the platelet plugs.

16.3.4.5.2 Tinea Faciei   This disease is a dermatophyte infection of glabrous skin. There is a history of photoexacerbation. The skin lesions are characterized by the lack of follicular plugging and true poikiloderma. The main three pathogens are *Tricophyton rubrum, Tricophyton mentagrophtyes,* and *Micrococcus canis.* In association with AIDS, tinea facialis is manifested as a superficial dermatomycosis produced by an infection of *Dermatophytes,* mimicking seborrheic dermatitis.

16.3.4.5.3 Pityriasis Rosea of Gibert   Pityriasis rosea (PR) is an acute eruption of the skin, most commonly in spring and autumn. The herald patch, e.g., a primary patch on the right scapula, precedes a general exanthematous eruption along the axis of the lesions, following the lines of cleavage in a "Christmas tree" distribution. The secondary lesions appear *in crops at intervals of a few days (maximum 10 days).*

The primary lesion may be at the site of cutaneous infection, flea bite, or smallpox vaccination. The pathogens are dermatophytes, fungi, *Staphylococcus albus,* hemolytic streptococci, and spirochetes.

There is a lasting immunity to PR, and consequently recurrences are rare. Among drugs able to induce PR are bismuth, barbiturates, gold, organic mercurials, metronidazole, D-penicillamine, isoretinin, and salvarsan. Lesion distribution corresponds to contact surface with garments.

16.3.4.6 Keratoacanthoma

Keratoacanthoma (molluscum sebaceum; molluscum pseudocarcinomatosum) is a benign, heavily keratinizing, rapidly evolving epithelioma situated mainly on the exposed areas of the skin; it simulates squamous epithelioma but undergoes spontaneous evolution. Sunlight is thought to be a factor, but chemical carcinogens are also believed to be responsible.

Keratoacanthoma resembles the Shope papilloma of unquestioned viral pathogenesis. The spontaneous healing of keratoacanthoma is similar to that seen in common warts, which are caused by a specific virus. Hybridization studies have shown the presence of human papilloma viruses types 9, 16, 19, 25, and 37, but the viral etiology remains uncertain. In immunosuppressed patients the development of keratoacanthomas is attributed to decreased rejection of tumors.

The initial rapid growth and the self-regression property of keratoacanthoma are thought to reflect and recapitulate the cyclic history of the hair follicle. The keratinous mass may build up to form a horn, or may ulcerate or exfoliate to form a large papillomatous mass. In pseudocarcinomatous infiltration, islands and columns of cells break away from the main body of the tumor and produce a rather regular, limited infiltration of the stroma, mimicking an early well-differentiated carcinoma.

16.4 SUGGESTED READING: ROSACEA*

We are presenting an appendix to this chapter (i.e., Acnea Rosacea) as an illustrative example particularly because of its relevance to the differential diagnosis of photoaggravated diseases and the difficulties encountered in the assessment of light involvement. Acnea Rosacea is an example underlining the many unknowns in the pathogenesis of photoaggravated disease, the actual stage of the disease in which such processes take place as well as the cellular mechanisms involved.

Rosacea bears a certain resemblance to acne vulgaris and can easily be confused with that disease by those untutored in dermatology. The failure to make that distinction is a serious mistake, because some therapies that are effective in acne may aggravate rosacea.

PATHOGENESIS

In contrast to acne, causative factors in rosacea are a matter of speculation. Many different etiologies have been proposed (Marks, 1989). No consensus has been established for any of these. However, some conjectures about the etiopathogenesis are most supportable than others. A hereditary component seems operative, but sound pedigree studies are lacking. A fairly strong case can be made that excessive exposure to sunlight plays an influential role. Accordingly, rosacea could properly be considered among the photosensitivity disorders. This might explain why rosacea is rare in black-skinned races.

There is convincing evidence that those who are most at risk for rosacea are light-colored, thinned-skinned, blue-eyed, freckled persons of Scotch–Irish (Celtic) extraction (Irvine and Marks, 1989). These individuals are classified as phototype I; they burn early and do not tan. The disease is most prevalent among Celts in sunny climates, and has been called the "curse of the Celts." Although no cure is in sight, the disorder can be satisfactorily managed.

CLINICAL MANIFESTATIONS

Rosacea is localized to the central area of the face, a distribution that is very helpful in differential diagnosis (Plewig and Kligman, 1993). The disease develops against a background of persistent flushing, upon which crops of papules and pustules later develop. Rosacea usually makes its appearance after adolescence, it peaks in adulthood, and lasts indefinitely. It sometimes culminates in grotesque granulomatous lesions. It is more common in women but is more severe in men. The prevalence of the milder forms in women may approach 10–15%, but good epidemiologic data are lacking. The hallmarks of rosacea are persistent erythema, telangiectasia, and follicular papulo-pustules (see Figure

* This section has been contributed by Albert M. Kligman, University of Pennsylvania, Philadelphia, Pennsylvania.

16.22). The disease evolves in well-defined stages, but the course is erratic and unpredictable. Severity varies enormously over a period of decades.

Stage I

All rosacea patients are "flusher-blushers." The central face suddenly gets diffusely red and takes 10 min or more to clear. At first the episodes of erythema are transient. With time, flushing becomes more frequent, of deeper hue, and

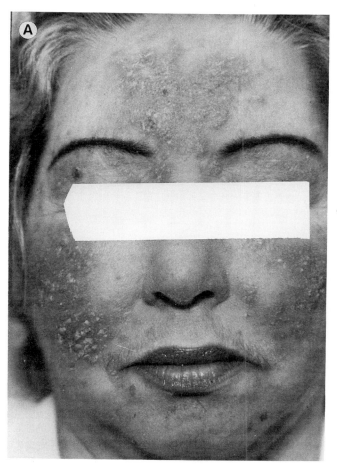

FIGURE 16.22 (A) Woman with acne rosacea. (B) Persistant erythema and telangiectasia in an adult. (C) Severe papulo-pustular rosacea on an erythematous background in a middle-aged male. (D) Rhinophyma in an older male with acne rosacea. Courtesy of Dr. Albert Kligman, School of Medicine, Department of Dermatology, University of Pennsylvania.

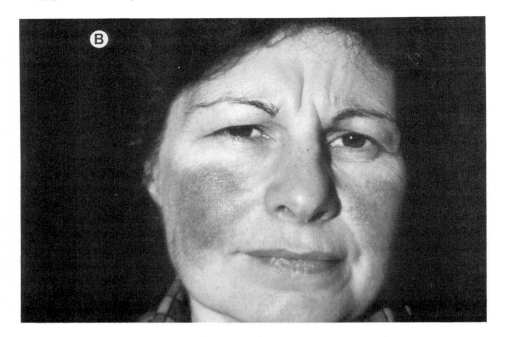

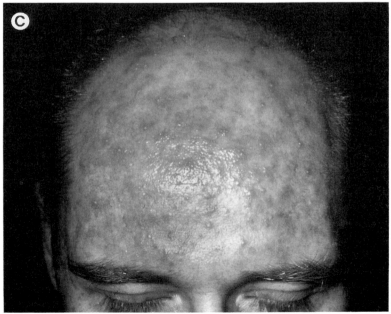

FIGURE 16.22 (Continued)

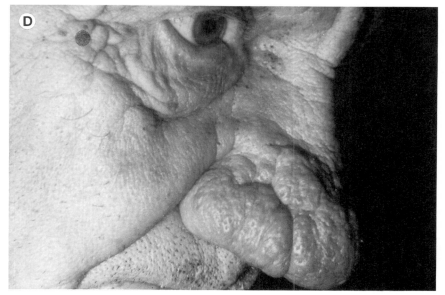

FIGURE 16.22 (Continued)

more persistent, often spreading to the neck and upper chest. Telangiectasias, which are permanently dilated venules, become superimposed on the red background, forming sprays of conspicuous vessels on the nose, cheeks, and glabella.

Heat, intensive exercise, and especially sunlight exacerbate flushing and deepen the erythema. Strong emotions can provoke prolonged bouts of long-lasting flushing; this is in contrast to people who blush briefly when embarrassed (Wilkin, 1983).

Patients with rosacea are thought to have sensitive skin that is vulnerable to a variety of chemical and physical stimuli. They seem to experience subjective discomforts, such as stinging and burning from skin-care products (sunscreens, moisturizers, and fragrances) that others use with impunity. Soaps, alcohols, abrasives, and peelers may produce these discomforts and in addition provoke a scaling dermatitis. Rosacea patients have to be cautioned against excessive use of skin-care products, especially cleansers.

Stage II

Papulo-pustules spring up on the erythematous landscape. These are strictly follicular in location and may involve the tiny invisible vellus follicles as well as the large-pored sebaceous follicles. These papules can be deep seated, hard, and surmounted by a yellowish, soft pustule. They customarily last for weeks but curiously do not leave scars, or leave only shallow ones. Pitted scars are from acne, not rosacea. Comedones (blackheads and whiteheads) are notably absent, a feature that distinguishes rosacea from acne.

Without protection against sunlight, the stigmata of photoaging (dermatoheliosis) emerge rapidly in early adulthood, namely, coarse and fine wrinkles and thickened yellowish, leathery, lax, dry skin. Premature aging may be seen as early as the twenties. The aggravating effect of excessive exposure to sunlight, especially in sunny climates is beyond dispute and is the basis for classifying rosacea in the category of a photosensitivity disorder, much like lupus erythematosis, in which sunlight is also an aggravating factor.

Crops of papulo-pustules may spread to the bald scalp, down the sides of the neck, and even onto the anterior and posterior trunk. These extra-facial locations are not rare but usually go unnoticed unless the entire body is examined.

Bacteriologic culture for pathogenic organisms yields nothing of interest. Only the resident microflora are recovered. Unlike acne, bacteria are not implicated in the causation of rosacea. Although the follicular pores are often enlarged, rosacea patients are generally not oily. Sebum output is in the normal range and apparently is of no importance in pathogenesis.

Stage III

A small number of patients, mainly men, go on, after many years, to develop a more serious and ugly manifestation. This takes the form of irregular and coarse thickenings in which widely dilated pores are conspicuous, containing a cheesy, oily material, easily expressed by the fingers. This enlargement of the skin's bulk leads to the well-known dramatic appearance of rhinophyma, a grotesque malformation of the nose. Enormous enlargement of sebaceous glands results in dozens of yellowish, small nodules on the cheeks and forehead. The coarsened, thickened skin is a consequence of a chronic diffuse inflammatory reaction, occupying the entire dermis. Nodular thickenings are extremely disfiguring and may eventually resemble the leonine facies of leprosy.

VARIANTS OF ROSACEA

The preceding description refers to classic, textbook rosacea. It is important, however, to recognize that rosacea has many variants. These often go unrecognized, to the detriment of the patient who then does not receive proper treatment.

Ophthalmic Rosacea

Eye involvement is much more common than generally supposed. A minimum of 20% of rosacea patients have appreciable ophthalmologic signs. The commonest expression is chronically inflamed lid margins (blepharitis) with scales and crusts, easily confused with seborrheic dermatitis. Photophobia, an intolerance of bright sunlight, is always present in ocular rosacea. Just as dermatologists too often fail to recognize the accompanying eye signs and symptoms, ophthalmologists, who might be the first to see these patients, often neglect to talk about or look at the skin. Eye involvement is independent of the severity of the skin disease, which, in fact, it may precede by years.

Other ocular manifestations include conjunctivitis, iridocyclitis, and, worst of all, keratitis, which may end in blindness. Rosacea patients should preferably be referred to an ophthalmologist, who is more likely to detect the early signs.

Steroid Rosacea

Topical corticosteroids, including the weakest ones, such as hydrocortisone, should never be used in a chronic disease such as rosacea (Leyden et al. 1974). Helpful initially, the inevitable outcome is a fierce exacerbation of the inflammatory lesions, namely, dense clusters of deep, tender papulo-pustules and even nodules on a flaming red, atrophic skin, traversed by networks of telangiectatic vessels. Withdrawal results in an almost intolerable aggravation of the papulopustular eruption. Weaning these patients from steroids is a vexation for both doctor and patient.

Rosacea Fulminans

Dermatologists formerly called this *pyoderma facial,* an unsuitable term because it is not an infection. It is now recognized as a ferocious expression of facial rosacea (Massa and Su, 1982), somewhat analogous to the worst form of acne, acne conglobata. Rosacea fulminans occurs only in young women, many of whom have no prior history of rosacea but who nonetheless can be classified as flusher-blushers on questioning.

Rosacea fulminans blows up into a horrendous inflammatory disorder with coalescent nodules, deep abscesses, and draining sinuses on a flaming-red background. Once seen it can never be forgotten. There is no agreement regarding causation. Nonetheless, there is commanding evidence that severe emotional stress, divorce, death of a loved one, or severe anxiety is antecedent to the blow-up. Patients often resist, giving an accurate account of a preceding emotional crisis. Empathetic questions will usually bring this out. This may be the archetypical pyschosomatic disease in all of dermatology. Rosacea is certainly colored and intensified by anxiety.

Phymas

Phyma is a Greek word for swelling or mass. The best known example is rhinophyma, which occurs exclusively on the nose of humans. This grotesque deformation is the end result of a diffuse chronic inflammatory process that provokes a severe fibroplasia, accompanied by enormous enlargement of sebaceous glands. Other phymas may occur and are usually misdiagnosed. These include gnathophyma (jaw), otophyma (earlobes), blepharophyma (eyelids), and otophyma (forehead). Another unusual manifestation of end-stage rosacea, usually missed, is a solid, hard, diffuse edema.

HISTOPATHOLOGY

The changes are characteristic but not diagnostic. The signs of chronic actinic damage are always present. These include a great proliferation of abnormal elastic fibers, ultimately degenerating into amorphous masses, accompanied by a great loss of collagen. Atrophy and atypia of the epidermis are also present. Although actinic damage does not cause rosacea, it is a constant that contributes to and parallels the severity of the disease. Phototype I Celts are susceptible both to sun damage and to rosacea.

Other nonspecific histologic signs include a diffuse lymphohistiocytic infiltrate, dilated vessels, fibroplasia with scarring, and sebaceous hyperplasia.

Demodex mites are almost always found in the follicular infundibila and sebaceous ducts, generally in small numbers. Their etiologic role is uncertain and controversial. When abundant, a contributory role cannot be ruled out. It is interesting that agents that inhibit *Demodex* seem to be therapeutically beneficial. Elemental sulfur is a venerable example.

The rupture of follicles, a frequent occurrence, releases hairs, sebum, bacteria, horn, and mites into the tissue, giving rise to varied, confusing pictures, including granulomas, foreign body giant cells, abscesses, etc.

TREATMENT

The aim of treatment is to keep the disease under reasonable control and to prevent progression to the disfiguring end stages. Treatment choices are varied and multiple and must be adapted to the ever-changing needs of individual patients.

Protection against all forms of chemical and physical trauma must be emphasized and brought to bear as soon as the diagnosis is made. The patient must scrupulously avoid frequent washing, astringents, alcoholic cleansers, abrasives, and peeling chemicals such as the increasingly popular α-hydroxy acids. Sunscreens are a must, preferably the newer ones that rely solely on highly micronized particles of titanium dioxide and contain no chemical sun filters (for example, benzophenones, which often cause stinging and burning). Titanium dioxide sunscreens are available at SPF values of 15 and above.

Topical Treatment

Topical antibiotics, such as those used in acne, are moderately effective when applied twice daily. Of these, erythromycin in a nonalcoholic vehicle is a good choice. Efficacy is greatest in papulo-pustular rosacea; it is not due to suppression of bacteria. Erythromycin seems to exert an antiinflammatory effect in some unknown way.

Topical metronidazole (Metrogel) is a new agent that is gaining popularity. It, too, is most effective in suppressing papulo-pustules by some unknown mechanism (Bleicher *et al.*, 1987).

Topical tretinoin, in low concentration, has recently been shown to be effective, though slower in bringing the disease under control. Telangiectasia seems to slowly disappear; this may not be as surprising as it at first seems, because retinoic acid is the preferred treatment for photoaged skin, the substrate in which rosacea evolves. Experts may wish to check by microscopic examination for *Demodex* in material extracted from the follicles. If high numbers are found, a trial with 5–10% benzoyl peroxide daily for 2–3 weeks is reasonable.

Telangiectasia can be satisfactorily obliterated by light electrocoagulation in the hands of an experienced dermatologist. For numerous large sprays on the nose, the tunable dye laser is the instrument of choice and gives splendid results, without scarring.

The surgical treatment of rhinophyma is very effective, using a variety of methods according to personal experience, including laser, dermabrasion, curettage, etc.

Systematic Treatment

Oral antibiotics are effective in controlling the severer forms of inflammatory rosacea. Tetracyclines seem to be more effective than erythromycins. These should be started at full doses, the equivalent of 1 g of tetracycline daily. Rosacea patients eventually figure out a maintenance dose after the disease has been brought under control. This may be as little as 250 mg of tetracycline daily.

Rosacea has its ups and downs and treatment should be accordingly adjusted. A combination of oral and topical therapy is advantageous as a starting maneuver.

The severest forms, such as rosacea fulminans, are best left to the specialists. Remarkable improvement can be achieved by a combination of oral isotretinoin and prednisone. These are potent drugs with many serious side reactions, and should not be employed by the inexperienced.

Bibliography

Arndt, K. A. (1993). Lichen planus. *In* "Dermatology in General Medicine" (T. B. Fitzpatrick, A. Z. Eisen, K. Wolff, I. M. Freedberg, and F. Austen, eds.), pp. 1134–1144. McGraw-Hill, New York.

Bell, L. P. (1983). "Lupus Patients Can Survive," pp. 52–63. Randen Press.

Behrman, R. E. (1992). Neonatal lupus phenomena. *In* "Nelson Textbook of Pediatrics," 14th ed (Behrman, R. E. ed., Kliegman, R. M. ed., Nelson, W. E., Vaughan, O. C. III), p. 627. Saunders, Philadelphia.

Bertolino, A. P., and Freedberg, I. M. (1993). Disorders of epidermal appendages and related disorders. *In* "Dermatology in General Medicine" (T. B. Fitzpatrick, A. Z. Eisen, K. Wolff, I. M. Freedberg, and F. Austen, eds.), pp. 671–709. McGraw-Hill, New York.

Bickers, D. R., Pathak, M., and Lim, H. W. (1993). The porphyrias. *In* "Dermatology in General Medicine" (T. B. Fitzpatrick, A. Z. Eisen, K. Wolff, I. M. Freedberg, and F. Austen, eds.), pp. 1854–1893. McGraw-Hill, New York.

Bjornberg, A. (1993). Pityriasis rosea. *In* "Dermatology in General Medicine" (T. B. Fitzpatrick, A. Z. Eisen, K. Wolff, I. M. Freedberg, and F. Austen, eds.), pp. 1117–1123. McGraw-Hill, New York.

Bleicher, P. A., Charles, J. H., and Sober, A. J. (1987). Topical metronidazole therapy for rosacea. *Arch. Dermatol.* **123**, 609–614.

Brugere-Picoux, J. (1982). "Biochimie du rumen—Aspect pathologiques," p. 48. Conference prononcée le 13 Mai 1982 aux Journées Nationales des Groupement Technique Vétérinaire (G.T.V.). d'Arles. G.T.V., 83-3-B-255.)

Burguess, G. H., Ghadially, R., and Ghadially, F. (1993). Keratoacanthoma. *In* "Dermatology in General Medicine" (T. B. Fitzpatrick, A. Z. Eisen, K. Wolff, I. M. Freedberg, and F. Austen, eds.), Vol. 1, pp. 848–855. McGraw-Hill, New York.

Buttery, J. E., Carrera, A. -M., and Pannall, P. R. (1990). Reliability of the porphobilinogen screening assay. *Pathology* **22**, 197–198.

Carr, R. I., and Jameson, E. J. (1986). Lupus Erythematosus (LE) A Handbook for physicians, patient and their families. *In* "Lupus Erythematosus," pp. 1–60. A publication of the Lupus Foundation of America, Inc., Rockville, Maryland. Patient Relations/Education Committee. In cooperation with The National Institute of Allergy and Infectious Diseases, National Institutes of Health Bethesda, MD.

Coffman, J. D. (1993). Cutaneous changes in peripheral vascular disorders. *In* "Dermatology in General Medicine" (T. B. Fitzpatrick, A. Z. Eisen, K. Wolff, I. M. Freedberg, and F. Austen, eds.), McGraw-Hill, New York.

Cotran, R. S., Kumar, V., and Robbins, S. L. (1989). Lupus erythematosus (LE). *In* "Pathologic Basis of Disease" (S. L. Robbins, ed.), 4th ed., p. 1107. Saunders, Philadelphia.

Crapley, T.G., and Fitzpatrick, T. B. (1993). Dermatologic diagnosis by recognition of clinical morphological patterns. *In* "Dermatology in General Medicine" (T. B. Fitzpatrick, A. Z. Eisen, K. Wolff, I. M. Freedberg, and F. Austen, eds.), p. 65. McGraw-Hill, New York.

Cripps, D. J., Gocmen, A., and Peters, H. A. (1980). Porphyria turcica. Twenty years after hexachlorobenzene intoxication. *Arch. Dermatol.* **116**, 46.

Dean, G., and Barnes, H. D. S. (1959). Porphyria in Sweden and South Africa. *S. Afr. Med. J.* **33**, 246 (cited in MacAlpine and Hunter, 1968b; Fitzpatrick, et al., 1993, p. 722).

DeLeo, V. A. (1992). Photocontact dermatitis. *In* "Topics in Clinical Dermatology Photosensitivity" (V. A. Deleo, ed.), pp. 84–141. Igaku-Shoin Medical Publishers, New York.

Deleo, V. A., Suarez, S. M., and Maso, M. J. (1992). Photoallergic contact dermatitis. Results of photopatch testing in New York, 1985 to 1990. *Arch. Dermatol.* **128**, 1513–1518.

Fitzpatrick, T. B., Eisen, A. Z., Wolff, K., Freedberg, I. M., and Austen, F., eds. (1993). "Dermatology in General Medicine." McGraw-Hill, New York.

Gibert, C. M. (1860). "Traité pratique des maladies de la peau et de la syphilis," 3rd ed., p. 402. Plon, Paris (cited in Fitzpatrick *et al.*, 1993, p. 1117).

Goldberg, A., and Rimington, C. (1962). "Diseases of Porphyrin Metabolism." Thomas, Springfield, IL (cited in MacAlpine and Hunter, 1968, p. 29).

Guttmacher, M. (1941). "America's Last King. An interpretation of the Madness of George III." Scribner, New York (cited in MacAlpine and Hunter, 1968b).

Harber, L. C. (1993). Abnormal responses to ultraviolet radiation: Drug-induced photosensitivity. *In* "Dermatology in General Medicine" (T. B. Fitzpatrick, A. Z. Eisen, K. Wolff, I. M. Freedberg, and F. Austen, eds.), pp. 1677–1689. McGraw-Hill, New York.

Harber, L. C., Kochevar, I. E., and Shalita, A. R. (1982). Mechanisms of photosensitization to drugs in humans. *In* "The Science of Photomedicine" (J. D. Regan and J. A. Parrish, eds.), pp. 323–347. Plenum, New York and London.

Hawk, J. L. M., and Magnus, I. A. (1979). Chronic actinic dermatitis in idiopathic photosensitivity syndrome including actinic reticuloid and photosensitive eczema. *Br. J. Dermatol.* **101**, Suppl. 17, 24 (in De Leo *et al.*, 1992, p. 1517).

Hawk, J. L. M., and Norris, P. G. (1993). Abnormal responses to ultraviolet radiation: Idiopathic. *In* "Dermatology in General Medicine" (T. B. Fitzpatrick, A. Z. Eisen, K. Wolff, I. M. Freedberg, and F. Austen, eds.), pp. 1661–1677. McGraw-Hill, New York.

Holbrook, K. A., and Wolf, K. (1993). *In* "Dermatology in General Medicine" (T. B. Fitzpatrick, A. Z. Eisen, K. Wolff, I. M. Freedberg, and F. Austen, eds.), p. 180. McGraw-Hill, New York.

Hurvitz, A. I. (1992). Lupus in the dog. *In* "Systemic Lupus Erythematosus" Sec. ed. (R. Lahita, ed.), pp. 195–202. Churchill & Livingston, New York, London, Edinburgh.

Ippen, H. (1982). Photodermatoses. *In* "The Science of Photomedicine" (J. D. Regan and J. A. Parrish, eds.), pp. 340–394. Plenum, New York and London.

Irvine, C., and Marks, R. (1989). Prognosis and prognostic factors in rosacea. *In* "Acne and Related Disorders" (R. Marks and G. Plewig, eds.), pp. 331–333. Dunitz, London.

Johnson, R. A., and Dover, J. J. (1993). Cutaneous manifestations of human immunodeficiency virus disease. *In* "Dermatology in General Medicine" (T. B. Fitzpatrick, A. Z. Eisen, K. Wolff, I. M. Freedberg, and F. Austen, eds.), pp. 2637–2689. McGraw-Hill, New York.

Jorizzo, J. L., and Sherertz, E. R. (1993). *In* "Dermatology in General Medicine" (T. B. Fitzpatrick, A. Z. Eisen, K. Wolff, I. M. Freedberg, and F. Austen, eds.), pp. 2058–2061. McGraw-Hill, New York.

Kauffman, L., Evans, I. K. E., Stevens, R. E., and Weinkove, C. (1991). Bone-marrow transplantation for congenital erythropoietic protoporphyria. *Lancet* **337**, 1510–1511.

Klimas, N. G., Patarca, R., Perez, G., Garcia-Morales, R., Schultz, D., Schabel, J., and Fletcher, M. A. (1992). Case Report: Distinctive immune abnormalities in a patient with procainamide-induced lupus and serositis. *Am. J. Med. Sci.* **303**, 99–104.

Kraemer, K. H. (1993). Heritable diseases with increased sensitivity to cellular injury. *In* "Dermatology in General Medicine" (T. B. Fitzpatrick, A. Z. Eisen, K. Wolff, I. M. Freedberg, and F. Austen, eds.), pp. 1974–1991. McGraw-Hill, New York.

Kushner, J. P. (1991). Laboratory diagnosis of the porphyrias. *N. Engl. J. Med.* **324**, 1432–1434.

Leyden, J. J., Thew, A. M. and Kligman, A. M. (1974). Steroid rosacea. *Arch. Dermatol.* **110**, 619–622.

Lyon, N. B. (1993). Geriatric dermatology. *In* "Dermatology in General Medicine" (T. B. Fitzpatrick, A. Z. Eisen, K. Wolff, I. M. Freedberg, and F. Austen, eds.), pp. 2961–2972. McGraw-Hill, New York.

MacAlpine, I., and Hunter, R. (1968a). "The Insanity of King George III: A Classic Case of Porphyria. Br. Med. Assoc., London.

MacAlpine, I., and Hunter, R. (1968b). Porphyria in the royal houses of Stuart, Hanover and Prussia. *In* "Porphyria: A Royal Malady," (I. MacAlpine, R. Hunter, C. Rimington, J. Brooke, and A. Goldberg, eds.), Br. Med. Assoc., London.

MacAlpine, I., Hunter, R., Rimington, C., Brooke, J., Goldberg, A., eds. (1968). "Porphyria: A Royal Malady." Br. Med. Assoc., London.

Magee, K. L., Rabini, R. P., Dovic, M., and Adler-Storthz, K. (1989). Human papilloma virus associated with keratoacanthoma. *Arch. Dermatol.* **125**, 1587 (in Ghadially and Ghadially, 1989).

Marks, R., (1989). Rosacea: Hopeless hypotheses, marvellous myths and dermal disorganization. *In* "Acne and Related Disorders" (R. Marks and G. Plewig, eds.), pp. 293–299. Dunitz, London.

Martin, A., and Kobayashi, G. S. (1993). Superficial fungal infections: Dermatophytosis, tinea, nigra, piedra. *In* "Dermatology in General Medicine" (T. B. Fitzpatrick, A. Z. Eisen, K. Wolff, I. M. Freedberg, and K. F. Austen, eds.), pp. 2420–2451. McGraw-Hill, New York.

Massa, M. C., and Su, W. P. D. (1982). Pyoderma faciale: A clinical study of twenty nine patients. *J. Am. Acad. Dermatol.* **6**, 84–91.

May, B. K., Bhasker, R. C., Bawden, M. J., and Cox, T. C. (1990). Molecular regulation of 5-aminolevulinate synthase. *Mol. Biol. Med.* **7**, 405–421.

McLean, D. I., and Haynes, H. A. (1993). Cutaneous manifestations of internal malignant disease. *In* "Dermatology in General Medicine" (T. B. Fitzpatrick, A. Z. Eisen, K. Wolff, I. M. Freedberg, and F. Austen, eds.), pp. 2229–2249. McGraw-Hill, New York.

Mills, J. A. (1993). Dermatomyositis. *In* "Dermatology in General Medicine" (T. B. Fitzpatrick, A. Z. Eisen, K. Wolff, I. M. Freedberg, and F. Austen, eds.), pp. 2148–2155. McGraw-Hill, New York.

Mosher, D. B., Fitzpatrick, T. B., Hori, Y., and Ortonne, J.-P. (1993). Disorders of melanocytes. *In* "Dermatology in General Medicine" (T. B. Fitzpatrick, A. Z. Eisen, K. Wolff, I. M. Freedberg, and F. Austen, eds.), pp. 903–995. McGraw-Hill, New York.

Page, E. H., and Shear, N. H. (1993). Disorders due to physical factors. *In* "Dermatology in General Medicine" (T. B. Fitzpatrick, A. Z. Eisen, K. Wolff, I. M. Freedberg, and F. Austen, eds.), pp. 1581–1592. McGraw-Hill, New York.

Pappas, A., Sassa, S., and Anderson, K. E. (1989). Porphyrias. *In* "The Genetic Basis of Inherited Disorders" (C. R. Scriver, A. L. Beaudet, W. S. Sly, and D. Valle, eds.), 6th ed., pp. 1301–1384. McGraw-Hill, New York.

Phillips, S. B., and Baden, H. P. (1993). Darier-White disease (keratosis follicularis) and miscellaneous hyperkeratotic disorders. *In* "Dermatology in General Medicine" (T. B. Fitzpatrick, A. Z. Eisen, K. Wolff, I. M. Freedberg, and F. Austen, eds.), pp. 547–552. McGraw-Hill, New York.

Plewig, G., and Kligman, A. M. (1993). "Acne and Rosacea." Springer-Verlag, Berlin.

Page, E. H., and Shear, N. H. (1993). *In* "Dermatology in General Medicine" (T. B. Fitzpatrick, A. Z. Eisen, K. Wolff, I. M. Freedberg, and F. Austen, eds.), pp. 1581–1592. McGraw-Hill, New York.

Ray, I. (1855). Insanity of King George III. *Am. J. Insan.* **12,** 1 (cited in MacAlpine and Hunter, 1968a, pp. 1–16).

Rennie, J. (1990). The body against itself. *Sci. Am.,* December, pp. 107–115.

Rock, A., and Moffat, J. L. (1956). Multiple self-healing epithelioma of Ferguson type: Report of a case of unilateral distribution, *Arch. Dermatol.* **74,** 525.

Rothfield, N. (1993). Lupus erythermatosus. *In* "Dermatology in General Medicine" (T. B. Fitzpatrick, A. Z. Eisen, K. Wolff, I. M. Freedberg, and F. Austen, eds.), pp. 2137–2148. McGraw-Hill, New York.

Rowe, L. D. (1989). Photosensitization problems in livestock. *Vet. Clin. North Am.: Food Anim. Pract.* **5**(2), 301–323.

Rubin, R. L. (1986). Enzyme-linked immunoabsorbent assay for anti-DNA and antihistone antibodies. *In* "Manual of Clinical Laboratory Immunology" (N. R. Rose, H. Friedman, and Fahey, J. L. eds.), pp. 744–749. Am. Soc. Microbiol., Washington, D. C.

Rubin, R. L. (1992). Autoantibody specificity in drug-induced lupus and neutrophil-mediated metabolism of lupus-inducing drugs. *Clin. Biochem.* **25,** 223–234.

Rubin, R. L., and Burlingame, R. W. (1991). Biochemical mechanisms of autoimmunity. *Biochem. Soc. Trans.* **19,** 153–159.

Rubin, R. L., Bell, S. A., and Burlingame, R. W. (1992). Autoantibodies associated with lupus induced by diverse drugs target a similar epitope in the (H2A-H2B)-DNA complex. *J. Clin. Invest.* **90,** 165–173.

Sato, K. (1993). Disorders of the eccrine sweat gland. *In* "Dermatology in General Medicine" (T. B. Fitzpatrick, A. Z. Eisen, K. Wolff, I. M. Freedberg, and F. Austen, eds.), pp. 740–753. McGraw-Hill, New York.

Schaunburg-Lever, G., and Lever, W. F. (1993). Familial benign pemphigus. *In* "Dermatology in General Medicine" (T. B. Fitzgerald, A. Z. Eisen, K. Wolff, I. M. Freedberg, and F. Austen, ed.), pp. 642–645. McGraw-Hill, New York.

Schelcher, F., Delverdier, M., Bezille, P., Cabanie, P., and Espinasse, J. (1991). Observation on bovine congenital erythrocytic protoporphyria in the blonde d'Aquitaine breed. *Vet. Rec.,* November 2, pp. 403–406.

Schreiner, E. C. W. (1993). Porokeratosis. *In* "Dermatology in General Medicine" (T. B. Fitzpatrick, A. Z. Eisen, K. Wolff, I. M. Freedberg, and F. Austen, eds.), pp. 565–571. McGraw-Hill, New York.

Schwartz, R. A., and Stoll, H. L. Jr. (1993). Epithelial precancerous lesions. *In* "Dermatology in General Medicine" (T. B. Fitzpatrick, A. Z. Eisen, K. Wolff, I. M. Freedberg, and K. F. Austen, eds.), pp. 804–821. McGraw-Hill, New York.

Short, P., and Adams, R. (1993). Neurocutaneous disorders. *In* "Dermatology in General Medicine" (T. B. Fitzpatrick, A. Z. Eisen, K. Wolff, I. M. Freedberg, and K. F. Austen, eds.), pp. 2249–2290. McGraw-Hill, New York.

Stanley, J. K. (1993). Pemphigus. *In* "Dermatology in General Medicine" (T. B. Fitzpatrick, A. Z. Eisen, K. Wolff, I. M. Freedberg, and K. F. Austen, eds.), pp. 606–615. McGraw-Hill, New York.

Syllabus (1991). Clinical Slide Collection on the Rheumatic Diseases. Prepared by the Visual Aid Subcommittee of the Professional Education Committee of the Arthritis Foundation. Atlanta, American College of Rheumatology

Waldenström, J. (1937). Studien uber die Porphyria. *Acta Med. Scand., Supple.* **82** (cited in MacAlpine and Hunter, 1968b).

Weston, W., and Lane, A. T. (1993). Neonatal dermatology. *In* "Dermatology in General Medicine" (T. B. Fitzpatrick, A. Z. Eisen, K. Wolff, I. M. Freedberg, and F. Austen, eds.), pp. 2941–2961. McGraw-Hill, New York.

Wilkin, J. K. (1983). Rosacea. A review. *Int. J. Dermatol.* **22**, 393–400.

Zucker-Franklin, D. (1993). Cutaneous manifestations of hematologic disorders. *In* "Dermatology in General Medicine" (T. B. Fitzpatrick, A. Z. Eisen, K. Wolff, I. M. Freedberg, and F. Austen, eds.), pp. 1993–2017. McGraw-Hill, New York.

17

Phototherapy of Neonatal Bilirubinemia and Vitamin D Deficiency

Previous chapters on light and health have dealt with the damaging effects of solar light and artificially produced UV irradiation. Beginning with this chapter we will now discuss the beneficial effect of light as used in phototherapy.

17.1 PHOTOTHERAPY OF NEONATAL BILIRUBINEMIA

17.1.1 Plasma Bilirubin Concentration as a Prognostic Index of Bilirubin Encephalopathy in the Neonate

Jaundice occurs in about half of all newborn infants; it is a result of the expression of high bilirubin concentration in the circulating plasma. Bilirubin, a tetrapyrrole pigment (Figure 17.1A) derived from the heme moiety of hemoglobin, is carried in plasma bound primarily to albumin. If the production of bilirubin exceeds the infant's ability to conjugate and excrete it, this pigment accumulates in the circulation, leading to a condi-

A

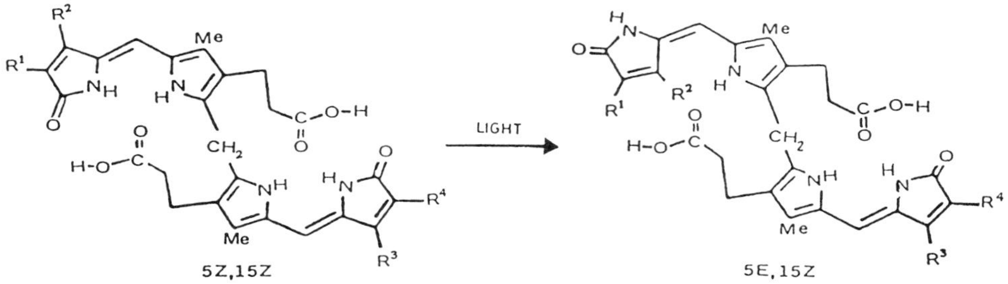

FIGURE 17.1A Structure of bilirubin and scheme of the photoconversion of bilirubin IX-α(z,2) to the EZ form. The photoisomerization affects the first ring on the left side of the figure. R^1,R^3,Me = methyl; R^2,R^4(CH=CH$_2$) = vinyl.

tion called hyperbilirubinemia. The pigment is then deposited in the skin and other tissues. The pathological consequences become serious if the amount of bilirubin in the plasma exceeds the available binding sites. Under such conditions, bilirubin is believed to circulate in a free (unbound or loosely bound) form. It then becomes possible for bilirubin to cross the hematoencephalic barrier. Bilirubin is particularly toxic to the cells of the central nervous system, especially in the basal ganglia, where it may produce irreversible brain damage, known as bilirubin encephalopathy. The term *kernicterus* is used to designate the most severe form of damage, with deafness, cerebral palsy, or death. Kernicterus can be established only at autopsy by the bilirubin staining (yellow staining) of basal nuclei.

Severe jaundice of the newborn must be treated promptly in order to prevent brain damage. *Phototherapy, i.e., irradiation of the infant with visible light, has proved to be an effective treatment.*

17.1.2 Prognostic Index of Bilirubin Encephalopathy

Bilirubin concentration in plasma (Figure 17.1B) is a deceptive guide to the prevention of brain toxicity, which can occur with plasma bilirubin levels as low as 15 mg / liter in full-term infants, and 90–100 mg / liter (or less) in sick, immature infants.

17.1.3 The Neurotoxic Form of Bilirubin

Unconjugated bilirubin (not its albumin-conjugated form) can cross the blood–brain barrier and become attached to phospholipids in the neurons, leading to cell damage. It is thought that bilirubin enters the brain cells, not because of its lipophilic properties, but because of its

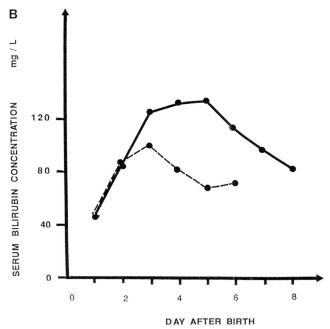

FIGURE 17.1B The effect of continuous phototherapy on the serum bilirubin concentrations of 35 neonates with physiologic bilirubinemia (---), compared with 37 matched control infants having the same condition but not treated (—). Mean peak bilirubin concentrations were lower and declined earlier in the treated group. Revised from Sisson, and Vogl (1982), p. 489.

attachment to membranes, as exemplified by the mitochondrial membranes. Reserve-albumin binding capacity and the saturation index of albumin are reliable guides to the effectiveness of therapy.

Very small amounts of blood can be studied by fluorometry. Concentrations of bound bilirubin and total bilirubin, bilirubin binding capacity of the blood, the partition of bilirubin among blood components, and the photoisomers of bilirubin in blood are determined in virtually one operation.

17.1.4 Bilirubin Photoproducts

The photoproduct described in Gunn rats is a true isomer of bilirubin IXa. Two photoisomer pairs produced anaerobically and excreted in rat bile are formed from irradiation of unconjugated bilirubin IXa. The photoisomerization is rapidly reversible (Figure 17.1A).

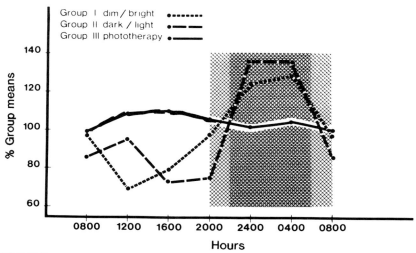

FIGURE 17.2 Under cycled light (both dim/bright light and total darkness/light) a circadian rhythm of plasma growth hormone is noted. With continuous phototherapy, the circadian rhythm is obliterated. From Sisson and Vogl (1982), p. 501. With permission of Plenum, New York.

17.1.5 Potential Hazards of Phototherapy

There are potential hazards associated with the kind of intense visible light irradiation used in the phototherapy of neonatal bilirubinemia:

1. A detrimental effect on platelet number.
2. Hemolysis following phototherapy.
3. Inhibition of erythrocyte glucose-6-phosphate dehydrogenase and glutathione reductase (indirectly by the reduction of riboflavin in red cells).
4. DNA cross-linkage and chromatid breakage as found in embryonic cultured mouse tissues.

In jaundiced infants the administration of phototherapy results in abnormal lactose tolerance tests. Phototherapy can cause lactase deficiency either directly on the intestinal lumen or, more plausibly, as an inhibitory influence of unconjugated bilirubin and/or its photoisomers on lactase production.

The circadian rhythm of human growth hormone (Figure 17.2), which is observed under cycled light, is eliminated with constant phototherapy (420–470 nm, 24 $\mu W/cm^2 \cdot nm$). Other biorhythms affected or distorted by phototherapy are blood calcium and glucose levels. Within a few hours of cessation of phototherapy, these abnormalities disappear.

In Gunn rat livers perfused with bilirubin (200 mg/l), 1 hr after irradiation at 420–470 nm, there was an almost twofold increase of cytochrome P450 and cytochrome b_5, i.e., there was a light-induced enhancement of drug metabolism by hepatic cells. Both bilirubin and its photoadducts have an inhibitory effect on immunoresponse.

17.1.6 Bronze Babies

The "bronze baby" syndrome is a complication of phototherapy. An intense grey–brown discoloration of the skin, serum, and urine, accompanied by anemia, occurs when phototherapy is used to reduce hyperbilirubinemia.

In this syndrome the spectral absorption curve of the serum shows maxima at 415 and 460 nm. The 415-nm maximum is attributed to free hemoglobin resulting from intravascular hemolysis or erythrocyte trauma from capillary sampling. The 460-nm peak is consistent with albumin-bound serum bilirubin. When a synthetic serum is prepared by addition of hemoglobin and bilirubin to an albumin solution in phosphate buffer, to yield the same hemoglobin/bilirubin concentrations as in the patient's serum, subtraction of the absorbancies of the synthetic serum from the absorbancies of patient's serum yields increased absorption between 300 and 500 nm, without distinct peaks. Normally the photodecomposition products of bilirubin formed *in vivo* are rapidly excreted, primarily in the bile and stool. In photoirradiated Gunn rats the pigments recovered from the bile exhibit a chromatographic behavior different from those recovered when bilirubin is photooxidized *in vitro* and subsequently injected.

The photodecay products of bilirubin show no toxicity, but the pigments of the bronze babies are toxic. The sera of infants exhibiting the bronze baby syndrome show absorption and emission spectra typical of porphyrins. Extraction and chromatographic separation of the porphyrin pigments show the presence of Cu^{2+}-uroporphyrin, Cu^{2+}-coproporphyrin, and Cu^{2+}-protoporphyrin.

Visible-light irradiation of synthetically prepared and bilirubin-added Cu^{2+}-porphyrins leads to the appearance of a brown color in the solution, coincident with an increase in absorbance in the near-ultraviolet and red spectral regions and a gradual decrease in the concentration of porphyrins. The same pattern occurs on irradiation of sera of infants with bronze baby syndrome. The pattern is intensified by addition of Cu^{2+} to the sera (it is likely that Cu^{2+} binds to ceruloplasmin).

These results suggest that the bronze color is due to the products of bilirubin-sensitized degradation of copper porphyrins. *The bronze baby syndrome is quite rare because its manifestation requires the simultaneous occurrence of high serum levels of porphyrins, of copper, and of bilirubin.*

In 13 cases of bronze baby syndrome produced by phototherapy, hepatic dysfunction was present in all of the infants. The infants were well, after their initial condition had improved. Bronzing disappeared within 2 months in all but one infant, but even with normalization of the hepatic function, spectroscopy demonstrated the presence of the pigment(s) at 1 year. Normal development was observed. The bronze color therefore seems to be a side complication brought out by phototherapy.

A case report of kernicterus (bilirubin encephalopathy associated in the worst cases with blindness and cerebral damage) occurring in association with a low bilirubin level in an infant with bronze baby syndrome suggests the possible toxicity of the accumulated pigment, but there was no bronze discoloration of the brain, and the bronze pigment in the skin was probably not the cause of the kernicterus.

17.2 THE PHOTOCHEMISTRY AND PHOTOBIOLOGY OF VITAMIN D

17.2.1 The Discovery of Vitamin D and Its Mechanism of Action

The first scientific description of a case of rickets appeared in 1645, by Glisson and Whistler. The disease reached epidemic proportions in Northern Europe and the United States with the urbanization of society and the consequent appearance of pollutants in the air, especially in areas of low-incident sunlight. Sir Edward Mellanby succeeded in reproducing the disease in dogs fed only oatmeal and maintained in the absence of sunlight. Because he could cure the disease with cod liver oil, known to contain the fat-soluble vitamin A, he concluded that the antirachitic activity of cod liver oil was due to vitamin A. McCollum demonstrated that the vitamin A activity of cod liver oil could be destroyed, leaving behind the antirachitic activity. The latter activity was attributed to another accessory food substance, vitamin D.

The confusion as to why both cod liver oil and UV light were effective in preventing rickets was resolved when it was demonstrated that *UV light induced the antirachitic vitamin* in fat-soluble portions of foods and in skin. Vitamin D_2 (a derivative of ergosterol, but not its photoproduct) was isolated and its structure was elucidated.

Windaus produced dehydrocholesterol synthetically and demonstrated its conversion to *another antirachitic vitamin (i.e., vitamin D_3)*. The function of vitamin D is to elevate plasma calcium and phosphorus to levels that support mineralization of bone. Vitamin D stimulates an active calcium transport process in the small intestine. *Rather than vitamin D directly functioning at the bone site, mineralization of bone in response to vitamin D results from elevation of blood plasma calcium and phosphorus concentrations.* Vitamin D functions neither in bone growth nor in mineralization if plasma calcium and phosphorus levels are maintained in the normal range. Instead, vitamin D plays an important role in bone resorption aspects of modeling and remodeling. Vitamin D, rather than stimulating the direct deposition of calcium in bone, actually stimulates the mobilization of calcium, a process that requires the presence of parathyroid hormone, with both the hormone and vitamin D working in conjunction to promote the mobilization of mineral from bone. By sequential hydroxylation in the liver and kidney, *vitamin D is converted to a hormone,* 1,25-dihydroxyvitamin D_3, that facilitates the movement of calcium and phosphorus in the appropriate directions, in order to prevent rickets and osteomalacia and to provide for normal regulation of calcium and phosphorus.

17.2.2 Nomenclature of Vitamin D

There are two active forms of vitamin D: vitamin D_2 and vitamin D_3. The term "vitamin D_1" is not much in use anymore. At first, when vitamin D was isolated by the irradiation of ergosterol, it was designated vitamin D_1. However, it was quickly realized that the product obtained was an impure mixture, and the term was dropped.

17.2.2.1 Vitamin D_2

When final purification of vitamin D from its parent ergosterol (a yeast sterol) was achieved, the product was named *ergocalciferol,* i.e., vitamin D_2 (Figure 17.3). Initially it was believed that vitamin D_2 was identical to the vitamin present in fish liver oils and was produced in the skin following exposure to sunlight. However, biologic evidence was presented that vitamin D_2, isolated from irradiated ergosterol solutions, was not the same compound that was isolated from UV-irradiated solutions of cholesterol.

UV-irradiated solutions of cholesterol have marked biologic activity in chickens, whereas irradiated solutions of ergosterol possess a minimum activity in this species.

Vitamin D_2 is not a direct photoproduct of ergosterol. Irradiation of ergosterol (*pro*vitamin D_2) at temperatures below 10°C results in a compound called *pre*vitamin D_2, which is thermally labile and undergoes a *thermally induced isomerization* of the triene system to form vitamin D_2.

17.2.2.2 Vitamin D_3

The parent compound of vitamin D_3 is 7-dehydrocholesterol (provitamin D_3; Figure 17.3). On UV irradiation, a 6,7-*cis*-hexatriene derivative, previtamin D_3, is obtained by ring opening. *Previtamin D_3 then undergoes a nonphotochemical, temperature-dependent isomerization to the more thermodynamically stable 5,6-cis isomer, vitamin D_3.* Vitamin D_3 is made in skin

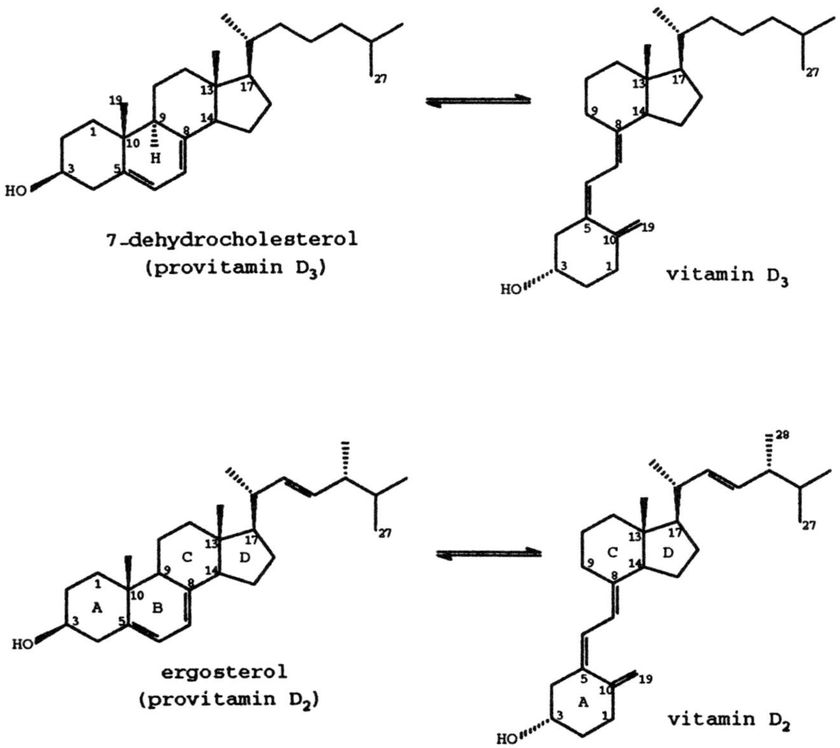

7-dehydrocholesterol
(provitamin D₃)

vitamin D₃

ergosterol
(provitamin D₂)

vitamin D₂

FIGURE 17.3 The structures of the parent compounds (7-dehydrocholesterol, ergosterol) and of vitamins D₃ and D₂.

during exposure to the sun. 7-Dehydrocholesterol, rather than ergosterol, is the parent compound present in the skin. Vitamin D₃, also called cholecalciferol, has equal antirachitic activity in the chick and the rat. *Vitamin D₃ is identical in structure to the vitamin D isolated from fish liver oils.*

When the designation vitamin D is used without a subscript, it means vitamin D₃ and/or vitamin D₂ interchangeably.

17.2.3 Chemistry of the Vitamin D Compounds

Among the nutritional forms of vitamin D, the steroidal side chains are ergosterol for vitamin D_2, cholesterol for D_3, 22,23-dihydroergosterol for D_4, sitosterol for D_5, and stigmosterol for D_6. Of these structures, the most important are D_2 and D_3, which are prepared from their respective 5,7-diene sterols. Vitamin D_3 is the natural form of vitamin D produced in skin. Vitamin D_2 is of historical importance as the cheapest source of vitamin D available and is still used to fortify foods for humans and domestic animals. Vitamin D_4, a chemical curiosity, has less than one-tenth the biological activity of D_3 in birds, and is only three-fourths as active as D_3 in mammals. Vitamins D_5 and D_6 are also chemical curiosities, not significant for use by humans.

All vitamin D compounds have a UV absorption maximum (see Chapter 5) at 265 nm, a minimum at 228 nm, and a molar extinction coefficient of 18,200 M^{-1} cm^{-1}. The intense UV absorption of vitamin D_3 renders it labile to light-induced isomerization. Its triene system is easily protonated, resulting in isomerization and giving rise to a compound known as isotachysterol, which is essentially devoid of biological activity. The lability of this triene structure has markedly limited the chemical approaches to modification of this molecule.

17.2.4 Biosynthesis of Vitamin D in Skin

The study of cholesterol biosynthesis shows that 7-dehydrocholesterol (provitamin D_3) accumulates in skin in abundant quantities (Figure 17.4). This compound had been previously isolated and identified as the intermediate in the photolysis of the $\Delta^{5,7}$-diene sterols that are converted to vitamin D_3 by UV irradiation. Absorption by 7-dehydrocholesterol of a quantum of light leads to an excited singlet state that may undergo ring opening between C-9 and C-10 to yield a 6,7-*cis*-hexatriene derivative, i.e., the intermediate previtamin D_3. Previtamin D_3 slowly equilibrates (i.e., thermal equilibration) to an equilibrium mixture, which at body temperature is 96% vitamin D_3 (the thermodynamically more stable 5,6-*cis* isomer of previtamin D_3) and 4% previtamin D_3. Previtamin D_3 can also absorb UV radiation and undergo either (1) ring closure, to form the parent compound (7-dehydrocholesterol) or its stereoisomer (lumisterol), or (2) isomerization of its triene chromophore to form tachysterol. On continuous irradiation, a quasi-photoequilibrium state is reached.

Vitamin D_2 is not a photoproduct of ergosterol. Irradiation of ergosterol at temperatures below 10°C results in a compound called previtamin D_2. Conversion of previtamin D_2 to vitamin D_2 is not the result of UV irradiation. The previtamin is thermolabile and undergoes a thermally induced isomerization of the triene system to vitamin D_2; this reaction has been characterized as an intramolecular hydrogen shift. When ergosterol is exposed to UV irradiation for prolonged periods, the resultant reaction mixture is devoid of biologic activity. There are a remarkable number of photoproducts of previtamin D_2 and tachysterol, called toxysterols—at least 13 reported so far. Under continuous irradiation, vitamin D_2 isomerizes to photoproducts called suprasterols, of which six major representatives are known. Both toxisterols and suprasterols are

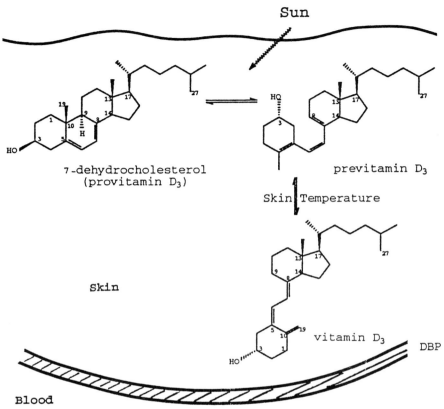

FIGURE 17.4 Biosynthesis in the skin of vitamin D_3 from 7-dehydrocholesterol; DBP, vitamin D binding protein.

believed to be devoid of biologic activity. To reach an equilibrium of 96% vitamin D_3 and 4% provitamin D_3 requires somewhere in the neighborhood of 48 hr at body temperatures.

The vitamin D binding protein (DBP), a 52,000-Da globulin, will not bind previtamin D_3 but will bind vitamin D_3. Thus, the previtamin will remain in the skin until it is converted to vitamin D_3, an especially attractive mechanism for the slow feed-in of vitamin D_3 into its metabolic and functional system.

The hormonal form of vitamin D_3, 1,25-$(OH)_2D_3$, accumulates in the nuclei of the malpighian layer of the skin. There is a receptor protein for this compound in such cells. It has been speculated that 1,25-$(OH)_2D_3$ might stimulate the production or accumulation of 7-dehydrocholesterol in the cells of the malpighian layer.

Of particular interest is the large accumulation of vitamin D_3 and to some extent D_2 in the livers of tuna and shark. Fish liver oils have been used for many years to treat and prevent rickets. Arguments exist that vitamin D is accumulated in the fish liver from the environment as a result of the food chain. The biogenesis of vitamin D in fish is not completely understood. There is evidence that 25-OH-D_3 and 25-$(OH)_2D_3$ may be formed in certain fish.

17.2.5 Functional Metabolism of Vitamin D

Following intravenous or oral administration of vitamin D_3 to animals, some 60–80% accumulates in the liver, but metabolites of D_3 do not remain in the liver even to a slight degree. Vitamin D_3 is hydroxylated on C-25 to 25-OH-D_3, predominantly in the liver, with some 25-hydroxylation taking place in intestine and kidney. *All known metabolism of vitamin D must progress through its 25-hydroxylated derivative.* A specific vitamin D 25-hydroxylase is located in the endoplasmic reticulum of hepatocytes. The 25-hydroxylase reaction has a K_m of 10^{-8} M for 25-OH-D_3 and requires NADH, a cytoplasmic factor, and molecular oxygen.

25-OH-D_3 must be metabolized further before it can carry out the functions of vitamin D_3 in the intestine, kidney, and bone. Together with the plasma binding protein, the 25-hydroxy derivative is transported to the kidney. The subsequent reaction takes place exclusively in the kidney of nonpregnant mammals, where 25-OH-D_3 is further hydroxylated in the 1α position to produce 1α-25-dihydroxyvitamin D_3 [1,25-$(OH)_2D_3$]. Cell-free preparations of kidneys from birds and mammals produce the dihydroxy vitamin, but anephric animals and mammals do not produce detectable amounts of this compound *in vivo,* when given injections of radioactive 25-OH-D_3. 25-Hydroxy vitamin D 1α-hydroxylase can be also found in placental tissue. This site is likely to produce 1,25-$(OH)_2D_3$ required by the fetus, but that cannot be proved at this point.

1,25-$(OH)_2D_3$ is metabolized very rapidly, whereas 25-$(OH)_2D_3$ is metabolized much more slowly, although more rapidly than vitamin D. The dihydroxy form is 10 times more active than vitamin D_3, and the monohydroxy form is twice as active as vitamin D_3. Anephric animals do not show a response to a physiologic dose of 25-OH-D_3, whereas 1,25-$(OH)_2D_3$ produces a response in the absence or presence of kidney. *Because 1,25-$(OH)_2D_3$ is produced in the kidney and has some of its functions in bone and intestine, it must be regarded as a hormone.*

The 1α-hydroxylase reaction that converts the monohydroxy vitamin D_3 into the dihydroxy form is perhaps the most significant enzyme in the metabolism of vitamin D. It is this hydroxylase that is regulated by the need for calcium or phosphorus and must therefore be regarded as the major regulated step in the metabolism of the vitamin.

In humans, 1,25-$(OH)_2D_3$ has a lifetime of approximately 2–4 hr in plasma and perhaps somewhat longer in target tissues. Thus, exogenous 1,25-$(OH)_2D_3$ must be administered very slowly. A possible alternative being experimentally studied is the use of minipumps to deliver the compound in small but steady concentrations.

A major metabolic product of 1,25-$(OH)_2D_3$ is calcitroic acid, a C_{23} carboxylic acid derivative found in the bile as a major excretory species of the active form of vitamin D. The first probable step in calcitroic acid formation is 23-oxidation of 1,25-$(OH)_2D_3$. Calcitroic acid is essentially without biological activity either in its free form or as its methyl ester. It is probably an inactivation product of the potent hormone.

Other pathways of vitamin D metabolism utilize 25-OH-D_3 as the substrate. A minor pathway is 26-hydroxylation to produce 25,26-$(OH)_2D_3$. 26-Hydroxylation is regarded as an interesting biological curiosity.

23-Hydroxylation with formation of 23S,25-$(OH)_2D_3$ produces an intermediate in the biosynthesis of a major metabolite of vitamin D, i.e., 25R-hydroxy vitamin D_3 26,23S-lactone. 23-Hydroxylation is probably followed by 26-oxidation, lactal formation, and oxidation to the lactone. This metabolite can be found in human plasma, especially under conditions of high vitamin D dosage. Neither its precursor nor the final lactone possesses measurable biological activity.

Another major metabolite of vitamin D found in human and animal plasma is 24R,25-$(OH)_2D_3$, which has less biological activity than its precursor, 25-OH_2, in the known systems that are responsive to vitamin D_3. The enzyme that catalyzes 24-hydroxylation is found in kidney, intestine, and cartilage. The hydroxylase is a cytochrome P450-dependent system. The 24-hydroxylation does not play a significant role in intestinal calcium transport or in the mobilization of calcium form bone, but it has been suggested that it is required for mineralization of bone, suppression of parathyroid hormone secretion, cartilage metabolism, and embryonic development.

It is noteworthy that when 1,25-$(OH)_2D_3$ is used as a sole source of vitamin D_3, delivered by a minipump parenterally, it produces normal bone mineralization through one entire generation. Thus, there is no evidence to support the idea that 24-hydroxylation plays a significant role in the function of vitamin D. Its absence produces no abnormalities or pathologies.

The conclusion is that, based on known pathways of vitamin D metabolism, *only 25-hydroxylation followed by 1-hydroxylation can be regarded as activation pathways.*

17.2.6 Photobiology of Vitamin D

The only differences between provitamin D_2 and provitamin D_3 (and their corresponding vitamins) are the double bond between carbons 22 and 23 and the methyl group on carbon 24. These small structural differences do not have a significant effect on the biological activity or metabolism of vitamin D_2 in humans.

The UV portion of the solar spectrum (290–315 nm) is responsible for the photochemical conversion of epidermal stores of 7-dehydrocholesterol (7-DHC) to previtamin D_3. Approximately 89% of the epidermal stores of previtamin D_3 reside in the stratum spinosum and stratum basale (malpighian layer). Although the dermis (i.e., sebaceous glands) and the epidermis contain about the same amount of 7-DHC per unit area, most of the UV radiation between 290 and 315 nm is absorbed by the epidermis, and hence very little previtamin D_3 is formed in the dermis.

Once previtamin D_3 is formed it immediately begins thermal equilibration to vitamin D_3. This process takes 2–3 hr to reach completion at physiologic temperatures in warm-blooded animals. The skin continues to synthesize and release vitamin D_3 into the circulation several days after exposure to sunlight.

Despite seasonal fluctuations of skin surface temperature by several degrees, *the previtamin–vitamin equilibration rate remains constant because it takes place at the dermo-epidermal junction, where the temperature is relatively constant.* Once vitamin D_3 is formed, the circulating vitamin D binding protein promotes its translocation from the dermo-epidermal junction into the circulation (Figure 17.4).

It has been speculated that heavily pigmented persons are protected from sun-induced vitamin D_3 intoxication because melanin efficiently absorbs the radiation that is responsible for vitamin D_3 production. It is further speculated that when people migrate away from the equator, those who are heavily pigmented are unable to make sufficient quantities of vitamin D_3 to maintain a healthy skeleton. The conclusion is that skin pigmentation regulates the transmission of solar UV radiation so that vitamin D photosynthesis is relatively constant and natural selection has favored black skin near the equator and light skin in areas distant therefrom.

Previtamin D_3 can either thermally isomerize to vitamin D_3 (without exposure to sunlight) or absorb a photon of UV radiation and isomerize to two biologically inert isomers, lumisterol and tachysterol. The amount of lumisterol and tachysterol that is formed in the skin depends on the spectral properties of radiation:

1. Up to 65% of 7-DHC may be converted to previtamin D_3 when the skin is exposed to narrowband UVB radiation (295 $\pm$5 nm).
2. A maximum of 20% of 7-DHC was converted to previtamin D_3 in a test group receiving simulated solar radiation.
3. Tachysterol is the major photoproduct of previtamin D_3 after narrowband irradiation, whereas lumisterol is the predominant isomer found in skin samples exposed to simulated solar radiation.

The reasons for these differences are in the power distribution of the radiation between 290 and 340 nm in the solar spectrum versus the absorption spectrum of 7-DHC and its photoisomers. The overlay of the curve of the action spectrum and that of the solar spectrum demonstrates the small portion of the solar UV spectrum that is involved with the formation of previtamin D_3.

The photoisomers lumisterol and tachysterol are not significantly bound by the vitamin D binding protein and they are not active in the stimulation of intestinal calcium transport or bone calcium mobilization. Probably these compounds remain in the skin to be sloughed off during the natural turnover of the skin, or to act as a chemical messenger to alert the epidermis that it is receiving excessive exposure to sunlight. *The unique interplay between thermal and ultraviolet energy is probably ultimately responsible for modulating the production of vitamin D in the skin.*

During the first 10–15 min of exposure to the sun at the equator during the early afternoon, a light-skinned Caucasian converts about 10–15% of his epidermal stores of 7-DHC to previtamin D_3. Previtamin D_3 is not increased by continued exposure to solar radiation, but is immediately converted to lumisterol and tachysterol as it is being generated from 7-DHC.

Bibliography

Cremer, R. J., Perryman, P. W., and Richards, D. H. (1958). Influence of light on the hyperbilirubinemia of infants. *Lancet* **1**, 1954.

DeLuca, H. F. (1984). The metabolism, physiology, and function of Vitamin D. *In* Vitamin D: Basic and Clinical Aspects" (R. Kumar, ed.), pp. 1–68. Martinus Nijhoff, Boston.

Holick, M. F. (1984). The photobiology of vitamin D_3 in man. *In* "Vitamin D: Basic and Clinical Aspects" (R. Kumar, ed.), pp. 197–216. Martinus Nijhoff, Boston.

Holick, M. F., MacLaughlin, J. A., Parrish, J. A., and Anderson, R. R. (1980). The photochemistry and photobiology of vitamin D_3. *In* "The Science of Photobiology" (J. D. Regan, and J. A. Parrish, eds.), pp. 195–218. Plenum, New York and London.

Jori, G., Reddi, E., and Rubaltelli, F. F. (1982). Bronze baby syndrome: Evidence for increased serum porphyrin concentration. *Lancet*, May 8, p. 1072.

Kopelman, A. E., Brown, S. R., and Odell, G. B. (1972). The "bronze" baby syndrome: A complication of phototherapy. *J. Pediatr.* **81,** 466–472.

Pettenazzo, A., Reddi, E., Granati, B., Camurri, S., Zaramella, P., and Rubaltelli, F. F. (1985). Cholestasis induced in Gunn rats as an experimental model of bronze baby syndrome. *In* "Primary Photo-Processes in Biology and Medicine" (R. V. Bensasson, G. Jori, E. J. Land, and T. G. Truscott, eds.), pp. 421–424. Plenum, New York and London.

Sisson, T. R. C. (1982). Advances in phototherapy of neonatal bilirubinemia. *In* "Trends in Photobiology" (C. Hélène, M. Charlier, T. Montenay-Garestier, and G. Laustriat, eds.), pp. 339–348. Plenum, New York and London.

Sisson, T. R. C., and Vogl, T. P. (1982). Phototherapy of hyperbilirubinemia. *In* "The Science of Photomedicine" (J. D. Regan and J. A. Parrish, eds.), pp. 477–509. Plenum, New York and London.

Tan, K. L., and Jacob, E. (1982). The bronze baby syndrome. *Acta Paediatr. Scand.* **71,** 409–414.

18

Photochemotherapy (PUVA Therapy) and UVB Phototherapy

Photochemotherapy combines irradiation therapy (phototherapy), generally UVA irradiation, with a topically or orally applied photosensitizer (e.g., a psoralen). The combination therapy, psoralen + UVA, is called PUVA.

18.1 LIGHT TREATMENT OF DISEASE: PSORIASIS, A KEY EXAMPLE

18.1.1 Psoriasis: The Disease

Psoriasis is a disease of unknown etiology, characterized by increased epidermal cell proliferation. It is identified by the symptomatic triad of hyperkeratosis (thickening of the stratum corneum), parakeratosis (presence of nuclei in the actively proliferating cells of the stratum corneum, which are normally anucleated), and acanthosis (thickening of the stratum spinosum in the epidermis) (Figure 18.1). The individual lesions are raised, red, circumscribed scaling plaques that tend to occur symmetri-

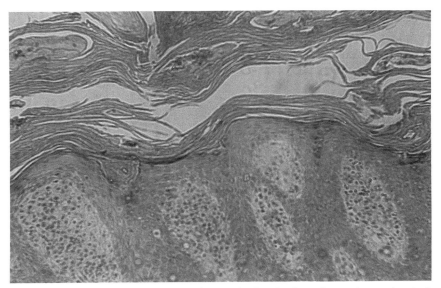

FIGURE 18.1   Histopathological image of psoriasis showing hyperkeratosis, parakeratosis, and acanthosis. Courtesy of Dr. George Elgart, Department of Dermatology, School of Medicine, University of Miami. Photomicrograph by Cahide Kohen.

cally over the body, with a predilection for elbows and knees. However, lesions may occur at any site. Psoriasis is associated with chronic, relapsing, variable erythrosquamous lesions.

Because of the typical appearance of the psoriasis plaques (Figures 18.2 and 18.3), the disease has been called (in the Turkish language) *sedeff hastalighi* (mother-of-pearl disease).

In United States the prevalence of psoriasis is 1%. In the Faeroe islands (61° north latitude) it is 2.8%, which may be explained by inbreeding and the absence of the "sunlight effect" that minimizes the symptoms of psoriasis. The incidence is low in West Africans, the Japanese, and North American blacks, and is absent in North and South American Indians. The disease has been observed in newborns and in a 108-year-old patient, although the peak incidence is in the third decade.

18.1.2 The Genetics of Psoriasis

There is an increased incidence of psoriasis among relatives of affected probands, and in the offspring of matings with one or both parents affected. There is a high concordance rate in monozygotic twins.

In psoriasis, with the increased antigenic human leucocyte antigen

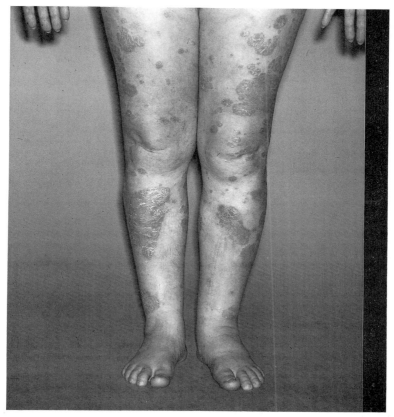

FIGURE 18.2   Psoriasis plaques in a patient before PUVA treatment. Courtesy of Professor Louis Dubertret, Department of Dermatology, Hopital Saint-Louis, Paris, France.

(HLA)-B17 complex, the onset of the disease is earlier and the symptoms are more serious (50% of immediate family members are affected).

18.1.3 Mediators in the Pathogenesis of Psoriasis

Leukotrienes (i.e., LTB4, a chemotactic factor for neutrophils) produce an infiltration of inflammatory cells and epidermal hyperplasia. 5-Lipoxygenase, which generates leukotrienes, is grossly increased in psoriasis, but multiple topical applications of LTB4 fail to simulate psoriatic lesions. Cytokines, growth factors, and proteases may be involved. Interleukin-8 is a chemoattractant of neutrophils and a stimulant of keratinocyte growth *in vitro*. Interleukin-8 stimulates neutrophils to produce LTB4.

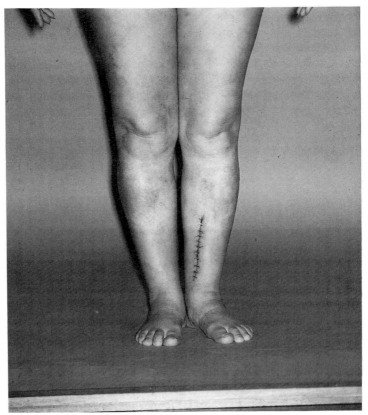

FIGURE 18.3 Same patient as Figure 18.2, post-PUVA therapy. Courtesy of Professor Louis Dubertret.

18.1.4 Characteristic Features of Psoriasis

The characteristic features of psoriasis (Figure 18.2) are (1) lesions with sharply demarcated borders, (2) noncoherent silvery scales and a glossy homogeneous erythema, (3) the Auspitz sign (i.e., when erythrosquamous scales are scraped, small blood droplets appear within seconds), and (4) following a nonspecific irritation, the appearance of lesions in areas not previously involved. The frequency of the Koebner effect (see definition in Chapter 16) increases when psoriasis requires multiple therapy to control the disease at an early age.

In contrast to the clear-cut hyperkeratosis/parakeratosis/acanthosis triad of psoriasis, in parapsoriasis there is slight hyperkeratosis, spotty parakeratosis, mild acanthosis, spongiotic dermatitis, and perivascular dermal lymphocytic infiltration.

18.1.5 UV Phototherapy of Psoriasis

18.1.5.1 UVB Phototherapy

18.1.5.1.1 Mechanism of Action   With UVB phototherapy, the skin usually returns to normal after 10–35 treatments. Remission may last from days to years, but the disease almost always eventually recurs. Exposure of skin to UV radiation causes a transient decrease in DNA, RNA, and protein synthesis. A rebound increase in synthesis of these macromolecules is accompanied by a proliferative state.

Psoriatic lesions are more sensitive to UV damage, perhaps because they are metabolically more active and replicate more actively. The therapeutic effect results from decrease of macromolecular synthesis in all psoriatic cells or there is a selective inhibition or killing of a smaller population of highly proliferative cells.

UV radiation may benefit psoriasis by (1) acting on lymphocytes and polymorphonuclear leukocytes, which take part in the pathophysiology of the disease; (2) photodegradative alteration of mediators (e.g., LTB4) required to maintain the hyperproliferative state; topical UVB-irradiated (290–320 nm) LTB4 causes a marked reduction in cutaneous erythema and transepidermal polymorphonuclear leukocyte (PMN) migration when compared with unirradiated and UVA-irradiated LTB4; and (3) altering the epidermal cell recruitment process from a resting to a proliferative phase; also, altering differentiation or action as a gene regulator or deregulator.

18.1.5.2 Action Spectrum for UVB Phototherapy of Psoriasis

The shorter wavelength UV radiation (254–280 nm) is far more erythemogenic than it is therapeutic (Figure 18.4), due to UVB absorption and scattering properties of psoriatic skin. In psoriasis, the proliferative compartment at the bottom of the epidermis is thicker than in normal skin. There is also an increased thickness of the stratum corneum. To reach the basal layer proliferative compartment, wavelengths shorter than 290 nm face two obstacles: (1) absorption by epidermal proteins as transmission decreases exponentially with thickness; (2) optical scattering, inversely related exponentially to wavelength, so that the shorter wavelengths have longer pathlengths, thus a greater chance of being absorbed on the way to the proliferative compartment.

Addition of the markedly erythemogenic UVC decreases the tolerance of normal skin to the more therapeutic UVB. UVC may have a net proliferogenic influence on the abnormal cells within the psoriatic plaques.

Daily doses of 1 minimal erythemal dose (MED) at 300–305 nm and of less than 1 MED at 313 nm are adequate for treatment of psoriasis.

Remittance spectra of psoriatic plaques and normal skin have been measured using an integrating sphere spectrophotometer before and immediately after application of mineral oil (refractive index, $n_D = 1.48$), isopropyl myristate ($n_D = 1.44$), and water ($n_D = 1.33$). A marked, broad-spectrum decrease in remittance of psoriatic plaques (originally exhibiting multiple air–skin interfaces) occurs within seconds after application of mineral oil or isopropyl palmitate.

18.1.5.3 The Goeckerman Regimen

This therapy for psoriasis is a combination of topical application plus UVB irradiation: crude coal tar applied to the skin plus a suberythemal dose of UVB. This treatment is superior to UVB radiation or tar alone. The topical agents may simply enhance transmission of UV radiation in the skin rather than having a direct effect. Liquor carbonis detergens, dithranol, or tar derivatives in gel base are more convenient to apply than crude oil tar.

The results of the therapy depend on treatment frequency, extent of body involvement, type of psoriasis, and patient motivation.

18.1.5.4 Action Spectra for Delayed Erythema: UVA and PUVA Photochemotherapy

UVA irradiation is markedly less erythemogenic than UVB irradiation (by three orders of magnitude). At doses close to 1 MED of irradiation at 334 or 365 nm, no significant improvement of psoriasis lesions is obtained. Using larger exposure doses (50–300 J/cm^2), primarily 365 nm is effective in clearing small psoriasis lesions. The MED for UVA irradiation is 20–100 J/cm^2.

The use of such large doses is impractical and possibly unsafe for patients undergoing whole-body irradiation. The addition of UVA spectra to UVB phototherapy in an exposure dose ratio of 20:1 does not noticeably improve UVB phototherapy. The selection of the most appropriate exposure source is based on (1) the action spectrum of acute phototoxicity of normal skin, (2) optical properties of the skin, and (3) spectral and geometric properties of available exposure sources (Figure 18.4).

Longer wavelengths are therapeutic at or slightly below erythema-producing doses; they penetrate more deeply into tissue and are more likely to affect abnormal blood vessels and cellular infiltrates, which is possibly important to the pathophysiology of psoriasis.

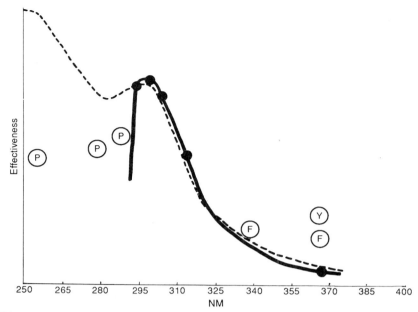

FIGURE 18.4 The proposed action spectrum for phototherapy of psoriasis obtained by plotting the effectiveness (reciprocal of threshold dose) on a log scale versus the wavelength in nanometers. The dashed line is the human erythema action spectrum. The solid line is the action spectrum for phototherapy of psoriasis based on several studies. Effectiveness (cm^2/J) is plotted on a logarithmic scale. The intercept of the human erythema spectrum with the ordinate corresponds to 10^2 cm^2/J; the effectiveness at the origin is 10^{-2} cm^2/J. (●) The reciprocal of the lowest daily dose that clears psoriasis; (○) the reciprocal of the highest daily exposure doses tried and found not to be effective. At the wavelengths indicated by the open circles and letters P, Y, F, the action spectrum of psoriasis must fall below these points. From Parrish (1982), p. 517, with permission of Plenum, New York.

Polychromatic radiation with a maximum output between 300–320 nm may represent an important compromise region for practical and effective phototherapy of psoriasis. This approach, called "selective ultraviolet phototherapy" (SUP), is more like solar radiation and utilizes wavebands that are transmitted deeper into the psoriatic tissue. In selected cases, SUP is as effective as PUVA, but PUVA is more effective for difficult cases of psoriasis (better penetration of UVA wavelengths compared to the SUP waveband). In all cases, PUVA is more effective in clearing psoriasis than UVA alone.

18.1.5.5 Heliotherapy of Psoriasis at the Dead Sea

At 1200 feet below sea level (location of the Dead Sea), solar radiation must travel the 1200-foot distance through the atmosphere before it reaches the earth. This results in a net spectral shift to longer wavelengths, appropriate for heliotherapy. Other advantages of heliotherapy at the Dead Sea are (1) the higher aerosol content of the atmosphere near the Dead Sea, accounting for biologically significant shifts in the terrestrial solar spectral power distribution, and (2) ease of alteration of optical properties of psoriasis plaques by soaking in the seawater, which may remove UV-absorbing substances from the skin.

18.1.6 PUVA Therapy of Psoriasis

PUVA is a very effective treatment of psoriasis (Figure 18.3). The action spectra for delayed erythema of normal skin and for phototherapy using PUVA are closely aligned in the UVA region. The most widely used psoralen derivative is 8-methoxypsoralen, also known as 8-MOP or methoxsalen. The quantity of UVA radiation required to elicit a minimal phototoxic dose is lowest between 1 and 3 hr after the ingestion of psoralens. Cutaneous photosensitization appears to last about 8 hr. There appears to be a general reciprocal relationship between psoralen serum level and UVA dose required to elicit a phototoxic response: peak serum levels are associated with the maximal period of sensitization following an oral psoralen dose. Metabolism and excretion of psoralens are altered in patients with severe impairment of hepatic or renal function.

A few minutes of sun exposure may result in blistering following topical application of 0.1 ml/5 cm² skin surface [8-MOP or trimethylpsoralen (TMP) in 0.1% solution]. Other furocoumarins used are 5-methoxypsoralen and certain alkyl-substituted angelicins (Figure 18.5).

When UVB therapy is combined with PUVA, the total cumulative UVA dose required is reduced by more than half, and the cumulative UVB is reduced to one-fifth of that required for UVB alone. Thus combined therapy may result in less cumulative phototoxic (or carcinogenic) insult to skin.

18.1.6.1 Psoralen–DNA photochemistry

18.1.6.1.1 General Mechanisms   The psoralens are known to intercalate between base pairs. Subsequent exposure to UVA radiation (320–

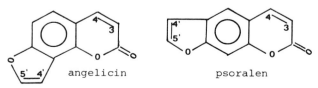

angelicin   psoralen

FIGURE 18.5 Chemical structures of angelicin (monofunctional furocoumarin) and psoralen (bifunctional furocoumarin). After Dall'Acqua (1982), p. 268. With permission of Plenum, New York.

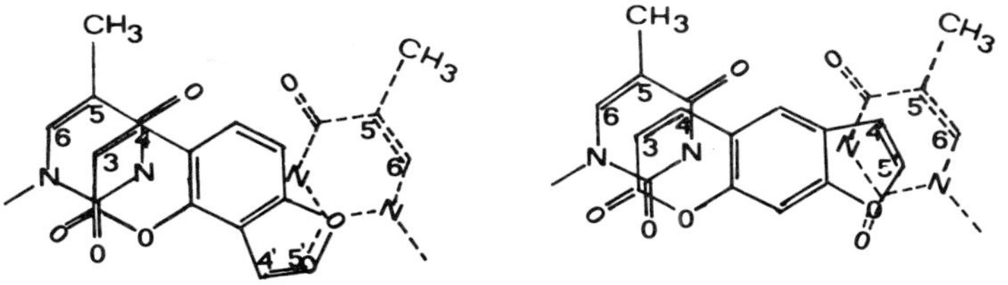

FIGURE 18.6 Representation of angelicin and psoralen intercalated. The angular structure of angelicin creates steric hindrance for biadduct formation, which, in contrast, is allowed by the linear structure of psoralen. When angelicin is intercalated in duplex DNA, only one of its two photoreactive sites can be aligned with the 5,6 double bond of one of the two pyrimidine bases involved in the intercalation. From Dall'Acqua (1982), p. 270. With permission of Plenum, New York.

400 nm) leads to covalent binding of the psoralen to DNA or RNA. In duplex DNA linear psoralens (Figure 18.6) intercalate in the following manner (see also Chapter 4):

1. The first step of the photoreaction after intercalation involves formation of a monofunctional adduct of psoralen with pyrimidine bases (thymine, cytosine) via the 4'-5' double bond of the furane ring of furocoumarin. The adduct formed absorbs at 360 nm and is fluorescent.

2. If two pyrimidines are adjacent at appropriate distances and on complementary strands, on subsequent absorption of a second photon by the monoadduct, the 3,4 double bond of the psoralen molecule forms a covalent linkage with the 5,6 double bond of the pyrimidine on the opposite strand, thus cross-linking the two strands of the DNA helix.

Photoaddition of psoralens to epidermal DNA leads to inhibition of DNA synthesis and cell division, and thus may account for their therapeutic effect in psoriasis. Psoralens can also bind to cytoplasmic structures, leading to photosensitization of plasma membrane components. The epidermal growth factor receptor and antigenic determinants are possible targets.

Use of the 360-nm excited fluorescence of psoralen adducts for *in situ* localization in cells may be misleading. UVA radiation enhances formation of Schiff bases (aminoiminopropenes), which emit in almost the same spectral region and may be improperly identified as psoralen adducts.

18.1.6.2 Monofunctional Psoralens

Compounds (Figure 18.5) forming only monofunctional adducts (Figure 18.6) have been developed to treat psoriasis. It is advocated that with photoactivation of a monofunctional agent, the photoadducts may be less mutagenic, especially if the target is not nuclear DNA but, rather, mitochondrial DNA. The risks of psoralen photochemotherapy can be minimized by utilizing those psoralens that preferentially form therapeutic rather than toxic photoproducts.

Angelicin (Figure 18.5) behaves as a pure monofunctional agent toward native DNA because of its angular structure (Figure 18.6). When angelicin is intercalated in duplex DNA, only one of its two photoreactive sites can be aligned with the 5,6 double bond of one of the two pyrimidines involved in the intercalation. One or more methyl groups have been introduced into the angelicin molecule in order to increase its photobinding capacity toward DNA.

The lesser mutagenicity of angelicin derivatives may be due to different enzymatic mechanisms of the DNA repair systems for removal of mono- and biadducts (as demonstrated in eukaryotic yeast). The yeast results are in accord with findings in cultured mammalian cells.

DNA interstrand cross-links, cell-kill, mutations, mitotic recombinations with crossing-over, carcinogenic effects, and direct phototoxicity are more evident with bifunctional than monofunctional furocoumarin. Using bifunctional furocoumarins, an efficient repair of monoadditions induced in nuclear DNA (increased survival) is accompanied by a low repair activity of lesions induced in mitochondrial DNA (induction of respiratory-deficient yeast, or *petite* mutants).

18.1.7 Management of PUVA Therapy and Risk/Benefit Ratio

18.1.7.1 Treatment Scheme and Light Dosimetry

18.1.7.1.1 Oral PUVA Therapy In psoriasis photochemotherapy, 8-methoxypsoralen (8-MOP) or 5-methoxypsoralen (5-MOP) is administered at a constant dose of 0.6–0.8 mg/kg in conjunction with a variable UVA dose. Approximately 2 hr after ingestion of 8-MOP or 5-MOP, UVA radiation is administered at the starting dose of 1 J/cm^2 (adjusted for skin type). The dose is increased stepwise in suberythematous amounts (2–4 treatments a week); 19–25 treatments are required. The resulting rapid pigmentation necessitates upward readjustment of the dose of light.

Overdosage leads to sunburn reaction, characteristically more delayed than with UVB radiation (24–48 hr posttreatment). Effects of therapy cannot be explained as being only due to DNA synthesis and mitosis arrest in response to biadducts of psoralen–DNA. However, PUVA appears to inhibit the chemotactic activity of psoriasis leukotactic factor, and may induce correction of immunologic abnormalities, such as depressed E-rosette formation by peripheral blood T lymphocytes.

18.1.7.1.2 Topical PUVA Therapy Topical psoralen photochemotherapy is effective but presents difficulties because of variations in percutaneous absorption of psoralen and the tendency of psoralen to cause irregular pigmentation around lesions.

Greater uniformity in topical application is obtained, and uneven pigmentation reduced, by utilizing a bath followed by exposure to UVA light (principally for the management of recalcitrant plaques, as a supplement to whole-body oral PUVA).

18.1.7.1.3 Combination Therapy Accelerated response to PUVA has been noted with combination therapies of PUVA and oral retinoic acid derivatives. The UVA dose at clearing is reduced by 30% or more with photochemotherapy in combination with retinoic acid. A combination therapy using PUVA and methotrexate has been required to treat pustular and erythrodermic psoriasis.

18.1.7.2 Skin Reactions and Side Effects of PUVA

The acute effects of PUVA on skin are (1) erythema, (2) melanogenesis, and (3) swelling, blistering, itching, pain, and desquamation (the UVB-induced sunburn reaction appears within 4–6 hr and peaks within 12–24 hr; the PUVA reaction appears around 24 hr and peaks 48–72 hr postirradiation, or even later). The PUVA dose–response curve for erythema is considerably steeper than that for UVB-induced erythema. Thus, a relatively small increment may cause an unexpectedly severe erythema.

Epidermal changes with PUVA occur later than with UVB-UVC radiation; fewer dyskeratotic cells are seen. Endothelial swelling and extravasation of red blood cells are seen with PUVA-induced erythema, and become more pronounced 7 days after exposure. A single exposure to PUVA enhances the activity of tyrosinase and there is an increase in the rate of transfer of melanosomes to adjacent keratinocytes. There is a change from the caucasoid pattern of melanosomes, aggregated in groups within membrane-limited vesicular bodies (melanosome complexes), to a pattern of singly dispersed melanosomes, characteristic in black skin. Gross pigmentation lasts from weeks to months.

18.1.7.3 PUVA-Induced Actinic Damage

The characteristics of actinic damage are (1) dry, wrinkled, inelastic, and leathery skin with irregular pigmentation, (2) atrophy of the epidermis, and (3) on microscope examination, replacement of the collagen of the upper dermis by masses of amorphous or granular material, slightly basophilic with hematoxylin–eosin staining (solar elastosis).

18.1.7.4 PUVA-Induced Carcinogenesis

PUVA adds to the cumulative damage from UV irradiation and enhances the risk of nonmelanoma skin cancer. Post-PUVA squamous cell carcinomas appear in unusual areas, i.e., those not normally sun exposed, and there is a strong association between the increased number of PUVA treatments and the risk of these tumors. Fair-skinned patients and patients with previous exposure to ionizing radiation are at highest risk.

After PUVA treatment, a decreased immunosurveillance mechanism may account for the rapid increase in the risk of cutaneous carcinoma.

18.1.7.5 Ophthalmologic Risks of PUVA

PUVA treatment may alter lens proteins to insoluble, higher molecular weight crystallines, with involvement of tryptophan photochemistry. Methoxsalen is found in the lens 12 hr after ingestion of a single therapeutic dose. In ducklings fed with the seeds of the *Ammi majus* and exposed to the sun for 4–5 hr a day, conjunctivitis was followed by mydriasis and severe retinopathy, but no cataracts.

18.1.7.6 Other Potential Long-Term Effects of PUVA

Long-term effects of PUVA include (1) bullous pemphigus, (2) increased incidences of lymphoma and leukemia in immunosuppressed individuals, (3) exacerbation of asymptomatic lupus erythematosus, and (4) skin contact allergy to 8-MOP.

18.2 MYCOSIS FUNGOIDES

Mycosis fungoides (MF) is a cutaneous T cell lymphoma that typically progresses through patch, plaque, and tumor phases of skin involvement (Figures 18.7–18.9). MF lesions frequently first appear on body regions that have not been exposed to sunlight, suggesting that solar UV radiation may exert an inhibitory influence on early disease. Response rates of 90% have been achieved in patients with pretumorous disease (Figures 18.8 and 18.9) utilizing PUVA.

Cutaneous T cell lymphoma (CTCL) initially appears in the skin and slowly progresses to involve the lymph nodes and viscera.

18.2.1 UVB Phototherapy of Mycosis Fungoides

Phototherapy with UVB light has been used in younger individuals, to avoid sterility that results from using chemotherapeutics. Out of 37 patients treated with UVB at the New York University Medical Center, 32 had histologically proved CTCL and 5 had parapsoriasis *en plaques,* the

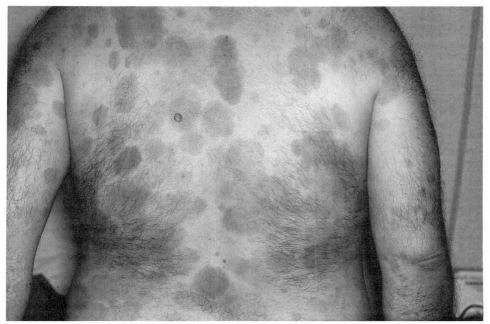

FIGURE 18.7 Mycosis fungoides. Thoraco-lumbar lesions, before PUVA treatment. Courtesy of Professor Louis Dubertret.

first stage of CTCL. The patients were categorized as either stage I (erythematous patches, no lymph node or organ involvement) or stage II (infiltrated plaques, no lymph node involvement). Phototherapy was started at 50–60% of MED, with increments of 50, 40, 30, and 20% of the previous dose at each of the next four treatments. The end point of increase was a generalized faint erythema 24 hr after exposure. Irradiation was three times per week, then twice, once on achieving remission, and then every other week on maintained remission of several months. The median duration of UVB therapy was 22 months.

Of patients in stage I, 25 achieved remission, 1 showed improvement, and 4 remained unchanged. Clearance of all cutaneous lesions for greater than 3 months was called clinical remission. The median duration of the complete response was 22 months (a range of 3 to 74 months). None of the patients with stage II lesions achieved a remission.

18.2.1.1 Mechanism of UVB Phototherapy

UVB radiation decreases the alloactivating and antigen-presenting capacity of the Langerhans cells, while increasing interleukin 1 and 6 production by keratinocytes, and also tumor necrosis factor. UVB radia-

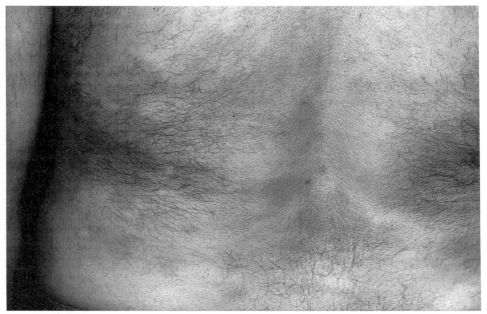

FIGURE 18.8   The same patient as in Figure 18.7, 15 years in remission after PUVA therapy. Courtesy of Professor Louis Dubertret.

tion suppresses human keratinocyte induction of the intercellular adhesion molecule at 1–24 hr, but induces expression at 48–96 hr. These alterations may serve an immunotherapeutic role controlling the progression of CTCL.

18.2.2 Home Phototherapy of Mycosis Fungoides

At the Cutaneous Lymphoma Clinic at Temple University, Philadelphia, 34 Caucasian patients with minimally infiltrated skin lesions and no extracutaneous involvement were selected for home phototherapy. Of the 34 patients, 31 had premyotic or early plaque phase MF. The other patients had parapsoriasis *en plaques* (chronic superficial dermatitis). The median age was 52 years.

Lesions tended to be located within the so-called bathing trunk region (lower trunk, buttocks, upper legs), and also on the breasts of female patients and around the axillar and upper inner arms (areas normally with little exposure to UV radiation). Patients were encouraged to administer home phototherapy (using four lamps emitting 280–320 nm) to covered regions primarily at suberythemogenic doses every other day. Maintenance alternate-day phototherapy was continued for 3–12

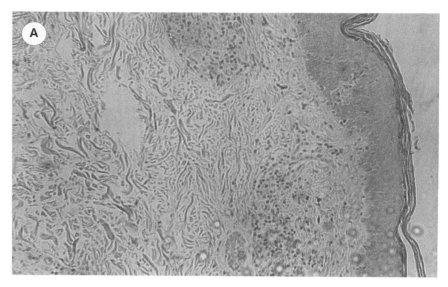

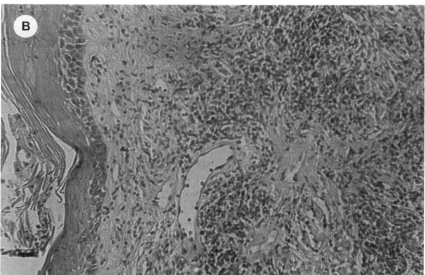

FIGURE 18.9 (A) Photomicrograph by Cahide Kohen of skin showing the early stage of mycosis fungoides. Note the mild infiltration of the dermis with tumoral lymphoid cells. Microscope slide, courtesy of Dr. George Elgart, Department of Dermatology, University of Miami, Miami, Florida. (B) Histopathological image of mycosis fungoides (cutaneous lymphoma), showing deep and massive infiltration of the dermis by malignant lymphoid cells. Courtesy of Dr. Elgart. Photomicrograph by Cahide Kohen.

months. Pretreatment biopsy specimens revealed in the upper dermis a moderately dense infiltrate of histiocytes and atypical lymphocytes, with hyperchromatic cerebriform nuclei, and similar cells diffusely into the lower hyperplastic epidermis. Posttreatment biopsy specimens revealed only a sparse perivascular lymphocytic infiltrate in the upper dermis.

For 19 of 31 early MF patients treated, a complete response followed and was maintained as long as phototherapy continued. The median time of complete response was 4 months, and the median of response duration was in excess of 18 months. Resumed UV therapy resulted in a second complete response for 4 of the 5 patients with recurrence off treatment.

In all instances, previously involved skin continued to show a mild perivascular infiltrate of morphologically normal mononuclear cells in the papillary dermis. In 5 patients, there was histological evidence of persistent MF despite clinical improvement of the skin.

18.2.3 Home Phototherapy in Early MF

In early MF, the key factors in patient selection for home phototherapy are poor tanning (melanization abrogates the effects of UV), early plaque-phase MF with infiltrates confined to the superficial dermis, and lesions on body areas accessible to phototherapy.

18.2.4 Adverse Effects of Phototherapy

Occasional adverse effects of phototherapy are basal cell carcinomas and actinic keratoses on sun-exposed areas. There have been no instances of squamous cell carcinomas.

18.2.5 Photophoresis: Extracorporeal Photochemotherapy of CTCL with 8-Methoxypsoralen and UVA

In treatment of the leukemic form of CTCL, the patient ingests 8-MOP. Two hours later 500 ml of blood is removed and centrifuged in the photophoresis apparatus. The white blood cell fraction is combined with saline and passed through a UV-transparent channel located between UVA lamps (optimal wavelength 330–350 nm). After irradiation, the white blood cells are recombined with other blood components and are returned to the patient.

This therapy has the following advantages: (1) lymphocytes are exposed to a drug activated *in situ* by UVA radiation, (2) due to the photoreactivity of 8-MOP, all of its activity is lost by the time the blood is returned to the patient, thus avoiding toxic effects to other organs, and (3) nonnucleated blood components are not significantly affected.

Erythrodermic patients with leukocyte counts of less than 15,000 are the best responders to photophoresis, with response also in patients with plaques (more than 15% of the body surface), circulating abnormal T cells, or lymph node involvement. Photophoresis results in an average 64% decrease of skin involvement and 88% loss of lymphocyte viability. The patients who respond initially receive monthly maintenance treatments. A case remaining free of skin lesions for over 4 years has been reported. The T cells that have been altered by the photosensitizer are recognized as "foreign" by the immune system, which therefore develop antibodies against these. Because of the polyclonal response of the immune system, not only T cells, but also the cancerous cells, are targeted by the elicited response for days after the photophoresis.

The optimal wavelengths (330–350 nm), away from the absorption maximum of 8-MOP at 300 nm, match the absorption spectrum of the 4′,5′ monoadduct, the precursor to the cross-link. Monoclonal antibodies that recognize the 4′,5′ monoadduct and psoralen–DNA cross-links are used to study adduct formation and removal in patient samples. 8-MOP photomodification of cell surfaces may lead to immunologically mediated events for the efficacy of photopheresis. Other targets of photomodification are unsaturated fatty acids and the epidermal growth factor receptor.

It is believed that the reinfusion of a malignant clone of T cells following their exposure to 8-MOP and UVA may induce an immune response by cells bearing antiidiotypic receptors recognizing the malignant clone, effectively resulting in "vaccination" of the host. Immunocompromised patients whose treatment is associated with direct destruction of the cells by photophoresis, rather than the induction of an immune response, require more frequent treatment.

18.3 VITILIGO

18.3.1 Characteristic Features

Vitiligo is a common acquired, possibly heritable, progressive depigmenting disorder, which often leaves patients visibly disfigured by the uneven coloration of the skin. It is a noncontagious disorder in which pigment cells are destroyed, resulting in irregularly shaped white patches on the skin (Figure 18.10); these may spread over most of the body or remain dormant for years.

The most susceptible race and sex are black and female. Vitiligo patches are due to the absence of melanocytes in small or large circumscribed areas of the skin. Melanocytes may be also affected in the inner ear, leptomeninges, and retinal pigment epithelium.

The spots of vitiligo occur where melanocytes are located. Common sites of pigment loss are exposed areas (hands, face, upper part of the chest), around body openings (eyes, nostrils, mouth, nipples, umbilicus,

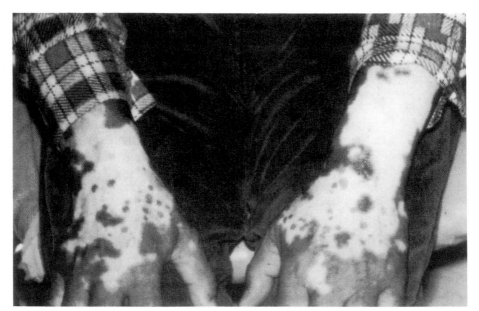

FIGURE 18.10   Vitiligo of the hands and arms in a black patient showing marked contrast between affected and normal skin. From R. M. Halder (1991). Copyright George V. Kelvin, Science Graphics, Great Neck, New York.

genitalia), body folds (arm pits, groin), sites of injury (cuts, scrapes, burns), hair (early graying of the hair and scalp, beard, or other areas), areas immediately surrounding pigmented moles, and the chorioid of the eye.

About one-third of all vitiligo patients say that other family members also have this condition. People with vitiligo face a greater risk of having: hyperthyroidism or hypothyroidism, pernicious anemia, Addison's disease, alopecia areata (round patches of hair loss), and uveitis.

In vitiligo, melanocytes cannot form melanin or their number decreases. Vitiligo is just one form of leukoderma (white skin). Severe trauma, like a burn, can destroy pigment cells, also resulting in leukoderma.

18.3.2 Mechanisms of Pigment Loss in Vitiligo

A combination of genetic, immunologic, and neurogenic factors is of major importance in causing most cases of vitiligo. Many people report pigment loss shortly after a severe sunburn. Others relate the onset of vitiligo to emotional trauma. Early graying of hair is part of vitiligo. Patients with vitiligo appear to have pigment cells that are genetically predisposed for destruction. These abnormal melanocytes are presumed to interact with the nervous and immune systems to bring about their destruction. Light-skinned people usually notice the pigment loss during the summer as the contrast between the vitiliginous skin and suntanned skins becomes accentuated. The degree of pigment loss can vary within each vitiligo patch, and a border of abnormally dark skin may encircle a patch of depigmented skin. Cycles of pigment loss followed by periods of stability may continue indefinitely.

Patients so totally depigmented that they are no longer bothered by contrasting skin color do not look like albinos because the color of their eyes and hair rarely changes.

18.3.3 Immune Processes in Vitiligo

Children with vitiligo have an increased incidence of autoimmune and/or endocrine disease, as well as premature graying in their immediate and extended family members. The human leukocyte antigen D (HLA-D) region genes encode cell surface antigens that present antigen to T cells and thus are involved in the regulation of the immune response. Specific HLA-D antigens have been implicated in the presentation of autoantigens to T cells and the induction of the subsequent autoimmune response that results in the production of autoantibodies and autoreactive cells responsible for pathogenesis.

In black patients, HLA-DR and HLA-DQ antigens are risk factors useful as a starting point for identification of the specific source of genetic susceptibility. Specific HLA genes serve as markers for the identification of high-risk patients; they may also identify a subgroup for immunosuppressive intervention sufficiently early in the course to prevent clinical disease.

In vitiligo subjects, mean total T lymphocytes and helper T cells are markedly depressed, natural killer cells are markedly elevated, and suppressor T cells are moderately elevated in comparison with control subjects. Thus, the regulation of cell-mediated immunity is defective in vitiligo. It is unknown whether these abnormalities are a direct cause or a result of vitiligo. Antibody-dependent cytotoxicity, utilizing killer cells with antimelanocyte antibodies, may be responsible for pigment cell destruction in vitiligo. Helper T cells may be reduced because of low levels or faulty production of T lymphocyte-stimulating factors in patients, or because of a serum factor that is toxic to helper T cells.

The cytotoxic effects of natural killer cells and peripheral T lymphocytes in patients with vitiligo need to be evaluated against normal melanocytes. It would be interesting to know the status of circulating T cell subsets and natural killer cells in patients who repigment, and also to study the same cells in patients with both malignant melanoma and vitiligo.

18.3.4 Basic Methods of Treatment

There are two basic methods of treatment: first, to try to restore the normal pigment (repigment), and second, to try to destroy the remaining pigment (depigment).

To achieve repigmentation, new pigment cells must come from the base of hair follicles, from the edge of the lesion, or from the patch of vitiligo, if depigmentation is not complete. The maximum amount of repigmentation that can be expected in any one spot in a year of treatment is an eighth to a quarter of an inch.

18.3.4.1 Repigmentation by PUVA Therapy

In repigmentation therapy, a patient is treated with PUVA. When psoralen drugs (trimethylpsoralen and 8-MOP) are activated by UVA light, they stimulate repigmentation by increasing the availability of color-producing cells on the skin's surface. A patient takes the prescribed dose orally 2 hr before lying in the sun or under artificial UVA light. The ideal time for natural sunlight is between 11 a.m. and 1 p.m. Treatment every other day is recommended (in the northern United States, from May to September). The winter rest period is desirable. Artificial UVA sources can be used all year (with a dermatologist's supervision). Patients with vitiligo should always protect their skin from sun exposure except during treatment time (sunscreen lotions and creams).

In patients with small, scattered patches of vitiligo, the lesions treated with topical psoralen can are exposed to sunlight. Such treatment makes a person very susceptible to burns and blisters following excessive sun exposure.

Candidates for repigmentation should meet the following requirements:

1. Patients should be at least 10 years old.
2. Pigment loss occurring for less than 5 years in patients over 20 years of age.
3. No sensitivity or allergy to sunlight.
4. Opportunity to be exposed to sunlight (1–2 hr, 2–3 times a week, for 2–5 years, in summer months). Pregnant women should not be treated because of potential harm to the fetus.

The response rate of topical PUVA therapy is 50–60%, with the best improvement occurring on the face and trunk and poorer results on distal extremities. PUVA treatment takes an average of 4 to 12 months of nonstop therapy.

Oral psoralen and topical psoralen seem to produce the best results. If 8-MOP is administered topically in low doses, there is less blistering and more likelihood of complete repigmentation. When treatment stops, the peripheral ring of hyperpigmentation slowly disappears: the repigmentation area blends with the normal skin and this new skin becomes permanent.

During the winter, most patients retain at least half of the repigmentation achieved in the summer months.

18.3.4.2 The PUVA Procedure in Vitiligo

18.3.3.2.1 Topical PUVA Therapy for Vitiligo PUVA is no panacea for the treatment of vitiligo. Genital areas are not treated with topical PUVA therapy because of the increased risk of phototoxic reactions. Once the patient maintains moderate asymptomatic erythema, no further increments are made in UVA dosage. Because phototoxicity peaks in 48 to 72 hr postirradiation, treatments are scheduled only once a week.

Repigmentation occurs in a perifollicular pattern. It is a gradual process and may require 4–12 months of continuous therapy. The new pigment should be regarded as permanent in the majority of patients.

Because topical 8-MOP is not absorbed through the skin in appreciable amounts, the risks of premature cataract, skin cancer, and liver toxicity, which may occur with oral psoralen, are virtually nonexistent.

18.3.4.3 Experimental Studies on Vitiligo Repigmentation by Transplants

Melanocyte transplants can be accomplished by a dermatologist. Pigment cells are taken from an unexposed, normally pigmented patch of the skin; they are grown in culture to large numbers, and are returned to a white patch on the patient.

18.3.4.4 Depigmentation Therapy

In a person with vitiligo occurring on over more than half of the exposed areas of the body, a trial for depigmentation of the remaining pigmented skin is very appropriate. The drug used for depigmentation is the monobenzylether of hydroquinone.

18.3.5 Khellin Photochemotherapy of Vitiligo

With psoralens, proper dosimetry of UVA radiation is a prerequisite to avoid phototoxic erythema reactions, because such reactions can lead to a Koebner effect. Khellin, with a chemical structure close to psoralen, behaves as a *monofunctional agent with respect to DNA and does not photoin-*

duce cross-links, which may explain its low photogenotoxicity plus lack of phototoxic erythema (even with irradiation up to 1000 kJ m^{-2}). The dose of khellin is kept constant at 100 mg. Khellin plus artificial UVA light appears at least as effective as PUVA. No long-term side effects are observed after 12 to 15 months of therapy. With no phototoxic skin reactions, khellin photochemotherapy can be considered safe for outdoor treatment with natural sunlight even in fair-skinned individuals. It therefore offers the opportunity of treating vitiligo outside of office hours and during summer vacation. At present, there exists no information regarding possible long-term risks of khellin plus UVA. The observation that it photobinds to DNA should, however, alert the photochemotherapist to the need for further investigations.

Bibliography

Averbeck, D. (1982). Photobiology of furocoumarins. *In* "Trends in Photobiology" (C. Hélène, M. Charlier, T. Montenay-Garestier, and G. Laustriat, eds.), pp. 295–305. Plenum, New York.

Bevilacqua, P. M., Edelson, R. L., and Gasparro, F. B. (1989). Extracorporeal photochemotherapy and UVA. *Spectrum* **2**(1), 6–8.

Bluefarb, S. M. (1959). "Cutaneous Manifestations of the Malignant Lymphoma," p. 173. Thomas, Springfield, IL, (cited in Milstein *et al.*, 1982).

Broq, C. (1902). Les parapsoriasis. *Ann. Dermatol. Syphiligr.* **3**, 433.

Chakrabarti, S. G., Halder, R. M., Johnson, B. A., Minus, H. R., Pradhan, T. K., and Kenney, J. A., Jr. (1986). 8-Methoxypsoralen levels in blood of vitiligo patients and in skin, ophthalmic fluids, and ocular tissues of guinea pig. *J. Invest. Dermatol.* **87**, 276–279.

Christopher, E., and Sterry, W. (1993). Psoriasis. *In* "Dermatology in General Medicine" (T. B. Fitzpatrick, A. Z. Eisen, K. Wolff, I. M. Freedberg, and K. F. Austen, eds.), pp. 489–514. McGraw-Hill, New York.

Dall'Acqua, F. (1982). Photochemical reactions of furocoumarins. *In* "Trends in Photobiology" (C. Hélène, M. Charlier, T. Montenay-Garestier, and G. Laustriat, eds.), pp. 267–277. Plenum, New York and London.

Dunston, G. M., and Halder, R. M. (1990). Vitiligo is associated with HLA-DR4 in Black patients. *Arch. Dermatol.* **126**, 56–60.

Edelson, R. (1988). Light-activated drugs. *Sci. Am.* **259**, 68–75.

Edelson R., Berger, C., Gasparro, F., Jegasothy B., Held, P. S., Wintroub, B., Vonderheid, E., Knobler, R., Wolff, K., and Plewig, G. (1987). Treatment of cutaneous T-cell lymphoma by extracorporeal photochemotherapy. *N. Engl. J. Med.* **316**, 297–303.

Eyre, R. W., and Krueger, J. C. (1984). The Koebner response in psoriasis. *In* "Psoriasis" (H. H. Roenigh, H. I. Maibach, eds.), p. 34. Dekker, New York (in Fitzpatrick *et al.*, 1993, p. 489).

Fitzpatrick, T. B., Eisen, A. Z., Wolff, K., Freedberg, I. M., and Austen, K. F., eds. (1993). "Dermatology in General Medicine." McGraw-Hill, New York.

Gilchrest B. cited in Bevilacqua, P. M., Edelson, R. L., and Gasparro, F. B. (1989).

Goekerman, W. H. (1925). The treatment of psoriais. *Northwest Med.* **24**, 299.

Grimes, P. E., Minus, H. R., Chakrabarti, S. G., Enterline, J., Halder, R., Gough, J. E., and Kenney, J. A., Jr. (1982). Determination of optimal topical photochemotherapy for vitiligo. *J. Am. Acad. Dermatol.* **7**, 771–778.

Halder, R. M. (1991). Topical PUVA therapy for vitiligo. *Dermatol. Nurs.* **3**, 178–180.

Halder, R. M., Walters, C. S., Johnson, B. A., Chakrabarti, S. G., and Kenney, J. A. (1986). Aberrations in T lymphocytes and natural killer cells in vitiligo: A flow cytometric study. *J. Am. Acad. Dermatol.* **14**, 733–737.

Halder, R. M., Grimes, P. E., Cowan, C. A., Enterline, J. A., Chakrabarti, S. G., and Kenney, J. A. (1987). Childhood vitiligo. *J. Am. Acad. Dermatol.* **16**, 948–954.

Halder, R. M., Pham, H. N., Breadon, J. Y., and Johnson, B.A. (1989). Micropigmentation for the treatment of vitiligo. *J. Dermatol. Surg. Oncol.* **15**, 1092–1098.

Henseler, T., Hönigsman, H., Wolff, K., and Christophers, E. (1981). Oral 8-methoxypsoralen photochemotherapy of psoriasis. European PUVA study: A cooperative study among 18 European countries. *Lancet* **1**, 853 (in Fitzpatrick *et al.*, 1993, p. 489).

Hönigsmann, H., and Ortel, B. (1965). Khellin photochemotherapy of vitiligo. *Photodermatology* **2**, 193–194.

Hu, C.-H. (1993). Parapsoriasis. *In* "Dermatology in General Medicine" (T. B. Fitzpatrick, A. Z. Eisen, K. Wolff, I. M. Freedberg, and K. F. Austen, eds.), pp. 1124–1128. McGraw-Hill, New York.

Kohen, E., Kohen, C., Reyftmann, J. P., Morliere, P., and Santus, R. (1984). Microspectrofluorometry of fluorescent photoproducts in photosensitized cells. *Biochim. Biophys. Acta* **805**, 332–336.

Le Vine, M. J. (1979). Components of the Goekerman regimen. *J. Invest. Dermatol.* **73**, 170.

Milard, B., Green, C., Ferguson, J., Raffle, E. J., and MacCloud, T.M. (1989). Study of the photodegradation of Leukotriene B4 by ultraviolet irradiation (UVB, UVA). *Br. J. Dermatol.* **120**, 145–152; from *Clin. Dig. Ser.* **1**, 216 (1989).

Milstein, H. J., Vonderheid, E. C., Van Scott, E. J., and Johnson, W. C. (1982). Home ultraviolet phototherapy of early mycosis fungoides: Preliminary observations. *J. Am. Acad. Dermatol.* **6**, 355–362.

Morliere, P., Honismann, H., Averbeck, D., Dardalhon, M., Huppe, G., Ortel, B., Santus, R., and Dubertret, L. J. (1988). Phototherapeutic, photobiologic and photosensitivity properties of Khelline. *J. Invest. Dermatol.* **98**, 720–724.

National Vitiligo Foundation, Inc. Tyler, Texas 757111.

Parrish, J. A. (1982). Phototherapy of psoriasis and other skin diseases. *In* "The Science of Photomedicine" (J. D. Regan and J. A. Parrish, eds.), pp. 511–531. Plenum, New York and London.

Parrish, J. A., Fitzpatrick, T. B., Taflenbaum, L., and Pathak, M. A. (1974). Photochemotherapy with oral methoxsalen and long wave ultraviolet light. *N. Engl. J. Med.* **291**, 1207–1211 (in Fitzpatrick, *et al.*, 1993, p. 489).

Parrish, A., Stern, R. S., Pathak, M. A., and Fitzpatrick, T. B. (1982). Photochemotherapy of skin diseases. *In* "The Science of Photomedicine" (J. D. Regan and J. A. Parrish, eds.), pp. 595–623. Plenum, New York.

Pathak, M. A., Haber, L. C., Seigi, M., Kukita, A. (1974). Photobiology and photochemistry of furocoumarins (psoralens). *In* "Sunlight and Man: Normal and Abnormal Photobiologic Responses" (M. A. Pathak, ed.), p. 25. Tokyo Univ. Press, Tokyo.

Ramsay, D. L., Lish, K. M., Yalowitz, C. B., and Soter, N. (1982). Ultraviolet-B phototherapy for early stage cutaneous T-cell lymphoma. *Arch. Dermatol.* **128**, 931–933.

Resnick, K. S., and Vonderheid, E. C. (1993). Home ultraviolet therapy of early mycosis fungoides: Long-term follow-up observations in thirty-seven patients. *J. Am. Acad. Dermatol.* **29**, 73–77.

Santella, R. M., Dharmaraja, N., Gasparro, F. P., and Edelson, R. L. (1985). Monoclonal antibodies to DNA modified by 8-methoxypsoralen and ultraviolet A light. *Nucleic Acid Research.* **13**, 233–244.

Stern, R. S., Laird, N., Melski, J., Parrish, J. A., Fitzpatrick, T. A., and Bleich, H. L. (1984). Cutaneous squamous cell carcinoma in patients treated with PUVA. *N. Engl. J. Med.* **310**, 1156–1161.

CHAPTER

19

Photodynamic Therapy

19.1 BASIC PRINCIPLES: MOLECULAR AND CELLULAR MECHANISMS OF PHOTODYNAMIC THERAPY

19.1.1 Principles

Photodynamic therapy (PDT) is based on the intravenous injection of tumor-localizing photosensitizers (generally porphyrin derivatives), followed by exposure of the tumor region to high fluence, usually of red light from a laser. PDT is effective only in cases wherein the whole tumor mass can be irradiated. Large tumors do not satisfactorily respond to PDT unless light is applied interstitially.

Unlike PUVA treatment, the effectiveness of which primarily relies on oxygen-independent photosensitization, PDT involves type 2 photodynamic reactions, which are oxygen dependent. Thus, hypoxic tumors do not satisfactorily respond to PDT. Another major difference with PUVA is that irradiation relevant to PDT is carried out with red to far-red radiations, making possible better penetration of light into the tumor (see

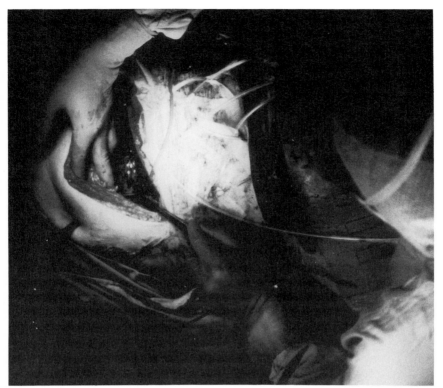

FIGURE 19.1   PDT being applied to a mesothelioma (postsurgery) from a laser source by interstitial fiber placement (adjuvant therapy). Courtesy of Professor Thomas J. Dougherty, Division of Radiation Biology, Roswell Park Cancer Institute, Buffalo, New York.

Chapter 13), the optimum being in the so-called "phototherapic window," in the 650–800 nm. New photosensitizers strongly absorbing in the 650–800 nm range are already under preclinical evaluation.

Although the tumor-localizing properties of porphyrins have been known for 70 years (Policard, 1924), the advent of lasers, fiber optics, endoscopy, and laparoscopy incited T. G. Dougherty to promote their use in phototherapeutic clinical applications (Figures 19.1 and 19.2). PDT is a safe treatment. The cytotoxic effect induced by singlet oxygen ceases when light is switched off, contrary to conventional chemotherapies and radiotherapy. Only the irradiated area is altered by PDT, with minimal systemic toxicity. Furthermore, PDT does not impede other treatments (surgery, chemotherapy, and radiotherapy). The interaction of the first excited triplet state of the sensitizer with oxygen produces singlet oxygen.

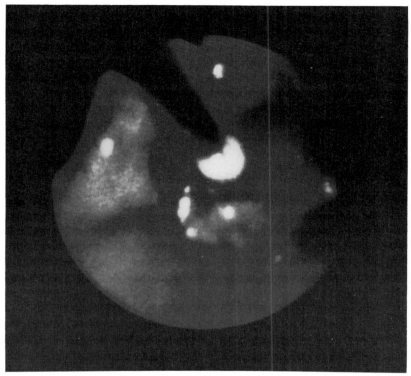

FIGURE 19.2 PDT treatment of bronchial tumor by interstitial fiber placement. Courtesy of Professor Thomas J. Dougherty.

The only drug toxicity encountered with Photofrin II, the presently used mixture of porphyrins, is a generalized photosensitivity that requires patients to avoid bright light for 30 days or more.

19.1.2 Tumor Selectivity of PDT

The preferential localization of anionic porphyrins and related dyes in tumors seems to be caused mainly by the special properties of tumor tissue:

1. Low pH consequent to high glycolytic activity, which favors the uptake and/or binding of dyes that increase in lipophilicity with decreasing pH.
2. Heightened low-density lipoprotein (LDL) receptor-mediated endocytosis activity of neoplastic cells.

3. Simple pooling of sensitizers in tumors due to leaky vasculature and poor lymphatic drainage.
4. The large interstitial space (stromal element) of tumor tissue (basis for nuclear magnetic resonance imaging), which may serve as a "sink" for injected substances.
5. Macrophages forming more than 50% of the total cell population (i.e., the important role of reticuloendothelial elements).

In the pancreas there is little or no difference in Photofrin II accumulation in normal and neoplastic tissue, but a large difference exists in treatment response. This may relate to the distribution of sensitizer among tumor compartments and/or the different sensitivities of normal and tumor-derived cells.

19.1.3 Transport and Delivery of Photosensitizers

19.1.3.1 Clearance of the Sensitizer

In mice, [^{14}C]Photofrin II is cleared from the blood with kinetics fitting a triexponential equation, with elimination half-lives of 4 hr, 10 days, and 36 days. Peak Photofrin II levels in tissue are reached 5–10 hr after injection (1% of injected material is in the circulation 24 hr after injection; 0.01% is still detectable at 75 days).

19.1.3.2 Delivery of Photosensitizers

Rapidly growing tumors and tissues acquire lipids by nonspecific endocytosis or by receptor-mediated endocytosis of low-density lipoproteins (LDLs). The high sensitivity of endothelial cells to PDT *in vitro* is due to a higher uptake of LDLs. However, several dyes that associate poorly with lipoproteins have been reported to be good tumor localizers. Photofrin can enter cells by pathways other than the LDL receptor pathway.

19.1.4 Mechanism of Action of Photosensitizers

19.1.4.1 Cells *in Vitro*

PDT affects mitochondria, lysosomes, the plasma and endoplasmic reticulum membranes, and to a lesser extent DNA. $^1O_2$, the main PDT-induced cytotoxic agent, can diffuse significantly less than 0.1 μm in cells during its lifetime. $^1O_2$ probably reacts at the sites of origin, but it may penetrate biological membranes. $^1O_2$ generated in the medium outside cells (e.q., from hematoporphyrin, which binds poorly to cells) is inefficient in inactivating cells as compared to $^1O_2$ generated from lipophilic and cell-bound sensitizers. Thus, the quantum yield for cell inactivation by cell-associated lipophilic Photofrin is 10 times larger than that for the hydrophilic dye tetraphenylporphine tetrasulfonate (TPPS$_4$). All these dyes have similar yields of $^1O_2$, so the focus is localization of the dyes in the cells and on $^1O_2$ reactions in membranes.

The PDT-affected plasma membrane (as indicated by depolarization) changes in terms of the morphologic structure of microvilli and the pattern of membrane proteins as seen by electron microscopy. Assays show leakage of enzymes to the extracellular medium. The damage to organelle membranes is comprehensive, as indicated by alterations of the endoplasmic reticulum (ER)-localized cholesterol-O-acetyltransferase and cytochrome P450, plus changes in mitochondrial and lysosomal membranes.

It is very hard to identify the main targets of PDT. Hematoporphyrin derivative (HPD) and Photofrin, a slightly modified version of HPD which differ in lipophilicity and intracellular localization patterns, have been used for studies of the mechanisms of PDT at the subcellular level. Fluorescence excitation spectra of lipophilic porphyrins in cells indicate close contact with proteins that are an important contributor to PDT-induced membrane damage (i.e., oxidation, cross-linking).

PDT enhances the transcription and translation of several oxidative stress genes (coding for a 70-kDa heat-shock protein similar to that induced by hyperthermia). This seems to be a cellular defense mechanism against oxidative stress.

19.1.4.2 Mechanism of Action of PDT in Tissues

19.1.4.2.1 Vascular Damage in Phototherapy In rodents, tumor eradication by Photofrin is dependent on the destruction of the microvasculature (Figure 19.3) at the tumor bed, which leads to tumor death by nutritional deprivation. The relative contribution of direct tumor cell photokilling may vary greatly with the photosensitizer and the treatment conditions.

Direct photodynamic tumor cell damage relies on the availability of oxygen in the tissue environment. It may be limited by two mechanisms: (1) rapid PDT-induced vascular occlusion and acute tumor hypoxia and (2) oxygen consumption by the photodynamic process, which may not be replenished as fast as required for continuing singlet oxygen generation. Vascular photosensitivity may depend on the level of circulating sensitizer and/or the binding of the sensitizer to vascular targets.

The earliest post-PDT vascular events involve damage to endothelial cells, reorganization of actin and tubulin elements, and cell rounding. A thrombogenic response may result from exposure of the vascular basement proteins through newly formed gaps: i.e., platelet activation and leukocyte adhesion to the endothelial lining vessels. The release of vasoactive eicosanoids (e.g., leukotriene B$_4$, thromboxane), vessel constriction,

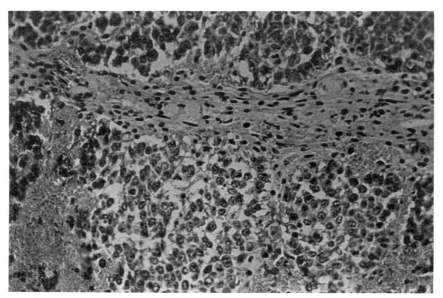

FIGURE 19.3 Histological image of mammary tumor subjected to PDT (3 hr after PDT). Blood vessels and cells have been destroyed. Courtesy of Professor Thomas J. Dougherty.

and alteration of permeability to macromolecules, combined with platelet aggregation, lead to vessel stasis.

Photofrin and its compounds, part of the new wave of photosensitizers (purpurins, chlorins, and phthalocyanines) such as mono-L-aspartyl-chlorin and monosulfonated zinc phthalocyanine act very similarly, causing vessel constriction and platelet aggregation leading to blood flow stasis. The disulfonated and tetrasulfonated zinc phthalocyanines do not appear to damage the microvasculature, even though using disulfonated zinc phthalocyanines does result in tumor destruction. The differences in response may be due to sensitizer location in the tissue or vasculature.

Dynamic capillaroscopy (DC) was adapted to observe acute photo-dynamic effects on the microcirculation of the mouse tail. Irradiation with white light or with red laser light caused rapid and long-lasting (longer than 5 min) cessation of capillary flow in animals sensitized with m-tetrahydroxyphenylchlorin (mTHPC). The Photofrin II absorption peak in the red is at ~625 nm; the mTHPC peak is at ~675 nm.

A comparative study of Photofrin and mono-L-aspartylchlorin (NPe6) was made on rat chondrosarcoma. Immediately after light treatment a decrease in cell viability was observed only in rats given NPe6. The data suggest that tumor cure using Photofrin and NPe6 is largely due to vascular stasis, but direct cell kill also is a factor.

The amount of HPD in tumor-associated macrophages is two to three times higher than that in the tumor cells 4–24 hr after administration. When macrophages are exposed to PDT they release tumor necrosis factor (TNF), which may contribute to tumor regression directly by inactivation of tumor cells or indirectly by leading to hemorrhagic necrosis of the vasculature.

Whereas different forms of PDT have been reported to result in immunosuppression, low doses of HPD and light given to a mixture of peritoneal macrophages and nonadherent cells (B and T lymphocytes) stimulated the Fc receptor-mediated ingestion activity of the macrophages. The latter phenomenon may be related to the aggregated nature of HPD whereas the light-dependent stimulation may be related to the formation of peroxidation products of membrane lipids.

19.1.5 Late Effects of Photodynamic Therapy

Prolonged skin photosensitivity in patients undergoing PDT can conceivably be abrogated by rapid photosensitizer clearance due to excretion and/or sensitizer breakdown. The second-generation photosensitizers meet the standards. For example, NPe6 causes no skin photosensitivity 24 hr postinjection in mice and guinea pigs, and it is particularly photodegradable.

Clearance of post-PDT residual sensitizer can be affected by (1) gradual photobleaching, (2) absorbing agents, diuretics, and modifiers of porphyrin metabolism, (3) interference from inflammatory effects, (4) singlet oxygen and free oxygen radical quenching carotenoids, and (5) photoprotection through sulfhydryl-containing compounds.

High levels of interleukin 1β, interleukin 2, and tumor necrosis factor-α are present in patients' urine up to 50 days after high-dose bladder PDT treatment. Cytokine release *in vivo* may represent a specific PDT response, or may be simply related to the degree of inflammation present.

19.2 A NEW APPROACH TO TREATMENT OF MALIGNANCY COMES TO TRIAL: A SELECTION OF EARLY CASE REPORTS WITH HEMATOPORPHYRIN DERIVATIVES

The first clinical trials of PDT were carried out with a mixture of porphyrin derivatives empirically formulated from animal tumor models and cultured cell studies. The mixture was called HPD or Photofrin I. An optimized version of Photofrin I was named Photofrin II™. It was used in the clinical applications described below. The following case reports are from a period when PDT therapy was in its infancy.

19.2.1 PDT of Lung and GI Tract Tumors

HPD was used as photosensitizer, with an argon dye laser yielding 630 nm emission. Therapy resulted in complete remission (CR), significant remission (SR; 60% or more of the original tumor volume disappeared), partial remission (PR; 20 to less than 60% of the tumor disappeared), or no remission (NR). The therapeutic results were as follows:

Lung cancer	10 CR, 3 SR, 0 PR–NR
Esophageal cancer	2 CR, 1 SR, 0 PR, 1 NR
Gastric cancer	7 CR, 6 SR, 1 PR, 1 NR

The indications for PDT are as follows: (1) precancerous lesions, (2) severe atypical squamous cell metaplasia of the bronchial epithelium, (3) inoperable early-stage cancer, and (4) in advanced cancer, preoperative PDT to reduce the extent of resection and preradiation therapy to decrease the required focal dose, in combination with chemotherapy. PDT is also used for recurrent tumors; intratumor PDT is for inoperable tumors using multiple quartz fibers.

Complete remission was not obtained in some of the early-stage cancer cases due to the laser beam not penetrating the entire lesion (unfavorable beam angle or tumor location) or due to obstruction of the laser beam by gastric folds and peristaltic movements of the esophagus and stomach.

A new fiber that provides a 360° field has been developed by Dr. Dougherty and his colleagues, a development that holds considerable promise.

19.2.1.1 HPD PDT of Endobronchial Lung Cancer

Waiting 24–72 hr postinjection in humans has been found optimal for obtaining preferential HPD distribution in tumors. HPD PDT was used to treat 60 patients with endobronchial tumors. Successful palliative therapy was achieved for extensive and obstructing endobronchial cancers with HPD irradiated at 630 nm (continuous argon laser, quartz fibers inserted into the biopsy channel of a flexible bronchoscope). Surface illumination of the tumor was either by forward directed output (FS) or by use of a scattering matrix applied to the fiber tip, to give an isotropic light distribution (CS). Full illumination of the lesion was obtained by sweeping the output over the lesion surface. Interstitial illumination was provided by direct point insertion into the tumor of a flat cut fiber (PI) or an isotropic cylindrical fiber (CI). For tumor tissue too firm to allow direct insertion, a long flexible-sheath endoscopic needle was used with a core fiber (PIN). Planned total light doses were 100 J/cm^2 for FS or CS, 200 for PI or PIN and for CI.

The purpose of phototherapy was to open up bronchi for ventilation, decrease shortness of breath, and prevent the retention of bacteria and secretions that may lead to pneumonia. The technique used resulted in very little bronchial inflammation or bleeding.

19.2.1.2 Endoscopic HPD−Laser Treatment of Gastric, Colorectal, and Bronchial Cancers

PDT was applied using HPD (Photofrin I) as photosensitizer, a rhodamine dye laser pumped by an argon laser (630 nm emission) as light source, and a 400-μm quartz fiber light guide passed through the channel of a rigid or a fiberoptic endoscope.

Patient selection was based on the following criteria: (1) advanced carcinoma obstructing the respiratory (trachea, main or lobar bronchus) or digestive (esophagus, stomach, large intestine) tract, (2) recurrent carcinoma (same locations), and (3) early carcinoma in high-risk patients (same locations).

The results, evaluated for metastasis (M+), were as follows:

Recurrent carcinoma of rectum, M+	60% regression
Sigmoido-rectal junction, M+	40% regression
Right main bronchus, advanced	50% regression
Left upper bronchus, early?	CR
Stomach antrum, early?	CR
Rectum, advanced, M+	70% regression

In the case of early cancer, complete remission was obtained. The tumor framework can be completely destroyed, because of adequate light penetration. PDT produced a constant response of tumor necrosis, with the only complication being hemorrhage (oozing or spurting).

19.2.2 PDT of Bladder Cancer

A tumor cell-positive urinary cytology, after macroscopic removal of all visible tumor, indicates persistance of foci that are responsible for the high rate of recurrence. Flexible fibers open the possibility of PDT by endoscopy.

PDT was given to 20 patients who had recurring superficial tumors after unsuccessful application of other treatments. Almost 80% of recurrent bladder carcinoma showed foci of dysplasia and carcinoma *in situ* within the bladder which cannot be detected by routine endoscopy. These multifocal growths could represent a reservoir for rapidly recurrent tumors. Patients with recurrent multifocal bladder tumors had been conventionally recommended to undergo cystectomy after failure of all other therapeutic procedures. These patients are good candidates for integral photodynamic therapy (PDT) providing homogeneous illumination of the whole bladder wall with therapeutically effective laser light.

The results from 17 patients who were treated by integral PDT and followed up comprehensively over at least 12 months are available. The phototherapy results were evaluated at 3 months intervals by endoscopy, cytology, bladder mapping and renal ultrasonography. The largest follow-up period is about 5 years. Preliminary results show complete destruction for 23 of the 28 tumors in above patients (83%). The failures in therapy can be explained by technical problems due to inhomogeneous light application.

For photosensitization of the patients the drug Photofrin II was used. It was injected intraveinously 67–72 hours before PDT treatment. In six patients treated with PDT no recurrence was found over the whole observation period up to nearly 5 years. Four patients remained free of tumor (12 and 14 months) after repeated transurethral resection (TUR) and Neodymium:YAG (ND:YAG) laser therapy following PDT. Four patients required a second photodynamic treatment. In six patients slight mucosal dysplasia persisted for at least 2.5 years. According to these preliminary results, PDT with strict patient selection is justified in the case of recurrent superficial bladder carcinoma. PDT enables a complete destruction of superficial bladder carcinoma, but it is accompanied by considerable transitory side effects: cystitis and bladder shrinkage, which take a long time to heal. In the long term, tumor recurrences in the bladder cannot be avoided. This is probably due to the continuous local onset of carcinogenesis by toxic substances in the urine. PDT dosimetry and tumor selectivity of photosensitizers need to be improved and suitable light applicators need to be developed. It should be noted that integral PDT is a method which is providing successful tumor destruction even for "worst case" patients.

19.2.3 PDT in Extensive Basal Cell Carcinoma of the Dorsal Skin

The entire dorsal skin area (18×21 cm^2) of multiple basal cell carcinoma (BCC) in a 75-year-old male was treated in 12 fractionated zones: in 5 fractions, an argon ion laser (all lines) was used; in 7 fractions, with the same irradiance, a dye laser was used. Multiple biopsy samples 2 months after the first treatment demonstrated no residual disease in areas treated, except at the left upper border of the area treated with the argon laser; a persistent small pigmented lesion was histologically confirmed.

A second treatment was performed on day 61 in 7 fractions. Irradiances were reduced to avoid side effects. Temperature measurements were done by means of thermocouples and thermographic maps. A maximum increase of 7°C was found at the surface level. Deep ulcerations, seen after the first treatment, were avoided in the second treatment. Two months after the second treatment the healing process was completed in all fractions, except for persistence of small pigmented lesions midway between two dye laser-treated sites and at the border between dye laser- and argon-treated sites.

19.2.4 PDT of Brain Tumors

Detectable levels of HPD accumulate in brain tumors, but HPD is effectively excluded by an intact blood–brain barrier. The modalities of PDT are as follows:

1. For inoperable deep tumors, the stereotactic implantation of one or more quartz fibers to provide argon–dye laser photoirradiation in conjunction with prior administration of HPD.
2. For recurrent tumors, PDT of the tumor bed after intravenous (and in some cases topical) administration of HPD.
3. For cystic or cavitary lesions, after intravenous and/or topical application of HPD, the cyst or cavity is filled with a diffusion medium and illuminated either with the laser–quartz fiber system or a high-intensity xenon arc lamp fiberoptic system.

Power densities higher than 200 mW of red light through a 0.6-mm quartz fiber will produce significant heating of tissue. There is less significant heating when light is delivered through a large-diameter (5.0-mm) fiberoptic system. If tissue heating is avoided, posttherapy cerebral edema can be avoided or minimized. Patients are managed with pre- and postoperative corticosteroid therapy. Results were rather poor, although mean lifetime was increased.

Developments in progress include new less toxic light–drug combinations with deeper penetration of tumor tissue; multiple-fiber lasers; systemic or topical methods to improve dye uptake by tumor cells; metabolic enhancers of the cytotoxic effect in the tumor tissue and quenchers of the photodynamic effect in normal tissue; combination of PDT with stereotactic CO_2 laser resection; interstitial and conventional radiation therapy; and chemotherapy, hyperthermia, and immunotherapy. The results of PDT were as follows (HPD 5 mg/kg, 72 hr):

Astrocytoma	8 dead after 2–37 months, 1 alive with recurrence (13 months), 2 alive after 13 months
Oligodendroglioma	1 dead after 3 months
Medulloblastoma	2 alive after 13–31 months
Ependymoma	1 dead, 1 alive after 13 months
Sella turcica craniopharyngioma	1 dead after 4 months
Left orbit rhabdomyosarcoma	1 dead after 2 months

Metastatic tumor   1 alive with recurrence
                                                    after 11 months, 1 alive
                                                    after 9 months

19.2.5 PDT in Retinoblastoma

HPD PDT may prove to be beneficial for the destruction of retinoblastomas and uveal melanomas. HPD fluorescence in experimental ocular amelanotic melanomas is consistent with porphyrin uptake by the tumor. Relatively large levels of HPD could be detected in the aqueous humor of humans, but not in the vitreous humor. Thus, vitreous seeding of the tumor will not accumulate sufficient levels of HPD; however, there may be leakage of HPD out of the tumors to the vitreous humor.

The total level of light reaching the retina (via transpupil illumination) will differ by at least a factor of 2 for infants with retinoblastoma (less than 2 years of age) and adults with chorioidal melanoma (averaging 60 years of age).

Tumor blanching, swelling, and hemorrhage are observed following PDT of patients with uveal melanoma and retinoblastoma. PDT has been applied to the treatment of chorioideal melanoma. In five cases, treatment was followed by complete remission in three patients, 80% regression in one, and 50% regression in one. In six other cases the treatment was not favorable. Posttreatment complications included skin photosensitivity, iritis, exudative retinal detachment, vitreous reaction and hemorrhage, chemosis, cataract, and chorioideal detachment.

19.2.6 Photodynamic Therapy of Diseases Other Than Tumors

PDT therapy has been applied to psoriasis, atherosclerotic lesions, lesions induced by herpes simplex virus, and neurosurgical lesions. The psoriasis lesions disappeared after 15 days of treatment in of 80% of the patients. HPD has the following advantages compared with 8-MOP treatment:

1. Only one injection of HPD (1 mg / kg body weight) is sufficient for the whole photodynamic treatment during 2–3 weeks.
2. A mutagenic effect seems excluded.
3. The excitation of HPD can be performed in the wavelength region 400–630 nm, where no dangerous side effects are included.

A major disadvantage may be the persistence of photosensitivity to bright sunlight for weeks following the HPD injection.

19.3 PDT: WHERE ARE WE TODAY?

Twenty years of research with PDT have shown the advantages and drawbacks of this new cancer treatment. With the advent of compact and easily handled diode lasers, PDT shows great promise in treating nodular tumors and growths in hollow organs. PDT has now been approved as a treatment for lung and stomach cancers in Japan and more recently in Canada in the form of "authorized Photofrin II treatment" for prophylaxis of papillary bladder cancers.

The following results have been reported using Photofrin II as the photoactive drug.

Early stage lung cancers. At Tokyo Medical College, results of 108 cases (83 superficial and 26 nodular tumors) have been published. The superficial tumors of 0.5 cm or less showed 89.4% Complete Remission (CR), whereas lesions larger than 2 cm (14 cases) yielded poor responses (only 28.5% CR). In the nodular group the best CR results were again obtained with lesions of less than 0.5% cm.

Early stage stomach cancers. The Tokyo group reported on 115 cases of early stage patients, dividing the patients into *mucosal* (i.e., superficial) and *submucosal* (*i.e.*, invasive beyond the mucosal layer) groups, with 85% CR for lesions of 1 cm or less in the former group and only 8% CR for lesions of comparable size in the latter group. In Osaka 100% CR was obtained on 15 patients with follow-ups of 2–19 months.

Early stage bladder tumors. The Tokyo group reported CR in 29 of 40 cases of superficial transitional cell carcinomas of the bladder, but found recurrence in 21 cases over 31 months. Five patients were disease-free for 5–84 months after PDT. For cervical lesions on 17 cases (11 carcinoma *in situ*, 6 dysplasias) 16 CR were reported.

In the U.S., preliminary results have been published on PDT (18 patients) vs observation of 16 patients having undergone resection in cases of superficial papillary bladder tumors. In the latter group recurrence was 81% vs only 39% for the PDT group. The average time of recurrence was 91 days and 394 days for the two groups respectively. Although these results are still preliminary and limited to a relatively modest number of cases, they are very promising, especially with the appearance of second-generation photosensitizers.

19.4 THE FUTURE OF PDT

19.4.1 Long-Wavelength Absorbing Photosensitizers

The best light penetration in tissues can be achieved with photosensitizers strongly absorbing in the range of 650–800 nm.

Photofrin II, although of a composition not completely agreed upon, is the main PDT sensitizer being used in the clinic. Its tumor accumulation and tumor demarcation properties are equal to or better than those of most of the proposed second-generation sensitizers.

Among the second-generation sensitizers, chlorins and phthalocyanines, chlorines have been most extensively studied. Skin sensitivity to sunlight will be a smaller problem with this group of dyes, which absorb minimally in the region of 400–600 nm. Chlorin derivatives may also compete favorably with Photofrin II™. Diporphyrins joined via methylene bridges and benzoporphyrin derivatives are efficient *in vivo* a few hours after injection, but are more rapidly cleared from tissues than Photofrin. Favorable optical properties and biodistribution patterns are possessed by purpurins, which require solubilizing or emulsifying agents such as cremophore, liposomes, or lipoproteins. The use of merocyanine 540 for purging autologous bone marrow grafts, sterilization of blood products, and inactivation of leukemia cells and herpes simplex virus type 1 is being investigated.

19.4.2 Photodynamic Therapy with Endogenous Protoporphyrin IX

5-Aminolevulinic acid (ALA) is a precursor of protoporphyrin IX (PPIX) in the biosynthesis of heme. Conversion into heme being slow, PPIX accumulates in *photosensitizing concentrations*, but only in certain types of cells and tissues exposed to ALA. *Tissue-specific photosensitization provides a basis for using ALA-induced PPIX for photodynamic therapy.*

ALA by intradermal injection can induce PPIX photosensitization in cells of the epidermis and its appendages, but not in the dermis. ALA in aqueous solution passes readily through abnormal keratin, but not through normal keratin, thus inducing photosensitization that is restricted primarily to the abnormal epithelium of actinic keratoses and superficial BCC or SCC (photoactivating light selectively destroys such lesions).

The ALA-induced PPIX fluorescence in various skin structures of mice correlates well with the degree of phototoxic damage. In the urinary bladder, the urothelium shows intense fluorescence, and the underlying muscle shows very little; in the uterus the endometrium is strongly fluorescent, but the myometrium is not. This permits the selective destruction of urothelium or endothelium cancers without danger of perforating the bladder or uterus.

The systemic administration of ALA does not induce fluorescence or photosensitization in the dermis of mice. By inference, in clinical trials, photosensitization induced by topical ALA should remain restricted to the epidermis and the epidermal appendages. *Such a tissue-specific photosensitization* would be ideal for the treatment of BCC and SCC.

In a male volunteer, the sun-damaged skin on the cheek, on the back of the hand, and on the scalp at the hairline became moderately photosensitized following exposure to ALA, with no detectable PPIX fluorescence in normal skin of the forearm even after 6 hr of exposure. With topical ALA (2 hr), areas of mild actinic keratose (but not adjacent skin) became both fluorescent and photosensitized. Following exposure to photoactivating light, there were no recurrences of the keratoses during the subsequent 18 months. However, the topical application of ALA was ineffective on verruca vulgaris, verruca plantaris, and seborrheic keratoses.

Standard treatment of BCC (Figure 19.4) and SCC involved the topical application of 20% ALA dissolved in a proprietary oil-in-water emulsion. After 3–6 hr (to allow penetration and synthesis of PPIX) the lesions and adjacent tissue were exposed through a long-pass colored glass filter (wavelengths higher than 600 nm). The total dose of the photoactivating light was varied from 15 to 150 mWh cm^{-2}.

Superficial BCC showed a complete response to topical ALA in 75% of cases (plus 7.5% partial responses) following a single exposure to light. Of the first 80 BCC cases treated, 72 showed a complete response when examined 2–3 months after treatment, 6 showed a partial response (area decreased by more than 50%), and two lesions were definite therapeutic failures. The partial responses occurred mainly with lesions exhibiting a central ulcerated area and an elevated rim covered by keratin. Every lesion that was relatively flat and covered by a rough keratin layer showed a complete response after a single treatment.

ALA-induced PPIX was used to treat six lesions diagnosed either as *in situ* or early invasive SCC. All showed complete response. However, two SCC lesions elevated 10 mm above the surface of the skin showed partial response after repeated weekly treatments. Actinic keratoses responded well to topical ALA PDT. Very extensive areas of sun-damaged skin on the face necessitated treatment of only a limited area at the time, in order to minimize the skin reaction. Although strong PPIX fluorescence and photosensitization were induced in percutaneous nodules of four cases with metastatic carcinoma of the breast, there was no effect on subcutaneous nodules. There was also no effect on the more diffuse deposits of malignant cells that characterize the inflammatory form of breast carcinoma. ALA in simple aqueous solution does not penetrate the keratin of normal skin. Even percutaneous nodules are difficult to eradicate because their periphery generally lies beneath the normal skin.

Compared to ALA therapy, with HPD it is not possible to administer a lethal dose of light to every part of the tumor without giving an overdose to at least some parts of the normal tissue within the treatment field. For example, if a superficial tumor of the skin involves the bridge of the nose, the tip, and both sides, the treatment field will consist of a complex set of curved surfaces that are very difficult to illuminate evenly because of the

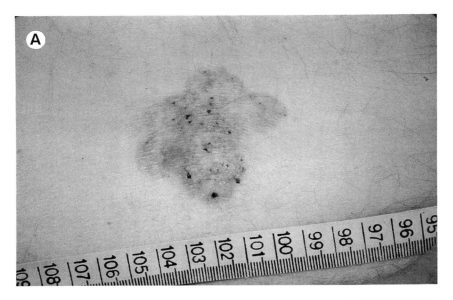

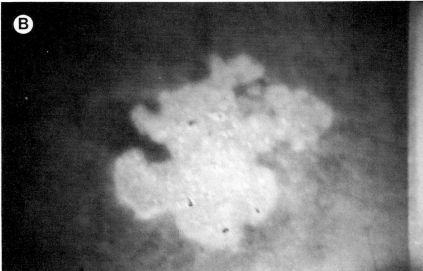

FIGURE 19.4   Basal cell carcinoma of human skin, treated with a cream containing 20% ALA 3 hr prior to PDT. (A) Photographed under white light; (B) photographed under UV light. The application of 150 mW cm^{-2} of red light (via a 35-mm slide projector lamp) during 15 min was effective in totally destroying the cancer. From slides prepared in the laboratory of Dr. James Kennedy, Department of Oncology, Queen's University, Kington, Ontario, Canada. Courtesy of Dr. Roy Pottier, Royal Military College of Canada, Kingston, Ontario, Canada.

variation in the cosine correction factor. Consequently, it is very difficult to avoid underexposing some parts of the treatment field while overexposing (and thus damaging) others. In the case of breast cancer, to ensure that the deep surface of each nodule receives a lethal dose of light, it would be necessary to give the skin surface a dose of light so large that even the relatively low concentration of HPD in the skin would be sufficient to produce serious damage.

In contrast, because ALA-induced PPIX photobleaches very rapidly, it is possible to overdose part of the treatment field in order to make sure that some other part will receive the normal dose, yet it is possible to do so without causing serious damage to normal tissues in the overdosed field. Also, it is possible to safely overdose normal tissues at the surface of the field in order to ensure that an adequate dose of light reaches the deep surface of a tumor.

A generalized but still quite tissue-specific photosensitization may be induced if ALA is administered either by subcutaneous or intraperitoneal injection or by mouth. Using irradiation via an endoscopic approach it should be possible to treat tumors arising from the mucosa of internal cavities (e.g., intestinal).

Bibliography

Amato, I. (1993). Hope for a magic bullet that moves at the speed of light. *Science* **262,** 32–33.

Bandieramonte, G., Marchesini, R., Zunino, F., Melloni, E., Andreola, S., Andreoli, C., Di Pitro, S., Spinelli, P., Fava, G., and Emanuelli, H. (1984). Hematoporphyrin-derivative and phototherapy in extensive basal-cell carcinoma of the dorsal skin. *In* "Porphyrins in Tumor Phototherapy" (A. Andreoni and R. Cubeddu, eds.), Plenum, New York and London, pp. 389–394.

Berg, H., Gollnick, F. A., Bohm, F., Meffert, H., and Sonnichsen, N. (1985). Photodynamic hematoporphyrin therapy of psoriasis. *In* "Photodynamic Therapy of Tumors and Other Diseases" (G. Jori and C. Perria, eds.), pp. 337–344. Libreria Progetto Editore, Via Marzolo, Padova, Italy.

Bruce, R. A., Jr. (1984). Photoirradiation for chorioideal malignant melanoma. *In* "Porphyrins in Tumor Phototherapy" (A. Andreoni and R. Cubeddu, eds.), Plenum, New York and London, pp. 455–461.

Doiron, D. R., and Balchum, O. J. (1984). Hematoporphyrin derivative photoradiation therapy of endobronchial lung cancer. *In* "Porphyrins in Tumor Phototherapy" (A. Andreoni and R. Cubeddu, eds.), Plenum, New York and London, pp. 395–403.

Dougherty, T. J., Potter, W. R., and Weishaupt, K. R. (1984). The structure of the active component of hematoporphyrin derivative. *In* "Porphyrins in Tumor Phototherapy" (A. Andreoni and R. Cubeddu, eds.), Plenum, New York and London, pp. 23–35.

Fingar, V. H. (1993). Mechanisms leading to vessel damage after photodynamic damage. *Photochem. Photobiol.* **57,** Suppl., 18S–19S.

Gomer, C. J., Murphree, A. L., Doiron, D. R., Szirth, B. C., and Razum, N. J. (1984). Preclinical examination of ocular photoradiation therapy. *In* "Porphyrins in Tumor Phototherapy" (A. Andreoni and R. Cubeddu, eds.), Plenum, New York and London, pp. 447–454.

Hayata, Y., Kato, H., Konaka, C., Ono, J., Saito, M., Takahashi, H., and Tomono, T. (1984). Photoradiation therapy in early stage of cancer cases of the lung, esophagus and stomach. *In* "Prophyrins in Tumor Phototherapy" (A. Andreoni and R. Cubeddu, eds.), Plenum, New York and London, pp. 405–412.

Henderson, B. W., and Dougherty, T. J. (1992). How does photodynamic therapy work? *Photochem. Photobiol.* **55,** 145–157.

Henderson, B. W. (1993). *Photochem. Photobiol.* **57,** Suppl., 17S–18S. Vascular photosensitization: its implications for photodynamic therapy.

Hilf, R. (1993). Consideration of cellular targets in photodynamic therapy. *Photochem. Photobiol.* **57,** Suppl., 19S.

Jocham, D., Stachler, G., Unsold, E., Chaussy, C., and Lohrs, U. (1984). Dye-laser-photoirradiation therapy of bladder cancer after photosensitization with hematoporphyrin derivative (HpD)—Basis for an internal irradiation. *In* "Porphyrins in Tumor Phototherapy" (A. Andreoni and R. Cubeddu, eds.), Plenum, New York and London, pp. 427–438.

Jori, G., and Perria, C. (1985). "Photodynamic Therapy of Tumors and Other Diseases." Libreria Progetto Editore, Via Marzolo, Padova, Italy.

Kennedy, J. C., and Pottier, R. H. (1992). Endogenous protoporphyrin IX, a clinically useful photosensitizer for photodynamic therapy. *J. Photochem. Photobiol., B: Biol.* **14,** 275–292.

Kennedy, J. C., Pottier, R., and Pross, D. C. (1990). Photodynamic therapy with endogenous protoporphyrin IX: Basic principles and present clinical experience. *J. Photochem. Photobiol., B: Biol.* **6,** 143–148.

Lampidis, T. J., Bernal, S. D., Summerhayes, I. C., and Chen, L. B. (1982). Rhodamine-123 is selectively toxic and preferentially retained in carcinoma cells *in vitro. Ann. N. Y. Acad. Sci.* **397,** 299–301.

Lampidis, T. J., Salet, C., Moreno, G., and Chen, L. B. (1984). Effects of the mitochondrial probe rhodamine 123 and related analogs on the function and viability of pulsating myocardial cells in culture. *Agents Actions* **14,** 751–755.

Laws, E. R., Wharen, R. E., Jr., and Anderson, R. E. (1985). Photodynamic therapy of brain tumors. *In* "Photodynamic Therapy of Tumors and Other Diseases" (G. Jori and C. Perria, eds.), pp. 311–316. Libreria Progetto Editore, Via Marzolo, Padova, Italy.

Lombard, G. F., Tealdi, S., and Lanotte, M. M. (1985). The treatment of neurosurgical infections by lasers and porphyrins. *In* "Photodynamic Therapy of Tumors and Other Diseases" (G. Jori and C. Perria, eds.), pp. 363–366. Libreria Progetto Editore, Via Marzolo, Padova, Italy.

Maziere, J. C., Morliere, P., Biade, S., and Santus, R. (1992). La photothérapie antitumorale: Bases biochimiques, application thérapeutique et perspective. *C. R. Seances Soc. Biol. Ses Fil.,* 186, 88–106.

Menezes da Silva, F. A. (1993). Non-invasive dynamic capillaroscopy of photodynamic effects on the microcirculation. *Photochem. Photobiol.* **57,** Suppl., 19S–20S.

Moan, J., and Berg, K. (1992). Photochemotherapy of cancer: Experimental research. *Photochem. Photobiol.* **55,** 931–948.

Monfreica, G., Martellotta, D., Bruno, G., Cariello, L., Zanetti, L., and Santoianni, P. (1985). An approach to the treatment of psoriasis with a porphyrin analogue. *In* "Photodynamic Therapy of Tumors and Other Diseases" (G. Jori and C. Perria, eds.), pp. 345–348. Libreria Progetto Editore, Via Marzolo, Padova, Italy.

Okuda, S., Mimura, S., Otani, T., Ichii, M., and Tatsuta, M. (1984). Experimental and clinical studies on HPD-photoradiation therapy for upper gastrointestinal cancer. *In* "Porphyrins in Tumor Phototherapy" (A. Andreoni and R. Cubeddu, eds.), pp. 413–421. Plenum, New York and London.

Policard, A. (1924). Étude sur les aspects offerts par des tumeurs expérimentales examinées à la lumière de Wood. *Biologie. Comptes Rendas* 91, 94–95.

Powers, S. K., Beckman, W. C., Brown, J. T., and Kolpack, L. C. (1987). Interstitial laser photochemotherapy of rhodamine-123 sensitized rat glioma. *J Neurosurg.* **67,** 889–894.

Siegel, K. A., Fingar, V. H., and Wieman, T. J. (1993). Mechanisms of tumor destruction using photofrin, HPPH, and NPe6. *Photochem. Photobiol.* **57,** Suppl., 20S.

Spinelli, P., Andreola, S., Marchesini, R., Melloni, E., Mirabile, V., Pizzetti, P., and Zunino, F. (1984). Endoscopic HpD-laser photoirradiation therapy of cancer. *In* "Porphyrins in Tumor Phototherapy" (A. Andreoni and R. Cubeddu, eds.), pp. 423–426. Plenum, New York and London.

Straight, R. C., Vincent, G. M., Hammond, E. H., and Dixon, J. A. (1985). Porphyrin retention and photodynamic treatment of diet induced atherosclerotic lesions in pigs. *In* "Photodynamic Therapy of Tumors and Other Diseases" (G. Jori and C. Perria, eds.), pp. 349–352. Libreria Progetto Editore, Via Marzolo, Padova, Italy.

Straight, R. C., Stroop, W. G., Spikes, J. D., and Dixon, J. A. (1985). Photodynamic inactivation of the herpes virus. *In* "Photodynamic Therapy of Tumors and Other Diseases" (G. Jori and C. Perria, eds.), pp. 353–358. Libreria Progetto Editore, Via Marzolo, Padova, Italy.

Van Leengoed, H. L. L. M., Schuitmaker, J. J., van der Veen, N., Dubbelman, T. M. A. R., and Star, V. M. (1993). Fluorescence and photodynamic effects of bacteriochlorin *a* observed *in vivo* in 'sandwich' observation chambers. *Br. J. Cancer* **67,** 898–903.

INDEX